# Structures 2018



Transportation Research Record: Journal of the Transportation Research Board

Volume 2672, Issue 41

ISBN: 978-1-5108-8602-5

**Printed from e-media with permission by:**

Curran Associates, Inc.
57 Morehouse Lane
Red Hook, NY 12571

**Some format issues inherent in the e-media version may also appear in this print version.**

Published by Sage Publications USA



ISBN (Print) 978-1-5108-8602-5
ISBN 2018 Printed Set (All Issues) 978-1-5108-7735-1

Printed by Curran Associates, Inc. (2019)



**Additional copies of this publication are available from:**

Curran Associates, Inc.
57 Morehouse Lane
Red Hook, NY 12571 USA
Phone: 845-758-0400
Fax: 845-758-2633
Email: curran@proceedings.com
Web: www.proceedings.com

Volume 2672 Number 41 December 2018

TRR
JOURNAL OF THE TRANSPORTATION RESEARCH BOARD

# Contents

## Structures 2018

### Articles

* Membership as of February 2019.

# TRB

# TRANSPORTATION RESEARCH BOARD

## CORE SPONSORS

**Transportation Departments of the 50 States and the District of Columbia**

***Federal Government***

U.S. Department of Transportation
- Federal Aviation Administration (FAA)
- Federal Highway Administration
- Federal Motor Carrier Safety Administration (FMCSA)
- Federal Railroad Administration (FRA)
- Federal Transit Administration
- National Highway Traffic Safety Administration (NHTSA)
- Office of the Assistant Secretary of Transportation for Research and Technology (OST-R)

U.S. Air Force Civil Engineer Center (AFCEC)
U.S. Coast Guard (USCG)
U.S. Department of Energy
U.S. Department of the Interior (DOI)
U.S. Environmental Protection Agency (EPA)

***Nongovernmental Organizations***

American Public Transportation Association (APTA)
Association of American Railroads (AAR)
California Air Resources Board

*The National Academies of*
SCIENCES • ENGINEERING • MEDICINE

The **National Academy of Sciences** was established in 1863 by an Act of Congress, signed by President Lincoln, as a private, nongovernmental institution to advise the nation on issues related to science and technology. Members are elected by their peers for outstanding contributions to research. Dr. Marcia McNutt is president.

The **National Academy of Engineering** was established in 1964 under the charter of the National Academy of Sciences to bring the practices of engineering to advising the nation. Members are elected by their peers for extraordinary contributions to engineering. Dr. C. D. Mote, Jr., is president.

The **National Academy of Medicine** (formerly the Institute of Medicine) was established in 1970 under the charter of the National Academy of Sciences to advise the nation on medical and health issues. Members are elected by their peers for distinguished contributions to medicine and health. Dr. Victor J. Dzau is president.

The three Academies work together as the National Academies of Sciences, Engineering, and Medicine to provide independent, objective analysis and advice to the nation and conduct other activities to solve complex problems and inform public policy decisions. The Academies also encourage education and research, recognize outstanding contributions to knowledge, and increase public understanding in matters of science, engineering, and medicine.

Learn more about the National Academies of Sciences, Engineering, and Medicine at **www.national-academies.org.**

The **Transportation Research Board** is one of seven major programs of the National Academies of Sciences, Engineering, and Medicine. The mission of the Transportation Research Board is to increase the benefits that transportation contributes to society by providing leadership in transportation innovation and progress through research and information exchange, conducted within a setting that is objective, interdisciplinary, and multimodal. The Board's varied activities annually engage about 7,000 engineers, scientists, and other transportation researchers and practitioners from the public and private sectors and academia, all of whom contribute their expertise in the public interest. The program is supported by state transportation departments, federal agencies including the component administrations of the U.S. Department of Transportation, and other organizations and individuals interested in the development of transportation.

Learn more about the Transportation Research Board at **www.TRB.org**.

Article

TRR
JOURNAL OF THE TRANSPORTATION RESEARCH BOARD

Transportation Research Record
2018, Vol. 2672(41) 1–10

DOI: 10.1177/0361198118798990
journals.sagepub.com/home/trr
SAGE

# Section Idealization of Corrugated Thermoplastic Pipe in AASHTO Design

**Brent J. Bass[1] and Jesse L. Beaver[1]**

## Abstract

The AASHTO Load and Resistance Factor Bridge Design Specifications (AASHTO LRFD) thermoplastic pipe design method requires corrugated pipe local buckling resistance to be determined based on corrugation effective area. The effective area may be determined through calculations or physical tests on sections of pipe. When determined through calculations, effective area is based on individual corrugation element (e.g., crest, valley, web) slenderness following methods published by the American Iron and Steel Institute (AISI) for cold-formed steel design. Cold-formed steel members are rolled from constant-thickness steel sheet and have cross-sections divided into elements by distinct corners. In contrast, corrugated thermoplastic pipe cross-sections have variable geometries with non-uniform thickness, elements without distinct corners defining their ends, and elements that may be rounded or have other beneficial features such as intermediate ribs or stiffeners. Applying the calculation method requires idealization of corrugation elements into flat plates of representative clear width and thickness. As corrugation geometries have evolved with the increased use of thermoplastic pipe, there has not been a thorough review of appropriate methods of idealization to ensure current geometries meet the intent of the design method. This paper reviews the existing AASHTO LRFD effective area calculation method, information from background documents upon which the AASHTO LRFD method was based (NCHRP reports 438 and 631), and relevant information from the AISI Specification for the Design of Cold-Formed Steel Structural Members; identifies important concepts for cross-section idealization; and provides recommendations for idealization of corrugation members with curves and intermediate stiffeners.

The AASHTO Load and Resistance Factor Bridge Design Specifications (AASHTO LRFD) (*1*) thermoplastic pipe design method requires local buckling resistance for corrugated pipes to be determined based on corrugation effective area. The effective area may be determined through calculations or physical tests on sections of pipe. When determined through calculations, effective area is based on the slenderness of individual corrugation elements (e.g., crest, valley, web; Figure 1) following methods published by the American Iron and Steel Institute (AISI) for cold-formed steel design in AISI S100-07 (*2*). The method was developed for cold-formed steel members typically rolled from constant-thickness steel plate and with cross-sections divided into corrugation elements by small radius, distinct corners. In contrast, corrugated thermoplastic pipe cross-sections have variable geometries with non-uniform thickness, elements without distinct corners to define their ends, and elements that may be curved for their entire length. Applying the AISI-based calculation method, therefore, requires idealization of these corrugation elements into flat plates of representative clear width and thickness. There may also be intermediate stiffeners, which, until recently, were not addressed by the AISI method in a way that could be implemented for thermoplastic pipe design.

As corrugation geometries have evolved with the increased use of thermoplastic pipe and more manufacturers developing corrugations, there has not been a published review of idealization methods to ensure current geometries meet the intent of the design method. Requirements of the AASHTO LRFD effective area and section idealization method are presented, followed by a review of NCHRP reports 438 (*3*) and 631 (*4*) that provided the basis of the current AASHTO approach, followed by sample corrugation idealizations with methodology based on research and updates to AISI, and the resulting effective areas. Three section idealization examples considered are: 1) a standard corrugation with flat crest, 2) the Example 1 corrugation modified to have a substantially curved crest, and 3) the same

[1]Simpson Gumpertz & Heger, Inc., Waltham, MA

**Corresponding Author:**
Address correspondence to Brent J. Bass: bjbass@sgh.com

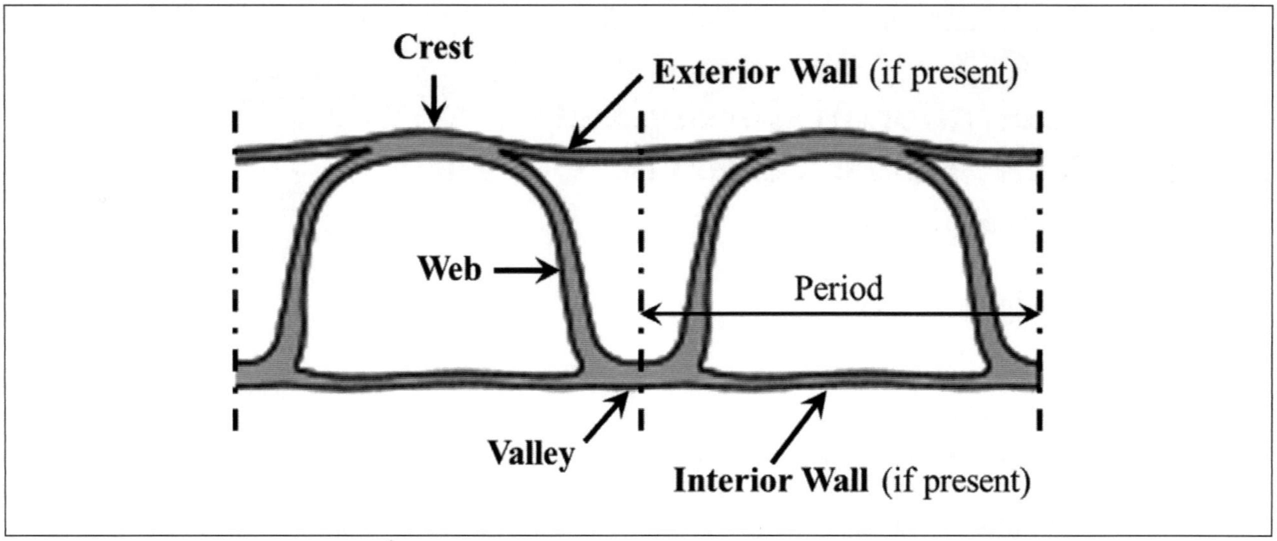


**Figure 1.** Sample corrugated wall with element terminology in bold.

corrugation from Example 1 but with a crest element with stiffener.

## AASHTO LRFD Section Idealization and Effective Area Calculation

### *AASHTO LRFD Requirements for Determining Effective Area by Calculation*

In AASHTO LRFD, effective area for local buckling of corrugated thermoplastic pipes may be determined by calculations in accordance with Section 12.12.3.10.1b. Key equations from this section are reproduced as equations 1 to 4 below, where terms with subscript $i$ are determined for each corrugation element $i$.

$$A_{eff} = A_g - \frac{\sum(w_i - b_{e,i})t_i}{\omega} \quad (1)$$

$$b_{e,i} = \rho_i w_i \quad (2)$$

$$\rho_i = \frac{1 - \frac{0.22}{\lambda_i}}{\lambda_i} \leqslant 1 \quad (3)$$

$$\lambda_i = \left(\frac{w_i}{t_i}\right)\sqrt{\frac{\varepsilon_{yc}}{k_i}} \geqslant 0.673 \quad (4)$$

where

$A_{eff}$ = effective area of pipe wall per unit length of pipe (in.$^2$/in.);

$A_g$ = gross area of pipe wall per unit length of pipe (in.$^2$/in.);

$w_i$ = total clear width of element $i$ between supporting elements (in.);

$b_{e,i}$ = effective width of corrugation element, $i$ (in.);

$t_i$ = thickness of element, $i$ (in.);

$\omega$ = spacing (or period) of corrugation (in.);

$\rho_i$ = effective width factor of corrugation element, $i$;

$\lambda_i$ = slenderness factor of corrugation element, $i$;

$\varepsilon_{yc}$ = factored compression strain limit for pipe material (in./in.); and

$k_i$ = plate buckling edge support coefficient for element $i$; $k$ = 4.0 for elements supported at both ends, and $k$ = 0.43 for elements supported at one end, such as freestanding ribs.

Commentary C12.12.3.10.1b notes that $k$ for these calculations is analogous to the effective length factor, $k$, in column buckling.

Note that the summation term in the numerator of Equation 1 is the sum of the ineffective width of each element, multiplied by the element thickness. When summed and divided by the period, this becomes the ineffective area per unit length of pipe, which is subtracted from the gross area per unit length to determine the effective area. The total clear width of the element between supporting elements, $w$, and effective width, $b_e$, are shown conceptually in Figure 2 (reproduced from AASHTO LRFD Figure C12.12.3.10.1b-1).

AASHTO LRFD 12.12.3.10.1b effective area calculations require corrugations to be idealized as straight elements. Each element is assigned a clear width based on the clear distance between the adjoining elements and a thickness based on the thickness at the center of the element. AASHTO LRFD Commentary C12.12.3.10.1b notes that the idealization should include the actual total area, that if an element such as the crest is curved, it should be idealized at the centroid of curvature, and that the idealization of the curved element need not touch the idealized webs. The Commentary references NCHRP Report 631 for guidance on other profile types (*4*).

### AASHTO LRFD Use of Effective Area

Effective area calculations, using the factored compression strain limit in Equation 4, assume every corrugation element is loaded to its strain capacity, $\varepsilon_{yc}$. In reality, the total section is not typically strained to capacity, therefore the AASHTO LRFD calculation method provides a conservative estimate of the effective area.

In AASHTO LFRD 12.12.3.10.1c–*Compression Strain*, the factored compressive strain from axial thrust, $\varepsilon_{uc}$, is checked against the material strain limit, $\varepsilon_{yc}$; $\varepsilon_{uc}$ is calculated from the total factored thrust demand acting on the effective area of the cross-section at $\varepsilon_{yc}$. In AASHTO LRFD 12.12.3.10.2b–*Combined Strain*, axial and flexural strains are combined and checked against a combined strain limit. The factored axial strain demand, $\varepsilon_{uc}$, is the same magnitude as determined in Section 12.12.3.10.1c, and is combined with flexural strain, which is typically calculated as the factored greatest outer fiber strain from a pipe deflected to its maximum allowed vertical deflection. The flexural strain is not dependent on an idealized or effective moment of inertia, and the effective area used for combined strain, being the same magnitude as for pure compression strain, still assumes all corrugation elements are strained to capacity.

### Noteworthy Items from Review of AASHTO LRFD Requirements

Recall that the terms to the right of the subtraction sign in Equation 1 are the ineffective area per unit length of pipe, which is subtracted from the corrugation gross area to determine $A_{eff}$. If the element clear widths and thicknesses are properly defined in the idealization, this process calculates an appropriate ineffective element area, and eventually an appropriate $A_{eff}$.

Commentary C12.3.10.1b notes the idealization should include the actual total area of the corrugation; however, this is not required by the specification or calculations, nor is it explained in sufficient detail to be applied consistently. The relative position of idealized elements and other differences between idealized and gross sections, such as neglecting corner areas by only idealizing appropriate element clear widths, have no bearing on calculated effective area. Properties like the idealized corrugation total area (including corners) and idealized corrugation moment of inertia (which would be affected by idealized element positions) are not used directly in the calculations; however, they may be useful for comparison with gross-section properties to ensure idealized areas are properly distributed to the idealized elements.

Although Commentary C12.12.3.10.1b specifies that a curved element should be idealized at the center of curvature, this location is not required by the specification or calculations. Furthermore, there is no specific guidance on how to set the clear width of the curved element or where to define the end of the intersecting element and beginning of the curved element.

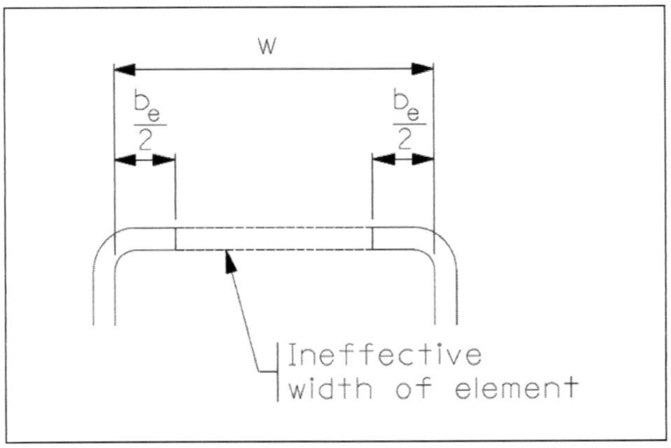


**Figure 2.** Effective width concept (reproduced from AASHTO LRFD C12.12.3.10.1b-1 [*1*]).

The clear width in Figure 2 extends beyond the start of the corner radius, to where the horizontal element intersects the supporting vertical elements; this design approach is conservative for a clear width, making it closer to a total width. The increased width will result in less effective area. The AISI effective area methods, upon which the AASHTO LRFD method is based, consider element clear width as the flat portion, excluding corner radii (*2*).

In the definitions following Equation 4, default values are provided for $k$, where $k = 4.0$ is for elements supported at both ends, and $k = 0.43$ is for elements with an unsupported end, such as a free-standing rib. In Commentary C12.12.3.10.1b, $k$ is likened to the column buckling effective length factor. For a plate buckling coefficient of $k = 4.0$, the analogy would be to a column with both ends pinned; for $k = 0.43$, the analogy would be, in theory, to a column with one end pinned and the other free. If the column were to include some intermediate bracing (or the corrugation element were to have an intermediate stiffener), it is possible to have values for $k$ that differ from 4.0 and 0.43. This concept will be investigated further below.

## Review of Research Leading to the AASHTO Effective Area Approach

### NCHRP Report 438

The first research adapting the AISI effective area approach to corrugated thermoplastic pipe design was published in 2000 by McGrath and Sagan in NCHRP Report 438 (*3*). The authors measured strains on pipe stubs (short chord lengths of pipe with parallel cut ends)

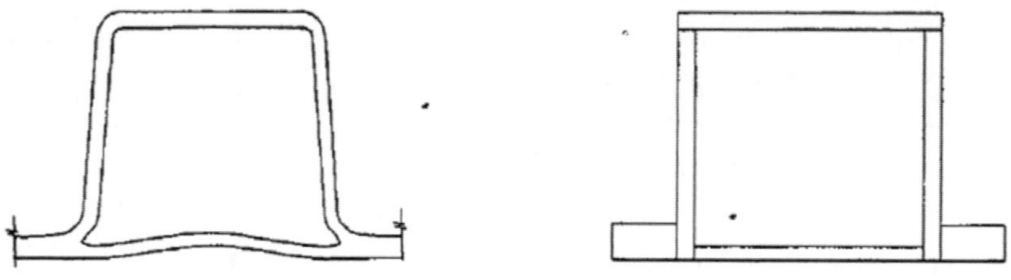


**Figure 3.** Sample profile idealization (reproduced from NCHRP Report 438, Figure B-2 [3]).

compressed between two parallel plates. They mounted strain gages on the crest, valley, and interior wall of the test specimen corrugations. The tests simulated the axial compression (hoop thrust) in an installed pipe. Measured test strains were compared with theoretical strains calculated by idealizing the sections, computing their effective areas, and using an estimated modulus of elasticity.

The report provided schematic idealizations for a typical corrugation with a crest, webs, valley, and interior wall, and for a honeycomb profile. The typical corrugation and its idealization from NCHRP Report 438 Figure B-2 is reproduced here as Figure 3.

The Figure 3 idealization has arbitrary element widths that do not correspond either to element total or clear widths. Element widths as shown would conservatively overestimate the clear widths of most elements, artificially reducing element effective areas and increasing the strain in the cross-section. However, the detailed calculations in NCHRP Report 438 Appendix D accurately account for the clear widths by subtracting, for example, the thickness of the two webs from the total width of the crest, as drawn in Figure 3, to arrive at the clear width of the idealized crest. In addition, the schematic idealization in Figure 3 has webs intersecting the crest and valley at right angles, whereas the nominal section has angled webs. Also note that the idealized valley thickness is not representative of the nominal section valley thickness. Idealization methods in NCHRP Report 438 were being developed for these initial strain comparisons; however, there was no standard procedure to arrive at idealized elements, particularly for corrugations with non-straight elements. The Figure 3 idealization is schematic for this cross-section, and properties derived from it require some interpretation (such as subtraction of appropriate intersecting element widths) to determine clear widths for effective area calculations.

Regarding stiffeners, NCHRP Report 438 Figure 9 shows a sample corrugation with two indentations in the crest that act as stiffeners. The report recommends increasing the area (i.e., thickness) of the idealized element to match the total area without otherwise accounting for the stiffeners.

NCHRP Report 438 concluded that the theoretical strain calculations accounting for effective width correlated well with physical test results.

## *NCHRP Report 631*

In NCHRP Report 631 (*4*), McGrath et al. provided updated recommendations for thermoplastic pipe design and materials, and included a detailed method for the stub compression test. The design and material requirements were adopted by AASHTO in the 2010 interim version of AASHTO LRFD (*5*), and the stub compression test method became AASHTO Test Method T-341 (*6*).

NCHRP Report 631 clarified the difference between the idealized widths as drawn in Figure 3 and the element clear width as needed for calculations. NCHRP Report 631 Figure 15, reproduced here as Figure 4, clearly identifies the clear width, $w$, of the idealized crest as differing from the total width as drawn, and appropriately shows the calculated effective width ($w_e$ in the figure).

NCHRP Report 631 Appendix B provides a general procedure for corrugation idealization. This procedure provides several requirements that are not actually necessary to perform the AASHTO effective area calculations. The Appendix B general procedure states that the corrugation is "idealized into rectangular elements that together have the same area, moment of inertia, element widths, element thicknesses, and distances to the inner and outer faces from the center of gravity."

Recall that the effective area calculations require only the corrugation gross area, idealized element clear widths and thicknesses, and element edge support coefficients. Total area of the idealized cross-section is not used and its requirement has been confusing to designers. The idealization can be developed with representative element clear widths and thicknesses without showing idealized element total widths and corner areas, or by showing

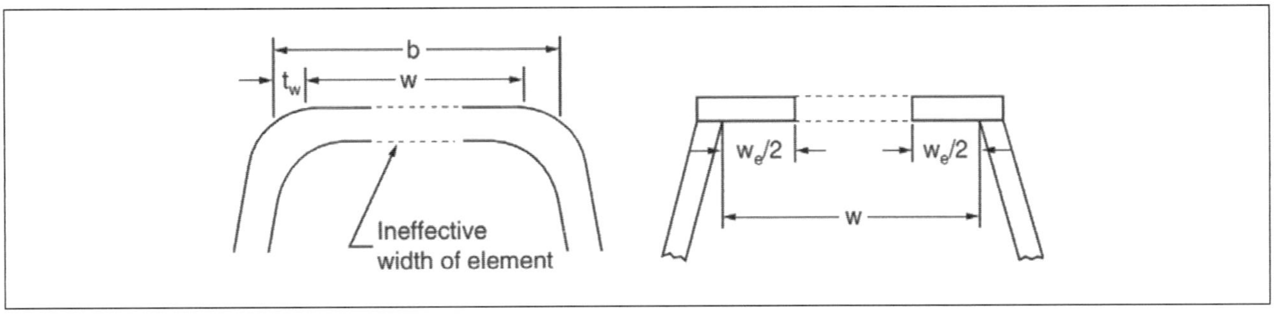


**Figure 4.** Illustration of crest element total, clear, and effective widths (reproduced from NCHRP Report 631, Figure 15 [*4*]).

corner areas separately to eliminate the need for subtraction of appropriate intersecting element thicknesses when entering into effective area calculations. Recall that the AASHTO LRFD effective area method subtracts the total ineffective area of all elements from the corrugation gross area, without use of the idealized section total area.

Other items, such as idealized moment of inertia and distances to the inner and outer faces from the center of gravity, are not required for evaluation of AASHTO LRFD compression strain or combined strain demands.

NCHRP Report 631 Appendix B also provides a recommended procedure for idealization of curved elements, such as an arch-shaped crest, as follows:

1. Define the start of the crest element where the webs are no longer straight,
2. Determine the centroid, thickness, and area of the curved crest,
3. Assign the thickness of the nominal, curved crest to the idealized element, and
4. Increase the element clear width so that the total area of the curved and idealized crests match, and place the centroid of the idealized crest at the centroid of the curved crest.

AASHTO LRFD Commentary C12.12.3.10.1b recommends a similar procedure.

Note, however, that a curved element with the same two end points as a shorter, straight element with like thickness through those end points will generally be less prone to local buckling than the shorter flat-plate element. The curvature forces the element into a higher buckling mode, thereby increasing its capacity. The steps for the Appendix B method do not take advantage of this increased capacity from curvature.

## Proposed Improvements to Corrugation Idealization Methods

### *Corrugations with Relatively Straight Elements*

Based on the requirements of the AASHTO LRFD effective area calculation method and other findings above, flat-element clear widths can be idealized directly, without idealizing element total widths or comparing gross and idealized section properties (e.g., total area and moment of inertia). This eliminates the bookkeeping of having to subtract appropriate intersecting element thicknesses when performing effective area calculations. If a comparison is desired between gross and total idealized section properties, corner areas may be drawn and included separately from the clear-width elements and used to determine the total idealized section properties. A sample corrugation and one method for its idealization are shown in Figure 5, with idealized element clear widths and thickness shown in green and corner areas shaded red.

In the Figure 5 sample idealization, the inside surface of the web was set to intersect the center of the arc between the web and crest, and the center of the arc between the web and valley. This results in the webs being angled inward slightly, in the directions of the curves. The outside surface of the web was located by offsetting it from the inside surface by the thickness at the center of the gross web. The ends of the inside and outside surfaces of the web were established by extending them to intersect the (horizontally extended or trimmed) inside surfaces of the crest and interior wall. All other elements were set to intersect the corners of the webs and their thicknesses were set to the thickness at the center of each gross element.

This idealization method falls somewhere between the AISI idealization method, which does not consider corner radii as contributing to clear widths, and the idealization figures from AASHTO LRFD.

### *Curved Corrugation Elements*

*Finite Element Analysis to Study Effect of Curvature.* The authors utilized three-dimensional finite element analysis (FEA) in ABAQUS (*7*) to study the effect of curvature on corrugation buckling. The FEA model considered a column of square plate elements representing a single corrugation element with 16 plate elements across its width, *w*, and with five different cases of curvature

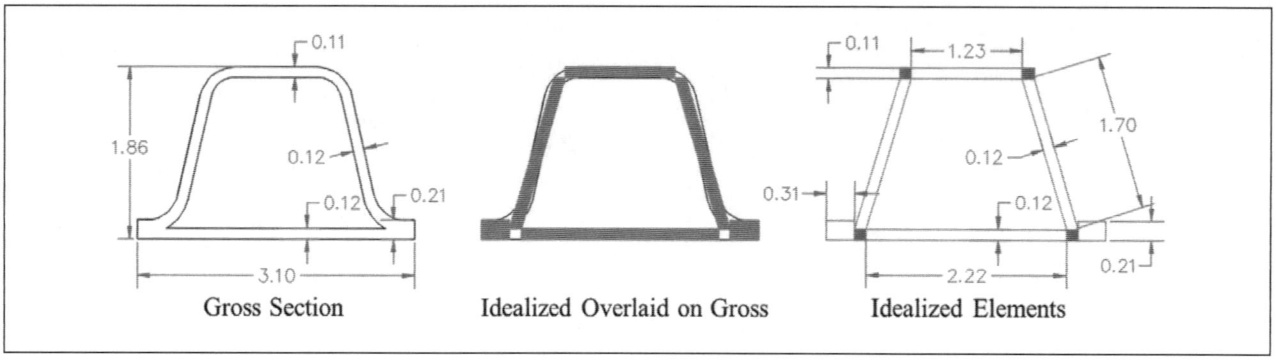


**Figure 5.** Sample corrugation idealization with flat crest (dimensions are inches).

considered, ranging from flat to a curve with radius equal to $w/2$ as shown in Figure 6. Boundary conditions included having translation fixed in three directions at the base and in the $X$- and $Y$-directions along the sides and at the top (orientation shown in Figure 6*b*). Width-to-thickness ratios included 5, 8, 13, and 20 with baseline thickness, $t = 0.125$ in. Nodes at the top of the model were loaded with a unit load (half the unit load for the two outside nodes), and an Eigenvalue analysis was performed to determine the first buckling mode scale factor. Material properties included an elastic modulus of 27,000 psi and Poisson's ratio of 0.4.

As the width, $w$, is fixed for all curvature cases (see Figure 6*a*), the cross-sectional area increases with curvature. To assess this effect on the Eigenvalue buckling load, separate analyses were performed on the $h = 0$ models with thickness changed to match the area of the cross-section in the models with $h = w/2$ and with $h = w/8$. As the purpose of the study was to compare relative theoretical performance between flat and curved elements while controlling variables other than curvature, element imperfections (such as mold lines or other anomalies that may be present in a physical corrugation) were not considered. Results from the analyses are shown in Figure 6*d* with the vertical axis normalized by $w$ for comparison across various width-to-thickness ratios.

Results show that increasing the curvature of a section has a significant effect beyond simply increasing the element area. For example, for $w/t = 8$ (circles), the buckling load is 6.5 times greater for the semi-circle ($h = w/2$) than for $h = 0$; similarly, for $w/t = 13$ (squares), the semi-circle buckling load is 10.7 times greater than for the flat plate. When thickness of the flat plate is adjusted to match total area of the semi-circle, the buckling loads for the semi-circles are 1.9 times and 2.9 times greater than the flat plates of like area for $w/t = 8$ and 13, respectively. For $w/t = 5$, which is a very stout element, the ratio is about 1.3.

Based on these results, to allow consideration of the benefit of curvature, the clear width of a curved element may be set as the distance between intersecting elements, and the idealized element thickness may be increased so its area matches the area of the curved element. This is a conservative approximation of complex results; more refined evaluation of the results and comparison with stub compression tests could produce a more exact, less conservative recommendation.

*Example Idealization with New Method for Curved Elements.* The cross-section from Figure 5 is redrawn with a curved crest in Figure 7 with the idealization method following the curved element idealization method from above. The web elements are shortened slightly to the start of the curved crest. The lower portion of the webs, valley, and interior wall remain as before. The clear width of the idealized crest is set to intersect the web corners, and its thickness is set to match the gross crest total area of 0.241 in.$^2$ (red area in Figure 7 Gross Section). The idealized crest thickness becomes 0.14 in. from the 0.241 in.$^2$ curved area divided by the 1.70 in. total width of the idealized crest. The idealized crest clear width is determined as 1.45 in. by subtracting the thickness of the two intersecting webs.

## *Stiffened Corrugation Elements*

*AISI Intermediate Stiffener Evaluation.* AISI first published a method to evaluate the ability of an intermediate stiffener to brace its parent element in Section B5.1, "Effective Widths of Uniformly Compressed Stiffened Elements with Single or Multiple Intermediate Stiffeners," of AISI S100-07 (*2*). Published in 2007, this method was not available during the development of NCHRP Report 438, which recommended against including effects of stiffeners aside from the increased total area.

When a corrugation element has an intermediate stiffener, two types of behavior are considered: 1) sub-element local buckling, in which the stiffener(s) divide the element into sub-elements that are evaluated for local

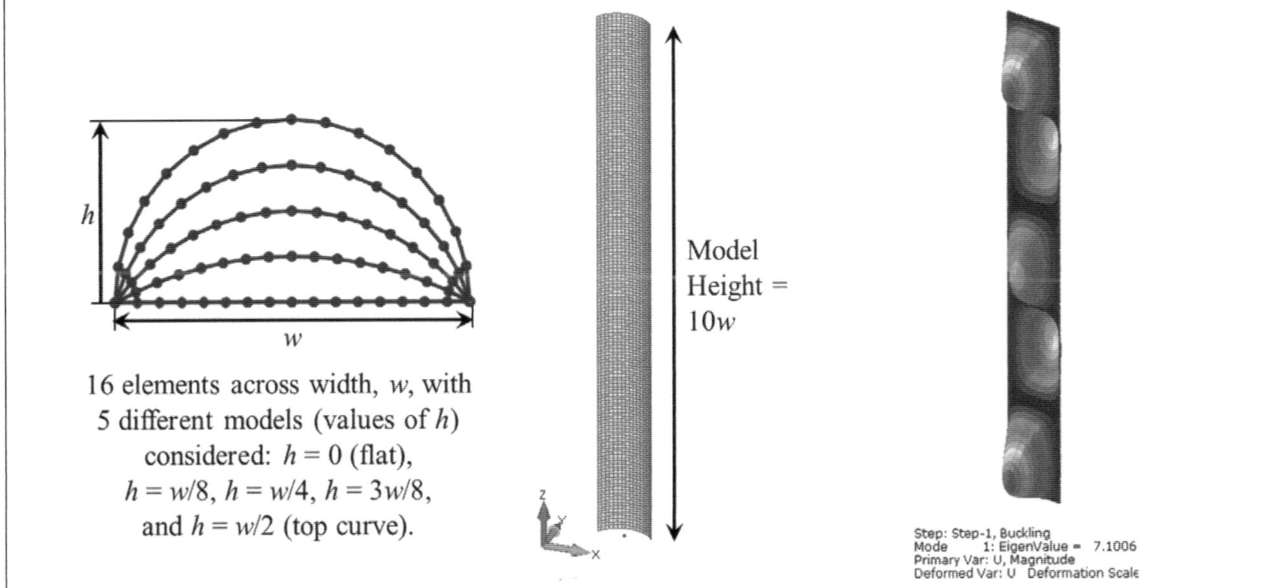


(a) Plan view, various models. (b) Isometric view of shell elements. (c) Eigenvalue result for $h = 0$, $w/t = 8$, and $t = 0.125$ in.

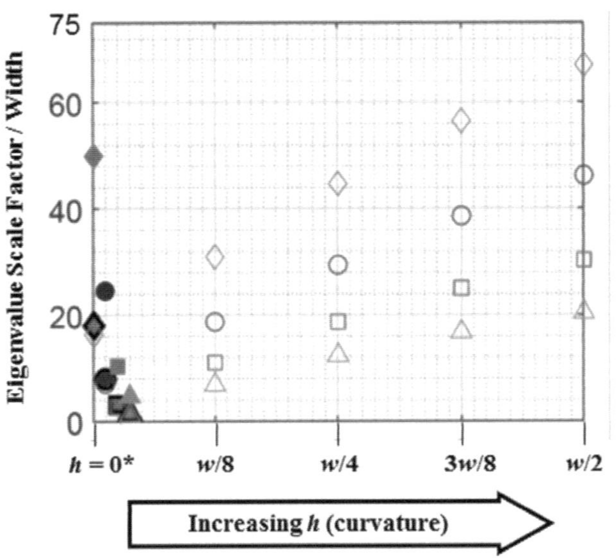


*Plot symbols for $h = 0$ are offset slightly along horizontal axis for clarity.

**Plot Legend**

| $w/t$ | $h$[1] | $t$ (in.) | $w$ (in.) | Model Height (in.) | Plot Symbol |
|---|---|---|---|---|---|
| 5 | 0 | 0.196[2] | 0.625 | 6.25 | ◆ (filled, grey) |
| | 0 | 0.130[3] | | | ◆ (grey, bold outline) |
| | 0 to $w/2$ | 0.125 | | | ◇ |
| 8 | 0 | 0.196[2] | 1.000 | 10 | ● (filled, grey) |
| | 0 | 0.130[3] | | | ● (grey, bold outline) |
| | 0 to $w/2$ | 0.125 | | | ○ |
| 13 | 0 | 0.196[2] | 1.625 | 16.25 | ■ (filled, grey) |
| | 0 | 0.130[3] | | | ■ (grey, bold outline) |
| | 0 to $w/2$ | 0.125 | | | □ |
| 20 | 0 | 0.196[2] | 2.500 | 25 | ▲ (filled, grey) |
| | 0 | 0.130[3] | | | ▲ (grey, bold outline) |
| | 0 to $w/2$ | 0.125 | | | △ |

1. See Figure 6a for $h$.
2. Thickness set to match area for $h = w/2$.
3. Thickness set to match area for $h = w/8$.

(d) Eigenvalue FEA results.

**Figure 6.** Overview and results of FEA to evaluate effect of curvature on element buckling.

buckling and the stiffener's ability to support the sub-elements is considered, and 2) distortional buckling of the stiffened element, in which the full element width including stiffener(s) buckles. The method determines edge support coefficients for the two types of behavior shown schematically in Figure 8 for an element with two intermediate stiffeners.

The AISI stiffener evaluation can be used to determine edge support coefficients for input into the AASHTO LRFD effective area equations 1 to 4, above. For plate

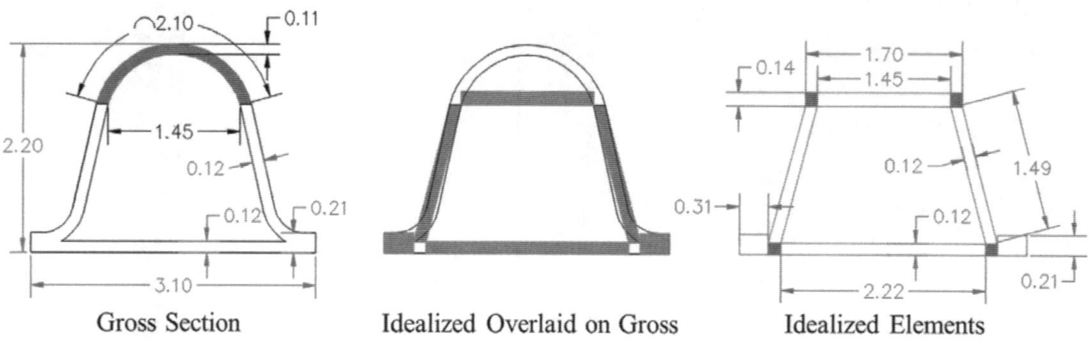


**Figure 7.** Corrugation idealization utilizing new curved element idealization method for crest (updated portion of corrugation compared with Figure 5 is shown shaded in gross-section view).

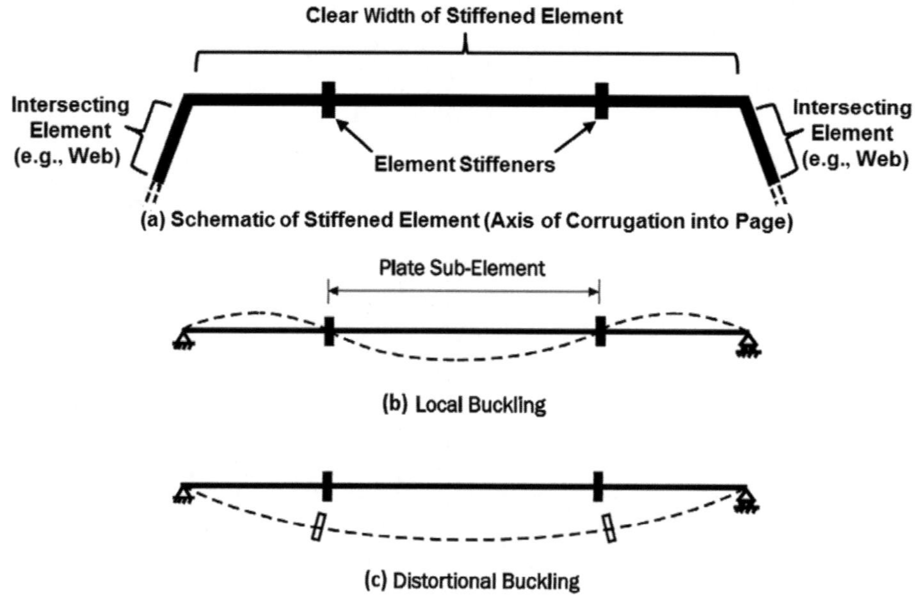


**Figure 8.** (*a*) Corrugation element with two intermediate stiffeners and intersecting elements, (*b*) local buckling in sub-elements adjacent to stiffeners, and (c) distortional buckling of full-width corrugation element including two intermediate stiffeners (adapted from AISI [2]).

sub-element local buckling, the edge support coefficient, $k_{stf}$, is determined based on stiffener and parent element geometry, and is determined as $k_{loc}$ from the AISI method Equation B5.1.1-1, divided by 4. This stiffener edge support coefficient, $k_{stf}$, will be 4.0 for one intermediate stiffener, indicative of the stiffener being adequate to support the sub-elements, similar to the element being supported by an intersecting element such as a web.

When considering distortional buckling of the full-width, stiffened element, stiffener geometric properties, including its moment of inertia about the mid-thickness of the parent element, are entered into the AISI calculations. A distortional buckling edge support coefficient, $k_d$, is determined by AISI Equation B5.1.1-2. The value of $k_d$ will generally be greater than 4.0 as the stiffener provides additional buckling stability compared with a full-width unstiffened element, which would have $k$ = 4.0 per AASHTO LRFD.

*Example Idealization with New Method for Stiffened Elements.* The gross and idealized sections from Figure 5 have been adapted by adding a stiffener to the crest in

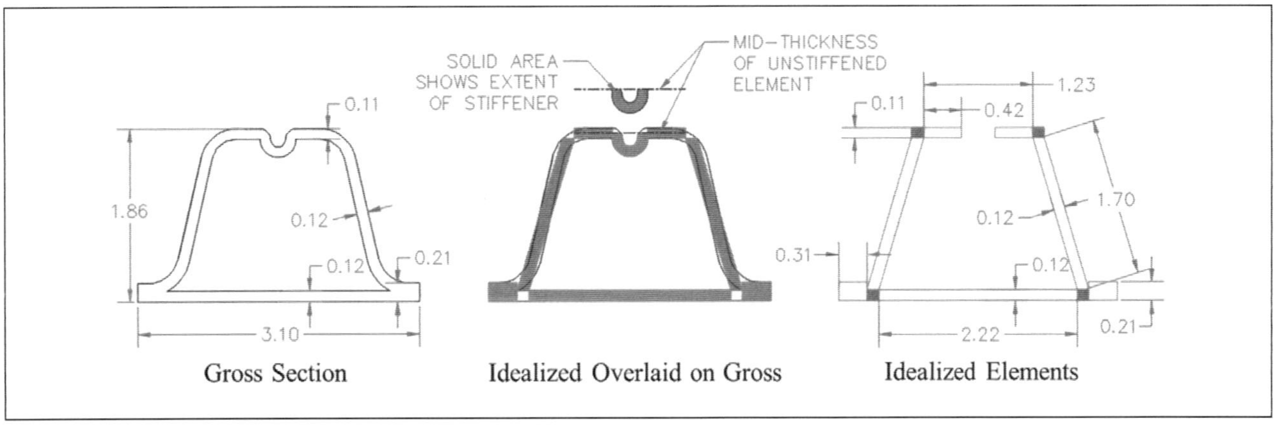


**Figure 9.** Corrugation idealization utilizing new stiffener evaluation method for crest with stiffener.

Figure 9. The stiffener and parent element geometric properties are entered into the AISI stiffener calculations. The Figure 9 stiffener is found to have an edge support coefficient, $k_{stf}$, of 4.0 for the 0.42 in. clear width sub-elements to either side of the stiffener. The distortional buckling edge support coefficient, $k_d$, is found to be 10.07 when considering distortional buckling of the 1.23 in. wide, stiffened crest element.

The effective area of the stiffened crest is calculated as the minimum effective area for distortional buckling or for sub-element local buckling. For distortional buckling, the total area of the crest of 0.14 in.$^2$ (clear width of 1.23 in. and thickness of 0.11 in.) with $k = 10.07$ results in an effective width and area of 1.27 in. and 0.14 in.$^2$, respectively (the area of the stiffener is neglected as it is at the center of the element and is first to become ineffective, but is accounted for in the increase in $k$ from the nominal 4.0 value). For sub-element local buckling, the total area considered includes the area of the stiffener (0.061 in.$^2$) plus the sub-elements to either side of the stiffener (2 * 0.42 in. * 0.11 in. = 0.092 in.$^2$), for a total area of 0.15 in.$^2$ This area is fully effective based on the stout width-to-thickness ratios of the sub-elements and restraint provided by the stiffener. Distortional buckling, therefore, is the controlling behavior and the crest effective area is 0.14 in.$^2$.

## *Comparison of Effective Areas for Corrugations in Figures 5, 7, and 9*

Effective areas for the three cross-section idealizations above were calculated assuming a limiting compressive strain, $\varepsilon_{yc} = 3.7\%$, for polypropylene (PP) and are presented in Table 1. Gross area for one period is presented, as is idealized area including red corners shown in the figure 5, 7, and 9 idealizations for comparison purposes.

All three cross-sections end up with a similar ineffective percentage of the gross area. The lowest ineffective percent of gross area occurs using the curved element idealization. This is partially attributed to increased effectiveness of the webs, which were shortened slightly in the Figure 7 idealization to end at the start of the curved crest, compared with the longer webs conservatively used in the other two idealizations.

## Summary

AASHTO LRFD effective width calculations and element idealization requirements were reviewed and found to have room for improvement and clarification. To have representative local buckling effective areas, idealized section clear widths and thicknesses must be representative of their gross-section element counterparts. Since the ineffective area of each element is summed and subtracted from the corrugation gross area, there is no requirement in the specification or calculations to determine idealized section total area and have it match gross area.

Background documents upon which the AASHTO LRFD effective width calculations are based include additional recommendations such as matching idealized and gross-section moments of inertia, centroids, and extreme fiber distances. In reality, these recommendations are not required by the specification or the calculations.

Results from a finite element study of curved elements show element curvature has a substantial benefit when evaluating effective area, allowing the element to reach a much higher load before buckling than a flat element through the same two end points. This is not considered in AASHTO LRFD or the contributing NCHRP reports. A conservative method to idealize curved elements as flat plates, based on the results of this finite element study, is to set the element clear width as the clear distance between intersecting elements, then increase the thickness of the flat idealized representation of the

**Table 1.** Nominal and Idealized Corrugation Properties for Corrugation with Flat Crest Including Effective Areas Assuming $\varepsilon_{yc}$ = 3.7% for PP

| Idealization method and figure number | Gross area, $A_g$ (in.$^2$) | Total idealized area,[(a)] $A_I$ (in.$^2$) | Idealized element | Clear width (in.) | Thickness (in.) | Ineffective width (in.) | Ineffective area (in.$^2$) | Ineffective area (% $A_g$) | Effective area (in.$^2$) |
|---|---|---|---|---|---|---|---|---|---|
| Flat crest idealization, Figure 5 | 1.00 | 1.00 | Flat crest | 1.23 | 0.11 | 0.36 | 0.33 | 33% | 0.67 |
| | | | Valley | 2*0.31 = 0.62 | 0.21 | 0.00 | | | |
| | | | Web | 1.70 | 0.12 | 0.67 | | | |
| | | | Interior wall | 2.22 | 0.12 | 1.16 | | | |
| Curved crest idealization, Figure 7 | 1.03 | 1.03 | Curved crest | 1.45 | 0.14 | 0.29 | 0.29 | 28% | 0.74 |
| | | | Valley | 2*0.31 = 0.62 | 0.21 | 0.00 | | | |
| | | | Web | 1.49 | 0.12 | 0.50 | | | |
| | | | Interior wall | 2.22 | 0.12 | 1.16 | | | |
| Stiffened crest idealization, Figure 9 | 1.02 | 1.02 | Stiffened crest | 1.23 | 0.11 | 0.39 | 0.34 | 33% | 0.68 |
| | | | Valley | 2*0.31 = 0.62 | 0.21 | 0.00 | | | |
| | | | Web | 1.70 | 0.12 | 0.67 | | | |
| | | | Interior wall | 2.22 | 0.12 | 1.16 | | | |

[a]Total idealized area includes red shaded areas in figure 5, 7, and 9 idealizations.

curved element so its area matches the total area of the curved element.

Updates to the AISI code include a method for evaluation of stiffened elements. This is accomplished by calculating edge support coefficients for the stiffened parent element considering distortional buckling, and for the sub-elements adjacent to the stiffeners considering sub-element local buckling. These edge support coefficients are able be used directly in the AASHTO LRFD equations to determine the controlling effective area for a stiffened element.

Finally, example corrugations were idealized, and effective areas calculated for a corrugation with flat crest, similar corrugation with curved crest utilizing the proposed curved member idealization method, and similar corrugation with a crest stiffener utilizing the AISI stiffener evaluation approach.

Though these methods are theoretically conservative, it may be possible to refine them through a more detailed evaluation of the results, which could include comparison of effective areas determined from stub compression with effective areas calculated by these methods.

## Author Contributions

The two authors, BB and JB, contributed to the study conception and design, analysis, interpretation of results, and draft manuscript preparation. Both authors reviewed the results and approved the final version of the manuscript.

## References


1. American Association of State Highway Transportation Officials. *AASHTO LRFD Bridge Design Specifications*, 7th ed. Washington, D.C.: AASHTO, 2014.
2. AISI S100-07. *North American Specification for the Design of Cold-Formed Steel Structural Members*. American Iron and Steel Institute, Washington, D.C., 2007.
3. McGrath, T. J., and V. E. Sagan. *NCHRP Report 438: Recommended LRFD Specifications for Plastic Pipe and Culverts*. Transportation Research Board of the National Academies, Washington, D.C., 2000.
4. McGrath, T. J., I. D. Moore, and G. Y. Hsuan. *NCHRP Report 631: Updated Test and Design Methods for Thermoplastic Drainage Pipe*. Transportation Research Board of the National Academies, Washington, D.C., 2009.
5. American Association of State Highway Transportation Officials. *AASHTO LRFD Bridge Design Specifications*, 6th ed., with 2010 Interim Updates. Washington, D.C.: AASHTO, 2010.
6. American Association of State Highway Transportation Officials. AASHTO T341-10 (2014), *Standard Method of Test for Determination of Compression Capacity for Profile Wall Plastic Pipe by Stub Compression Loading*. Washington, D.C.: AASHTO, 2014.
7. ABAQUS 3D Experience R2016x. *Dassault Systemes SIMULIA Corp*. Johnston, R.I., 2016.


*The Structures Section (AFF00) peer-reviewed this paper (18-06430).*

Article

TRR
JOURNAL OF THE TRANSPORTATION RESEARCH BOARD

# Development of Live Load Model for Strength II Limit State in AASHTO LRFD Design Specifications

Transportation Research Record
2018, Vol. 2672(41) 11–23

DOI: 10.1177/0361198118756874
journals.sagepub.com/home/trr
SAGE

**Peng Lou[1], Hani Nassif[1], and Paul Truban[2]**

The AASHTO LRFD Bridge Design Specifications defines Strength II limit state for agencies to consider the load combination by owner-specified special design vehicles, evaluation permit vehicles, or both. The configuration and characteristics of permit vehicles vary from state to state. In addition, the code calibration process performed in 1994 for the development of the live load factors was applied only to the Strength I limit state. In New Jersey, the design permit vehicle was not developed based on actual permit records or weigh-in-motion (WIM) data. Recently, with the development of permit-issuing management and WIM technology, there is a need to evaluate the effectiveness of design permit vehicles. This study aims to develop a live load model for the assessment of Strength II limit state for New Jersey Department of Transportation (NJDOT). Five years of permit vehicle records are provided by NJDOT for the development of the live load model. The distribution of Gross Vehicle Weight is best described as the Generalized Extreme Value distribution. Load effects are simulated for different span lengths. The mean and standard deviation (SD) of the 75-year maximum loads are predicted using different extrapolation approaches. The results show that NJDOT Design Permit Vehicle provides stable mean and SD of bias ratios at 75-year level. In comparison with the current AASHTO live load factor of 1.35, the averages of the bias ratios at the 75-year level are found to be 1.31, 1.23, and 1.16 for the positive moment, shear, and negative moment, respectively.

Overweight vehicles are commonly considered to affect both the structural safety and deterioration of bridge structures. In the 1960s, Roy Jorgensen and Associates (*1*) provided summaries of permit practices in the US, examining the discrepancy in permit issuance among different agencies, such as permit limits, application and issuance methods, routing issuance definitions, and types. The Transportation Research Board (TRB) (*2*) and Federal Highway Administration (FHWA) (*3*) initiated the work of seeking widely accepted provisions and procedures for permit issuance. Lately, Farrar et al. (*4*) in 2014 performed a scan report under NCHRP 20-68A about the advances in state Department of Transportation (DOT) superload permit process and practices.

One of the major issues of operating overweight vehicles is the deterioration of bridge structures. Because of the increasing number of permits issued, state DOTs have already initiated studies to help quantify the economic impact of overweight vehicles. Nassif et al. performed a comprehensive review of overweight studies conducted by various agencies in North America, including Arizona, New York State, Connecticut, Indiana, Louisiana, Ohio, Wisconsin, and South Carolina (*5*). Lou et al. presented studies about the impact of overweight vehicles on deteriorations of bridge decks and superstructures with a focus on New Jersey data (*6,7*). They found that the deterioration of concrete decks and prestressed concrete girders are highly affected by overweight vehicles, while steel girders are not as sensitive.

Another important aspect of operating permit systems is the structural safety of the bridge network. AASHTO LRFD defines Strength II limit state for agencies to consider the load combination relating to the use of bridges by owner-specified special design vehicles, evaluation permit vehicles, or both (*8*). Therefore, different design permit vehicles are used by different transportation agencies to consider regional-specific overweight vehicles. From the design manuals and specifications mandated by agencies, the Strength II limit state usually controls the design (*9–11*), while some

[1]Rutgers Infrastructure Monitoring and Evaluation (RIME) Group, Department of Civil and Environmental Engineering, Rutgers, the State University of New Jersey, Piscataway, NJ
[2]Bureau of Freight Planning and Services, New Jersey Department of Transportation, 1035 Parkway Avenue, Trenton, NJ

**Corresponding Author:**
Address correspondence to Peng Lou: penglou@rutgers.edu

states do not typically use the Strength II limit state (*12*). In addition, some states also incorporate the design permit vehicle in service limit state, as these heavy permit vehicles might cause serviceability problem for bridges, such as cracking and deflection. For example, NYSDOT and NJDOT use their design permit vehicle for "Service III" limit state as a design load case in addition to the AASHTO HL93 live load (*10,11*). A recent study conducted by Wassef et al. recommended a live load factor of 1.0 instead of 0.8 for Service III limit state in designing prestressed concrete bridges (*13*). Despite the importance of considering the permit vehicles during bridge design and evaluation, very few studies were found to calibrate or validate the local design permit vehicles. Early in 1987, Ghosn and Moses presented a study of the calibration of bridge rating equations for overload permit (*14*). Fu and Moses developed an overload checking method for highway bridges based on limited data of permit overloads from Ohio DOT (*15*). Furthermore, Fu and Hag-Elsafi covered several aspects of permit vehicles including the development of live load models, reliability models, and proposed permit load factors for overload checking (*16,17*). However, these studies focused mainly on the evaluation of highway bridges instead of design. A complete picture of the load effects caused by permit overweight vehicles was not fully investigated either.

In the United States (LRFD AASHTO) and Canada (Ontario Highway Bridge Design Code), Nowak started the work on developing the truck live load models based on truck survey data collected in Ontario in 1975 (*18,19*). Extreme 75-year loads were determined by extrapolation of the upper tail of the distribution of load effects. The results constituted the major development of live load models for Strength I limit state. Wassef et al. performed the calibration of service limit states for concrete bridges for AASHTO LRFD (*13*). Weigh-in-motion (WIM) data were used to derive the live load effect on bridges. Extremely heavy trucks ranging from 0.01% to 0.22% of the truck population were taken out of the database when considering service limit states.

During the last two decades, with the development of permit-issuing management and WIM technology, there is a need to evaluate the effectiveness of current design permit vehicle and develop a validated permit vehicle based on available data. This study presents an approach for the development of live load model for design permit vehicle. Design permit vehicles currently used by various transportation agencies are summarized. More than 208,487 Permit vehicle registration records from 2010 to 2015 are provided by NJDOT. The data are used as a basis for load extrapolation. The permit data are filtered for the oversize vehicles using Federal Bridge Formula. The distribution of Gross Vehicle Weight (GVW) is best described as a Type II Extreme Value distribution (Fréchet). Load effects, including simple span positive moment, simple span shear force, and continuous span negative moment are calculated for different span lengths ranging from 20 ft to 200 ft with an increment of 20 ft. The results can be served as an input for the load side when performing reliability calibration.

## Permit Vehicles from Various Agencies

Figure 1 shows the permit vehicles currently used by New Jersey, Pennsylvania, New York State, Wisconsin, California, Connecticut, and Virginia. Wisconsin and Virginia do not require a Strength II limit state during the design, but they do require bridges to be checked for the permit load rating using their permit vehicles. New Jersey and New York State not only require Strength II limit state but also require the use of permit vehicles in service limit states. State DOTs usually developed their state permit vehicle based on the envelope of load effects created by issued routine permit vehicles (*20,24*). However, this does not share the same philosophy of reliability-based LRFD design.

## Data Processing and Results of Permit Vehicle Records

In this section, 208,487 permit vehicle registration records are used. Figure 2 shows the number of the permits issued for each month from September 2010 to July 2015. The data do not include full-month data for August 2010 and August 2015; therefore, these two months are not included in Figure 2. From the figure, Januarys and Februarys usually have the lowest number issued. Overall, the number of permits has an average annual increase of 7.66%. Figure 3 shows the distribution of vehicles based on the number of axles. Five, six, seven, and eight-axle vehicles are found to be dominant, which constitute 94.06% of total records. Figure 4 shows the prevailing vehicle types from the permit record. The GVW and wheelbase of the maximum vehicle are 848 kips and 271.7 ft, respectively. Figure 5 shows the density distribution of GVW of collected permit vehicles along with the Normal, Lognormal, and Generalized Extreme Value (GEV) distribution approximations. To show the tail of the distribution, the Cumulative Density Function (CDF) is also plotted in Normal Probability Paper (NPP). Kolmogorov–Smirnov and Anderson Darling goodness of fit tests were performed for 63 types of distribution, such as GEV, Pearson 5, Pearson 6 lognormal, Beta, and Wakeby distribution. The GEV distribution is found to be the best fit from both tests. For the fitted GEV distribution, the location parameter is $\mu = 105.4$, scale parameter is $\sigma = 21.28$, and shape parameter is $\xi = 0.04526$, which indicates this is a Type II Extreme Value distribution, also known as Fréchet distribution.

## Live Load Effects and Load Extrapolation

Before simulating the load effects, Bridge Formula B is utilized to filter out the oversize vehicles. In addition, there are only 516 records (0.25%) that have a GVW over 250 kips

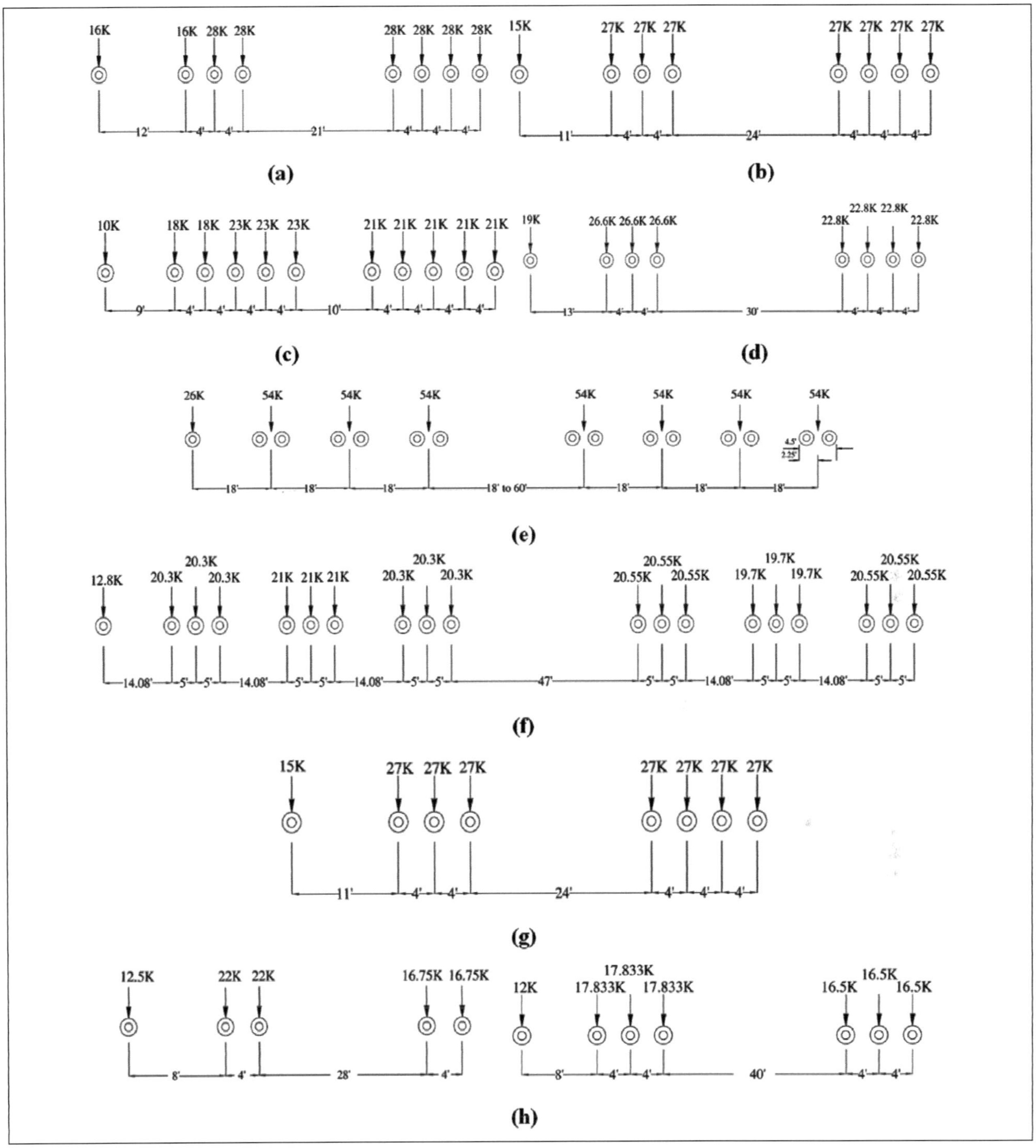


**Figure 1.** Permit vehicles from various agencies. (*a*) NJDOT design permit vehicle (*11*). (*b*) Pennsylvania design permit vehicle, P-82 (*20*). (*c*) NYSDOT design permit vehicle (*10*). (*d*) Wisconsin permit vehicle, Wis-SPV (*21*). (*e*) California design permit vehicle, P15 (*9*). (*f*) Connecticut department operating vehicle (design), P380 (*22*). (*g*) Connecticut department operating vehicle (design), P204 (*22*). (*h*) Virginia BP-90 and BP-115 (*23*).

and these vehicles are usually superload vehicles, which were distinguished from routine permits. The definition for superload varies from state to state. Some of the states (i.e., Idaho, South Dakota, and Wisconsin) do not use this term. For states that use the term superload, the definition can be based on dimensions, or weight, or a combination of both (*4*). Each permit vehicle record is simulated using simple beam analysis. Three types of load effects,

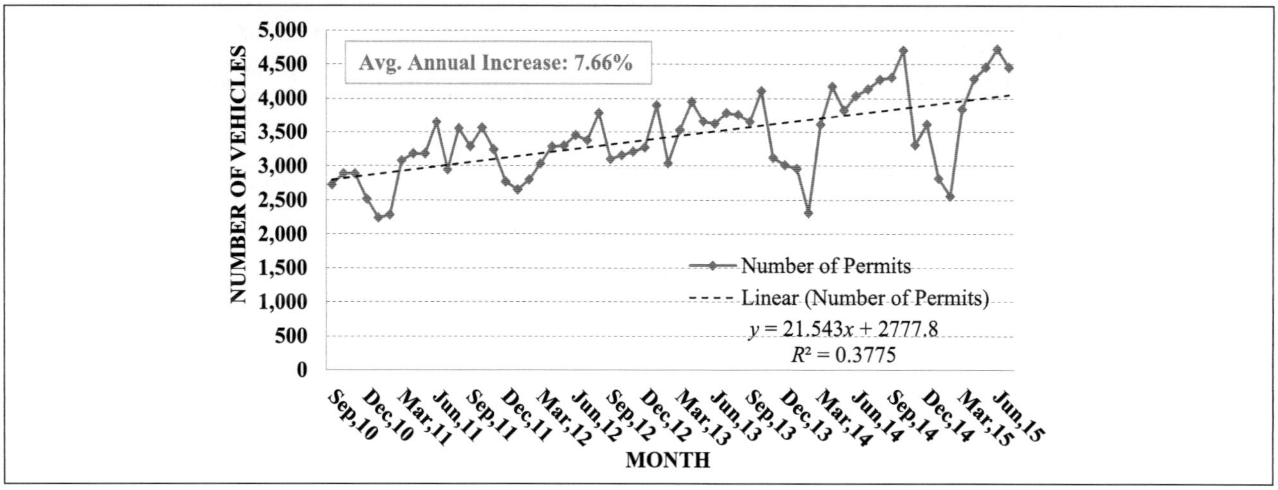


**Figure 2.** Number of permit issued from September 2010 to July 2015.

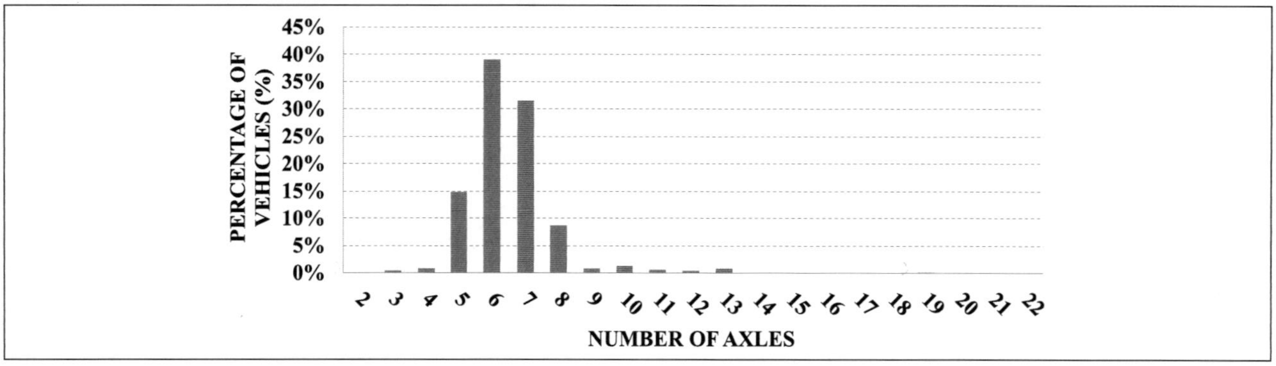


**Figure 3.** Number of axles for permit vehicles.

including maximum positive moment in simply supported beam (simple span), maximum shear force in simple span, and maximum negative moment in two equal span continuous beam, are calculated from simple beam analysis with different span lengths ranging from 20 ft to 200 ft with an increment of 20 ft. Selected results are shown in Figure 6 as examples. However, when developing or validating the design permit vehicles, one should not penalize all bridges for few extremely heavy superloads, which in some cases might even not be operated on the bridge network. Therefore, in this study, only permit vehicles that are below 250 kips are considered for the live load extrapolations. A total of 202,033 permit vehicle records are used in the future analysis.

When performing live load extrapolation for Strength I limit state, Nowak et al. used only the up tail of the whole truck population and assumed it followed normal distribution as only heavy trucks matter in that case (*24*). However, in this study, the permit data could be assumed as the up tail of the entire vehicle population. Therefore, all data points are used in this study to extrapolate the maximum load effects in 75 years. Extrapolation of the simple span positive moment for short span (40 ft), medium span (120 ft), and long span (200 ft) are shown as examples in Figure 7(*a*), (*b*), and (*c*), respectively. Two types of extrapolation approaches are used. The first approach is to fit all truck data into three distributions—normal, lognormal and GEV distribution—described here as approximation approach herein, while the other approach is to only fit the data point in NPP to the linear and lognormal line, described as plot fit approach. Nowak et al. use the plot linear fit approach (*24*). The first approach first assumes the distribution, then the characteristic parameters of the assumed distribution are calculated using all bias ratio data. The distribution with calculated parameters is then plotted in the NPP. This approach applies same weight to each vehicle, therefore is more stable as regards the extrapolation results. It is less sensitive to extreme loads as the number of extreme loads is very few. The second method, which is more affected by the outliers, only fits the line to the plots. Both Plot Lognormal Fit and Lognormal Approximation give close extrapolation results and describe the entire distribution very well. The coefficient of determination is higher than 0.90 in all cases for Plot Lognormal Fit, and higher than

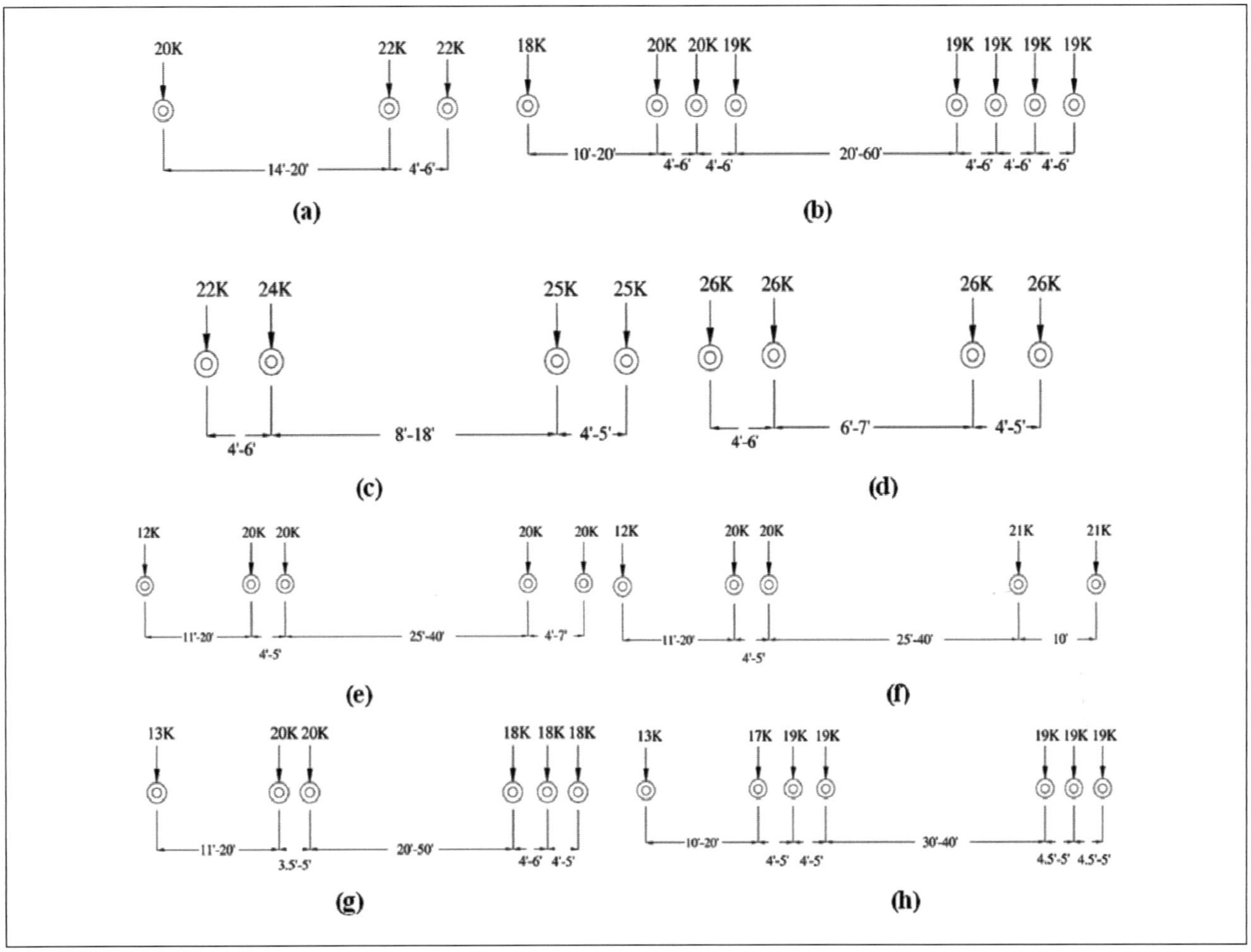


**Figure 4.** Prevailing permit vehicle configuration from the records: (*a*) three axles (499 out of 917, 54.4%); (*b*) eight axles (15,571 out of 17,991, 86.5%); (*c*) four axles Type I (802 out of 1834, 43.7%); (*d*) four axles Type II (122 out of 1834, 6.7%); (*e*) five axles Type I (15,303 out of 30,977, 49.4%); (*f*) five axles Type II (3186 out of 30,977, 10.3%); (*g*) six axles (66,737 out of 81,376, 82%); (*h*) seven axles (51,057 out of 65,760, 77.6%).

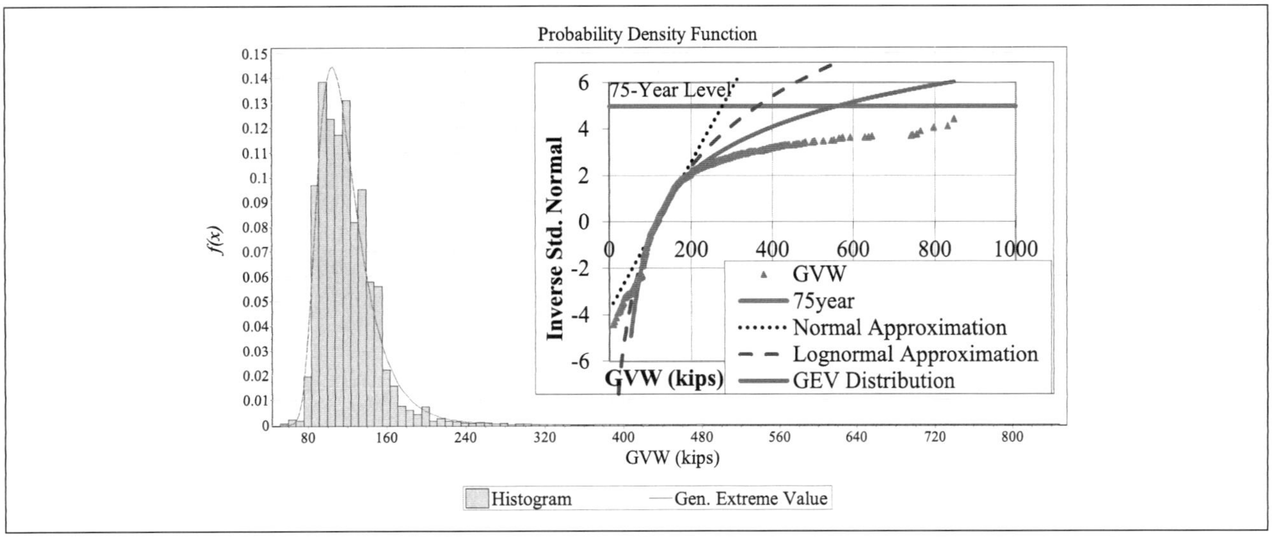


**Figure 5.** GVW distribution of permit records.

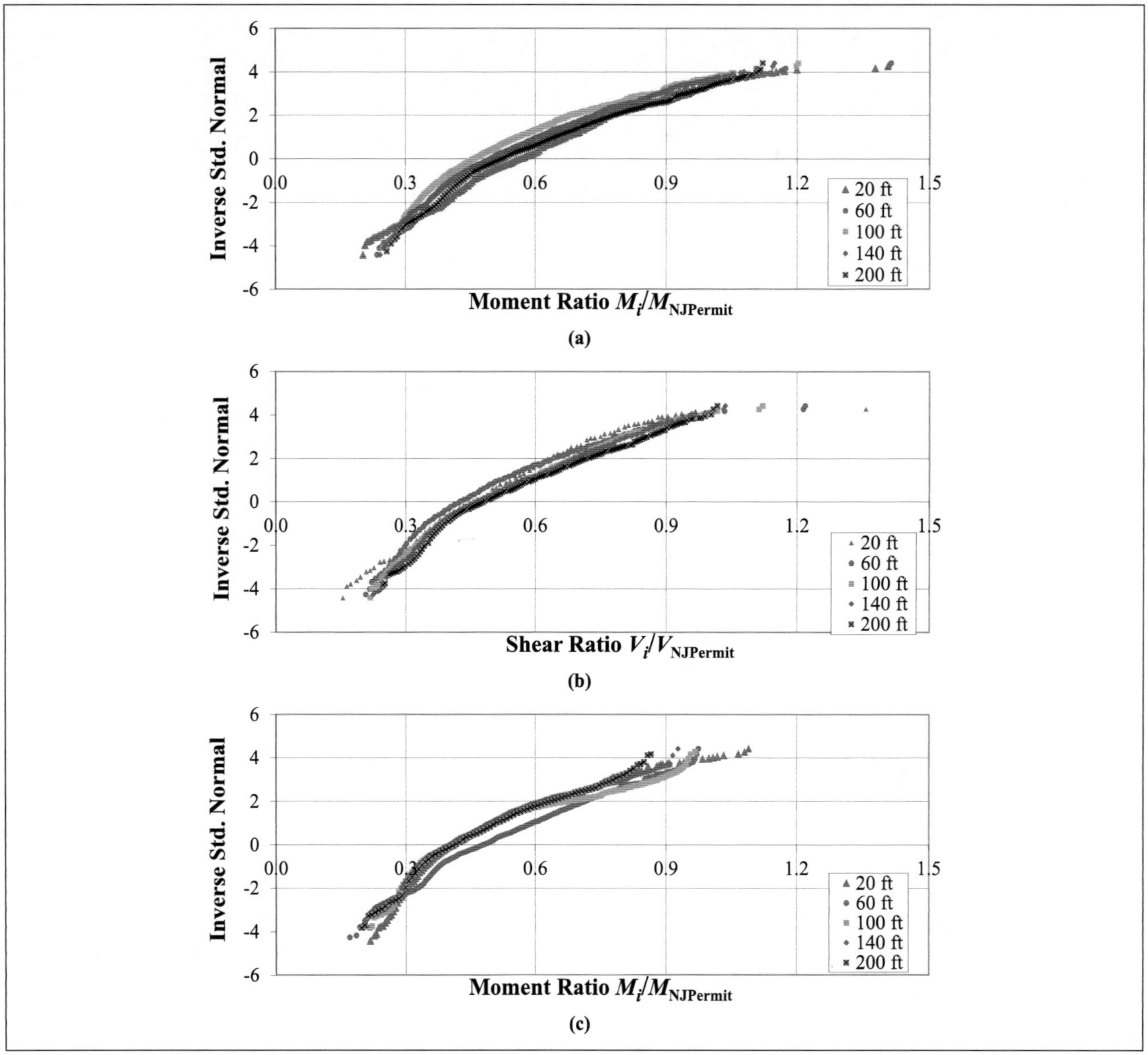


**Figure 6.** Load effects caused by permit records: (*a*) simple span positive moment; (*b*) simple span shear; (*c*) continuous span negative moment.

0.97 in all cases for Lognormal Approximation. The Plot Linear Fit and Normal Approximation are inferior in describing the distribution. Although the GEV Type II distribution describes the GVW distribution well, the GEV distribution approximation does not describe the up tail of load effect well. This study preferred the Lognormal Approximation and Plot Lognormal Fit. Another reason for the preference is that these two extrapolations give very similar results when using either 5-years permit data or single year data (2011), as shown in Figure 7.

For the following analysis results, the Lognormal Approximation is used, as for some cases where some extreme heavy outliers exist Plot Lognormal Ft would give fluctuated results. Figure 8 gives the mean and standard deviation (SD) of bias ratios for the predicted 75-year maximum loads for different permit vehicles. In those states that have multiple design permit vehicles, the maximum load effects of multiple vehicles are used to calculate the bias ratio. For the New Jersey design permit vehicle, the means and SD of bias ratios are consistent for different span lengths. This means the current NJDOT Design Permit Vehicle represents the local permit loads well, and there is no need to change the configuration of the design permit vehicle. The mean of bias ratio varies between 1.0 and 1.5, and the SD are about 0.25. Considering the current live load factor, 1.35 for design permit load, in Strength II limit state, the NJDOT Design Permit Vehicle

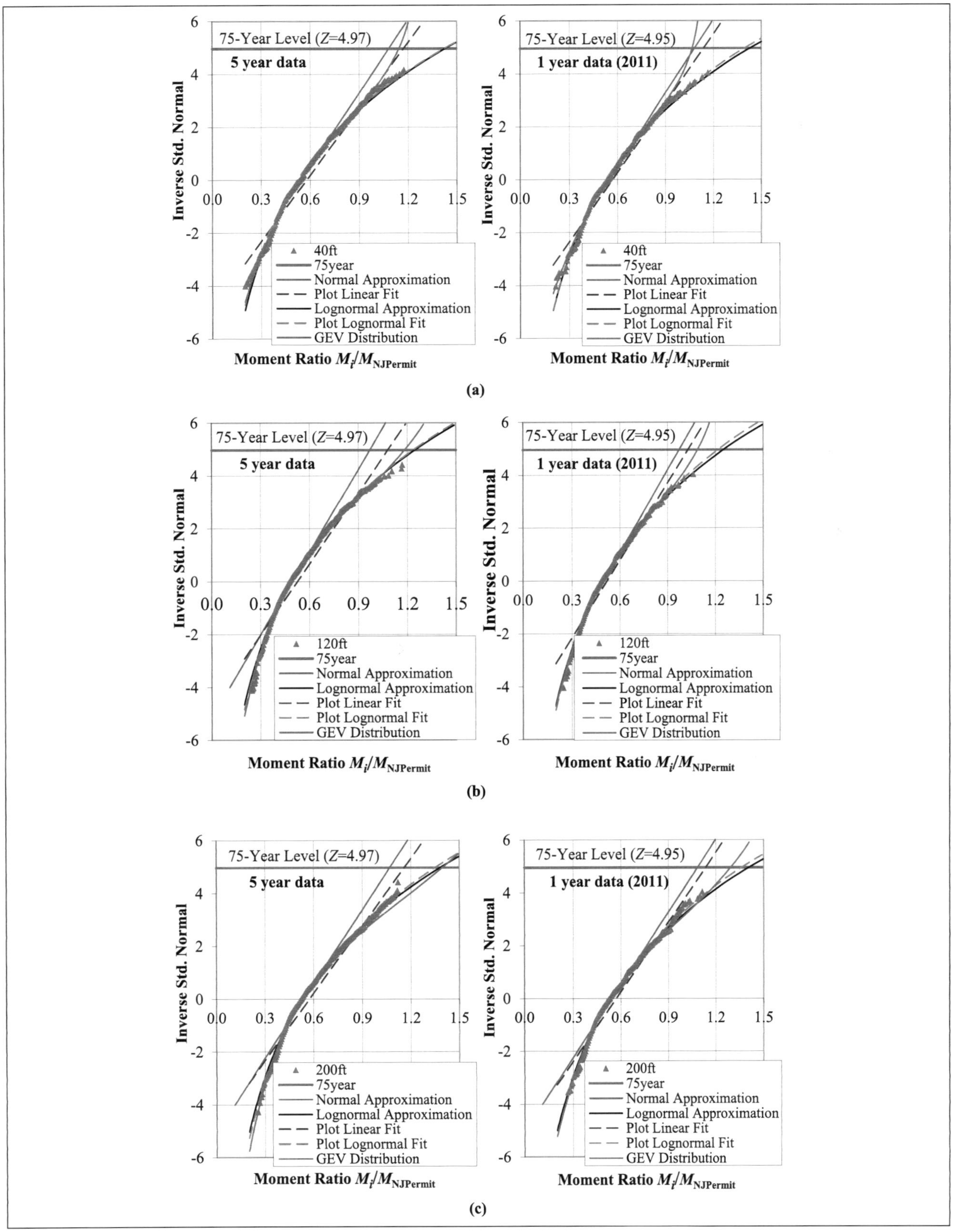


**Figure 7.** Extrapolation of load effects: (*a*) short span (40-ft) single span positive moment; (*b*) medium span (120-ft) single span positive moment; (*c*) long span (200-ft) single span positive moment.

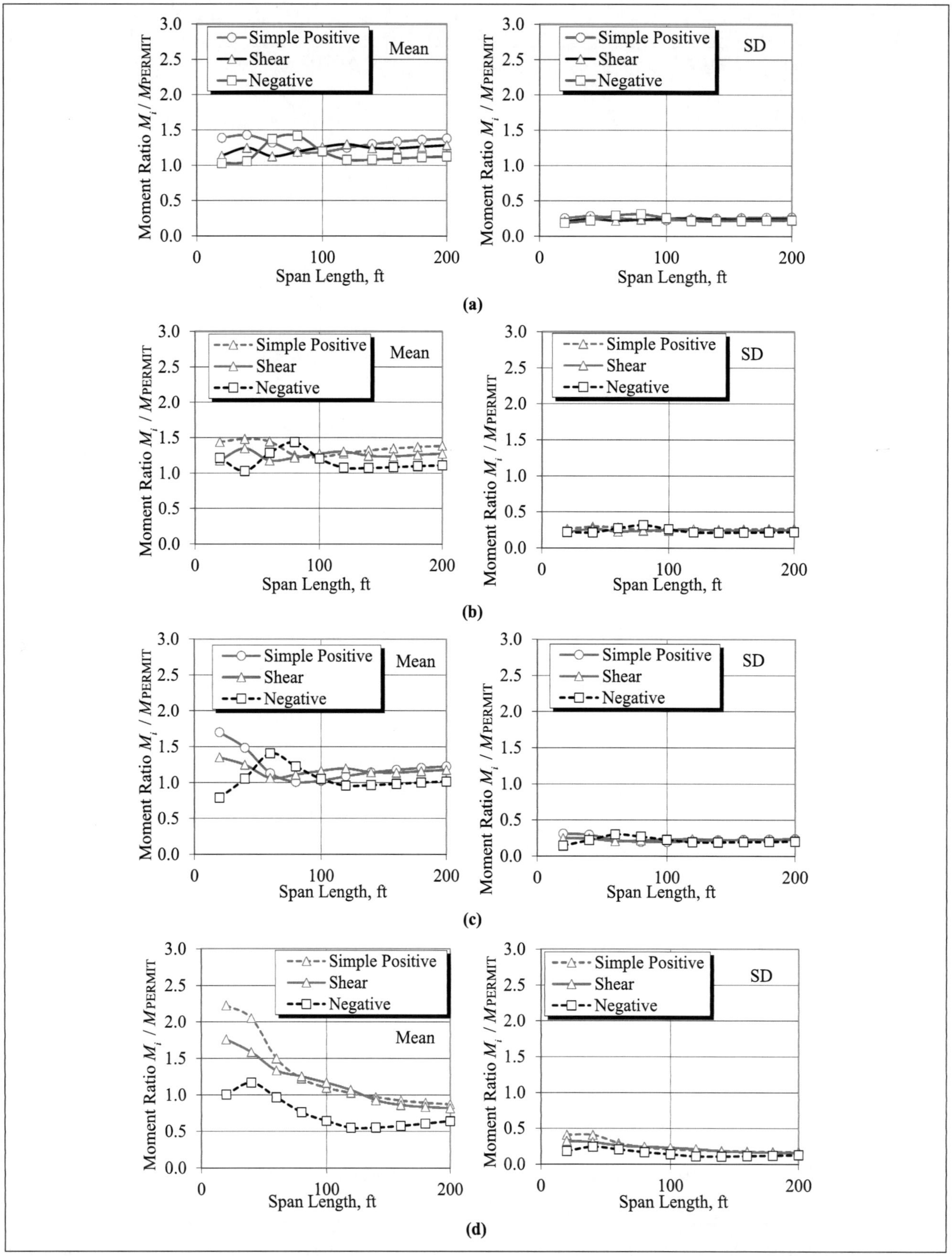
Simple Positive
Shear
Negative
Mean
SD
Moment Ratio $M_i$ / $M_{PERMIT}$
Span Length, ft
0
100
200
0.0
0.5
1.0
1.5
2.0
2.5
3.0
(a)
(b)
(c)
(d)

(continued)

**Figure 8. (continued)**

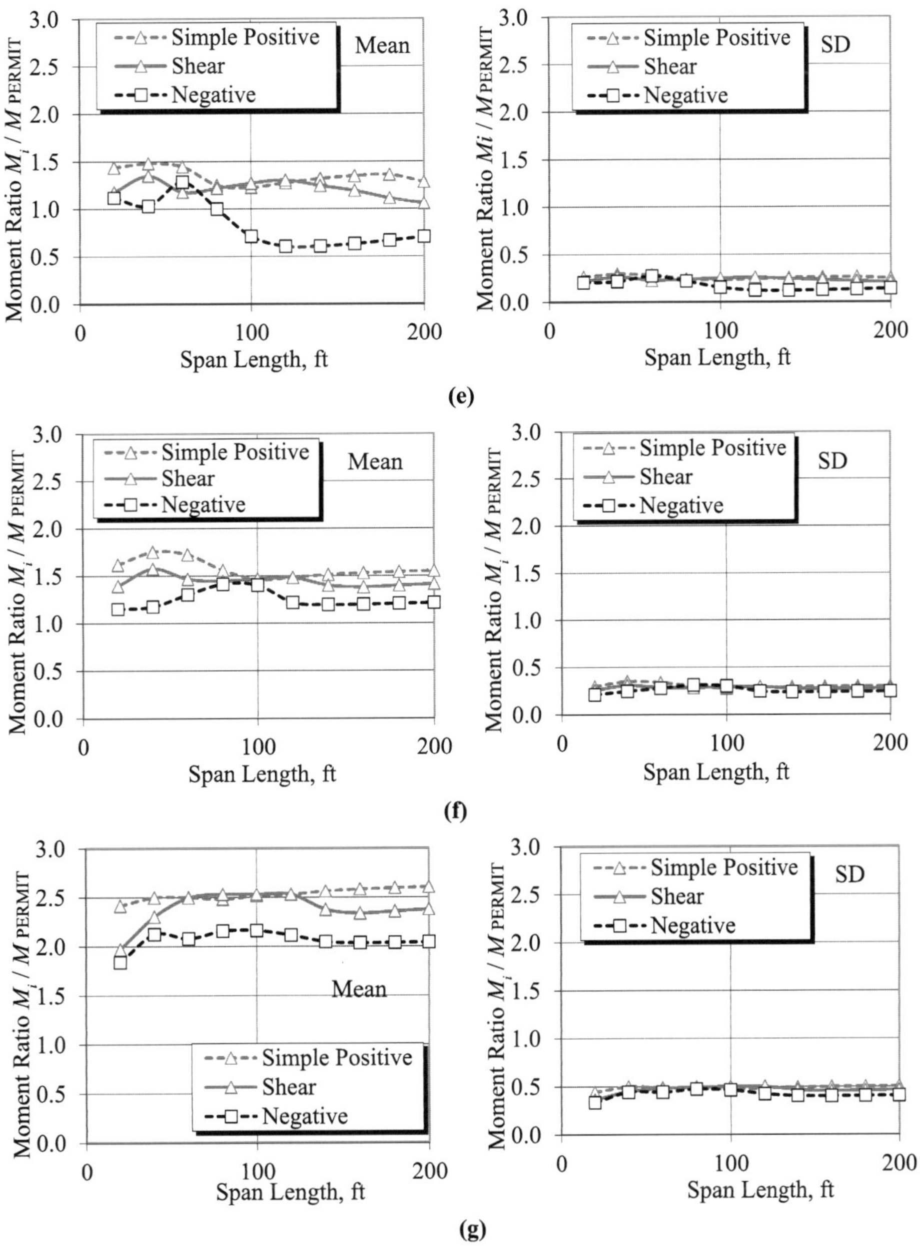


**Figure 8.** Bias ratios for different permit vehicles: (*a*) New Jersey; (*b*) Pennsylvania P-82; (*c*) New York State; (*d*) California P15; (e) Connecticut (*f*) Wisconsin; (g) Virginia.

gives close bias ratios based on the results. However, if the load extrapolation results are inconsistent over different span lengths, such as shown in Figure 8(*d*), a new configuration should be proposed to represent the permit load better. If the bias ratio over span length shows consistent results but with bias ratio higher than 1.35, such as shown in Figure 8(*f*), the agency might consider increasing the weight of its design permit vehicle while keep the configuration the same. In addition, it was found the coefficient of variations (CoV), which is about 20%, is higher than in previous calibration works (*13,18,19,24*). Therefore it is hard to judge whether the design permit vehicle is effective without performing a complete reliability calibration. The Pennsylvania design permit vehicle also gives stable results ranging between 1.0 and 1.5. Among the other permit vehicles, New York State and Wisconsin give relative stable bias ratios except for short span. California gives a higher bias ratio, which is greater than 2.0, for short span, then the bias ratio decreases below 1.0 as span length

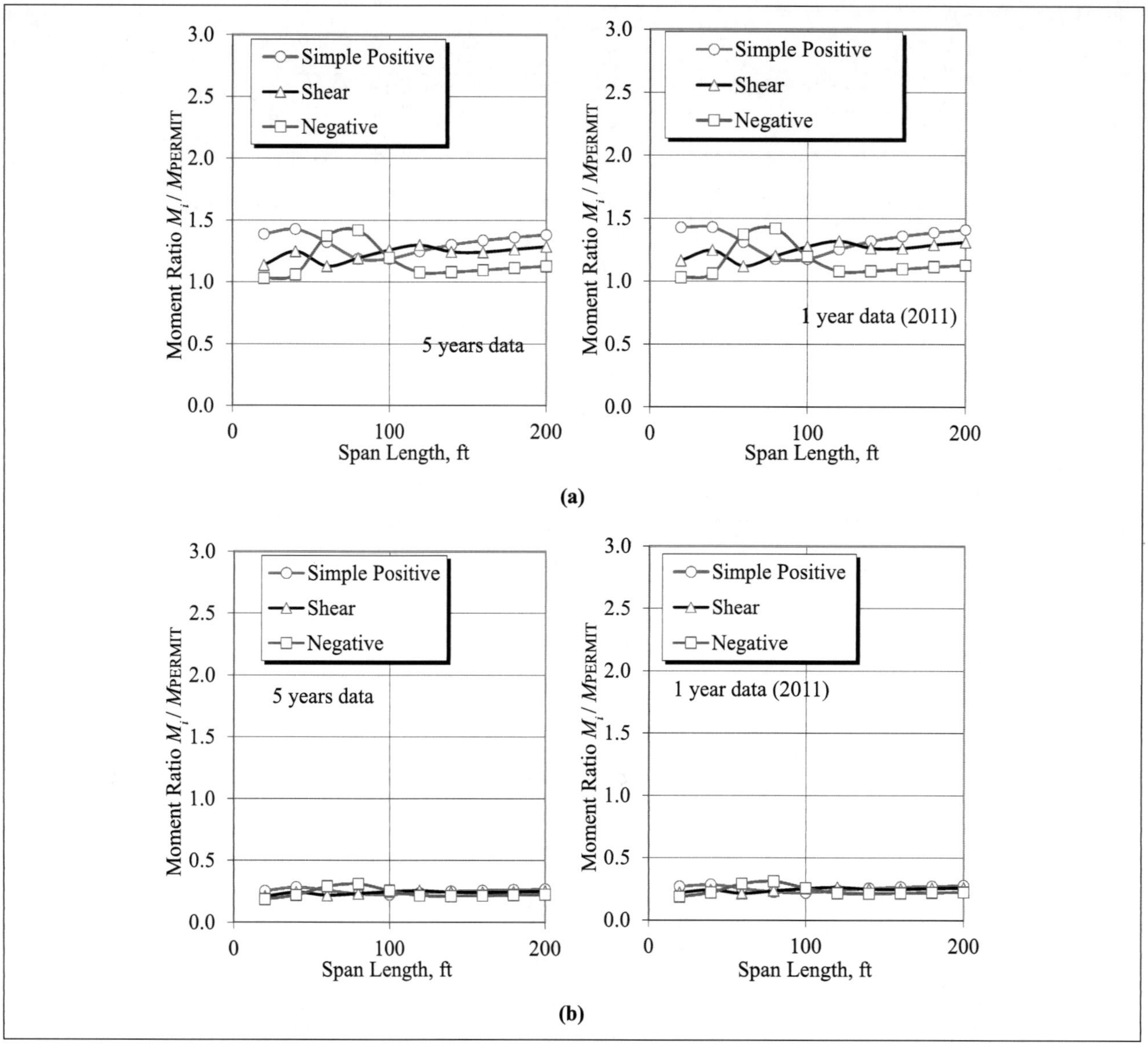


**Figure 9.** 75-year maximum extrapolation using 5-year and 1-year data: (*a*) mean; (*b*) SD.

increases. Connecticut permit vehicles give stable results except for the long span negative moment. The Virginia design permit vehicle is relatively light compared with others, as the bias ratios are mostly between 2.0 and 2.5. Assuming these permit vehicles represent the local permits, permit vehicles in New Jersey seem to have a much higher axle weight than other states, as short spans are usually control by the axle weight. It is important to mention that permit vehicles are regionally specific; both the weight distribution and configuration vary from state to state. Therefore, the design permit vehicle profile should be different based on local permit vehicles, and this is why the bias ratios for different design permit vehicles give large differences.

Figure 9 shows the extrapolation results both using both 5-year data and 1-year data (2011) using the same extrapolation method, Lognormal Approximation. The results are almost the same and the difference is within 5%. This comparison shows the extrapolation method used is stable and gives reliable results.

## Statistical Parameters for Strength II Limit State and Service Limit States

For service limit states, we are not concerned about the 75-year design life maximum load effects. Figure 10 shows the mean of bias ratio at the 75-year, 10-year, 5-year, and 1-year level. Meanwhile, the statistical parameters, including mean and CoV of bias ratio at the different levels, are presented in Table 1, Table 2, and Table 3 for the simple span positive moment, simple span shear, and continuous span negative moment, respectively. This can serve directly as

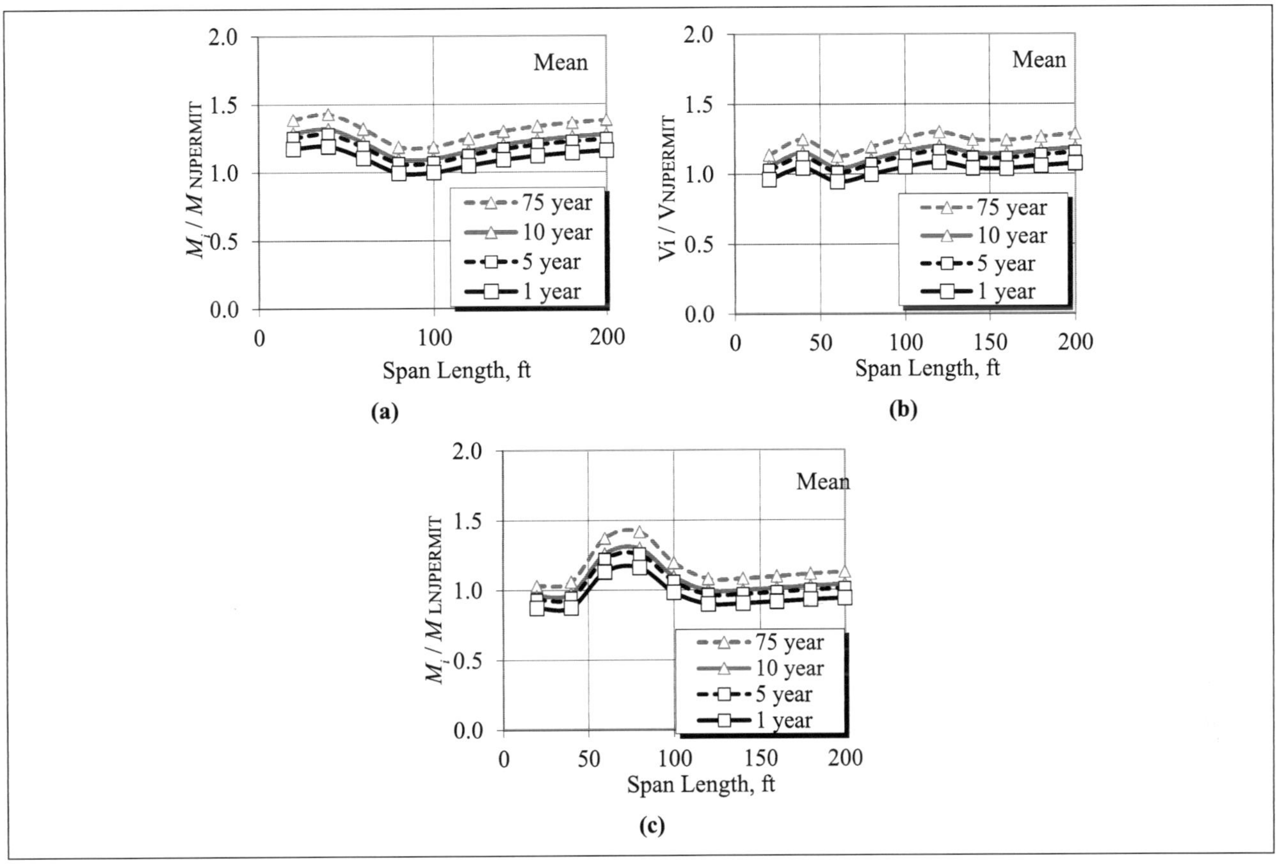


**Figure 10.** Bias ratio for 75-year, 10-year, 5-year, and 1-year maximum load: (*a*) simple span positive moment; (*b*) simple span shear; (*c*) continuous span negative moment.

**Table 1.** Statistical Parameters of Live Load for Simple Span Positive Moment

| | 75 year | | 10 year | | 5 year | | 1 year | |
|---|---|---|---|---|---|---|---|---|
| Span length (ft) | λ | CoV | λ | CoV | λ | CoV | λ | CoV |
| 20 | 1.39 | 18% | 1.29 | 18% | 1.25 | 18% | 1.17 | 18% |
| 40 | 1.43 | 20% | 1.32 | 20% | 1.28 | 20% | 1.19 | 20% |
| 60 | 1.32 | 20% | 1.22 | 20% | 1.18 | 20% | 1.10 | 20% |
| 80 | 1.19 | 19% | 1.10 | 19% | 1.07 | 19% | 0.99 | 19% |
| 100 | 1.19 | 19% | 1.10 | 19% | 1.07 | 19% | 1.00 | 19% |
| 120 | 1.25 | 19% | 1.15 | 19% | 1.12 | 19% | 1.05 | 19% |
| 140 | 1.30 | 19% | 1.20 | 19% | 1.17 | 19% | 1.09 | 19% |
| 160 | 1.33 | 19% | 1.23 | 19% | 1.20 | 19% | 1.12 | 19% |
| 180 | 1.36 | 19% | 1.26 | 19% | 1.22 | 19% | 1.14 | 19% |
| 200 | 1.38 | 19% | 1.28 | 19% | 1.24 | 19% | 1.16 | 19% |
| Max | 1.43 | 20% | 1.32 | 20% | 1.28 | 20% | 1.19 | 20% |
| Min | 1.19 | 18% | 1.10 | 18% | 1.07 | 18% | 0.99 | 18% |
| Average | 1.31 | 19% | 1.21 | 19% | 1.18 | 19% | 1.10 | 19% |

inputs for the reliability calibration of Strength II limit state and service limit states. It is important to mention that the average value of the mean of bias ratio at the 75-year level is 1.31, 1.23, and 1.16 for the positive moment, shear, and negative moment, respectively. These results are within the current live load factor for Strength II limit state, 1.35.

## Conclusions

This paper presents a study of the validation and development of design permit vehicle for the NJDOT Structural Design Manual. Permit vehicle registration records are used to perform live load extrapolation. The major conclusions are summarized as follows:

**Table 2.** Statistical Parameters of Live Load for Simple Span Shear

| | 75 year | | 10 year | | 5 year | | 1 year | |
|---|---|---|---|---|---|---|---|---|
| Span length (ft) | λ | CoV | λ | CoV | λ | CoV | λ | CoV |
| 20 | 1.14 | 18% | 1.05 | 18% | 1.03 | 18% | 0.96 | 18% |
| 40 | 1.25 | 20% | 1.15 | 20% | 1.12 | 20% | 1.04 | 20% |
| 60 | 1.13 | 19% | 1.04 | 19% | 1.01 | 19% | 0.94 | 19% |
| 80 | 1.19 | 20% | 1.10 | 20% | 1.07 | 20% | 1.00 | 20% |
| 100 | 1.26 | 20% | 1.16 | 20% | 1.13 | 20% | 1.05 | 20% |
| 120 | 1.30 | 20% | 1.20 | 20% | 1.16 | 20% | 1.08 | 20% |
| 140 | 1.24 | 20% | 1.15 | 20% | 1.12 | 20% | 1.04 | 20% |
| 160 | 1.24 | 19% | 1.15 | 19% | 1.11 | 19% | 1.04 | 19% |
| 180 | 1.26 | 20% | 1.17 | 20% | 1.13 | 20% | 1.06 | 20% |
| 200 | 1.28 | 20% | 1.19 | 20% | 1.15 | 20% | 1.07 | 20% |
| Max | 1.30 | 20% | 1.20 | 20% | 1.16 | 20% | 1.08 | 20% |
| Min | 1.13 | 18% | 1.04 | 18% | 1.01 | 18% | 0.94 | 18% |
| Average | 1.23 | 19% | 1.14 | 19% | 1.10 | 19% | 1.03 | 19% |

**Table 3.** Statistical Parameters of Live Load for Continuous Span Negative Moment

| | 75 year | | 10 year | | 5 year | | 1 year | |
|---|---|---|---|---|---|---|---|---|
| Span length (ft) | λ | CoV | λ | CoV | λ | CoV | λ | CoV |
| 20 | 1.03 | 18% | 0.96 | 18% | 0.93 | 18% | 0.87 | 18% |
| 40 | 1.06 | 21% | 0.97 | 21% | 0.94 | 21% | 0.87 | 21% |
| 60 | 1.37 | 21% | 1.26 | 21% | 1.22 | 21% | 1.13 | 21% |
| 80 | 1.42 | 22% | 1.30 | 22% | 1.26 | 22% | 1.16 | 22% |
| 100 | 1.19 | 21% | 1.10 | 21% | 1.06 | 21% | 0.98 | 21% |
| 120 | 1.08 | 20% | 0.99 | 20% | 0.96 | 20% | 0.90 | 20% |
| 140 | 1.08 | 20% | 1.00 | 20% | 0.97 | 20% | 0.90 | 20% |
| 160 | 1.10 | 20% | 1.01 | 20% | 0.98 | 20% | 0.91 | 20% |
| 180 | 1.11 | 20% | 1.03 | 20% | 1.00 | 20% | 0.93 | 20% |
| 200 | 1.13 | 20% | 1.04 | 20% | 1.01 | 20% | 0.94 | 20% |
| Max | 1.42 | 22% | 1.30 | 22% | 1.26 | 22% | 1.16 | 22% |
| Min | 1.03 | 18% | 0.96 | 18% | 0.93 | 18% | 0.87 | 18% |
| Average | 1.16 | 20% | 1.06 | 20% | 1.03 | 20% | 0.96 | 20% |

1. The distribution of Gross Vehicle Weight is best described as a Type II Extreme Value distribution (Fréchet);
2. Two types of extrapolation approaches, approximation and plot fit, are used for live load extrapolation at different year levels. Lognormal Approximation and Plot Lognormal Fit are found to describe the entire distribution very well. The Lognormal Approximation is used for the results since it provides the most stable for different span lengths;
3. Based on permit data, New Jersey and Pennsylvania permit vehicles give the most stable bias ratios for various span length;
4. The average value of the mean of bias ratios at the 75-year level is 1.31, 1.23, and 1.16 for the positive moment, shear, and negative moment, respectively.

The results are within the current live load factor, 1.35, for Strength II limit state. However, the effectiveness of the design permit vehicle remains unknown without performing a complete reliability calibration, as the CoV of bias ratio differs from the previous calibration works.

## Future Work

Reliability calibrations for Strength II limit state and service limit states using the design permit vehicle are needed to evaluate the effectiveness of design permit vehicles. Other aspects of design including dynamic impact, girder distribution factor for permit vehicles, and multiple presence events need to be investigated.

### Acknowledgments

This study is sponsored by NJDOT. The help and support from Andrew Ludasi in Bureau of Freight Planning and Services, Xiaohua "Hannah" Cheng, Ph.D., P.E and Eddy Germain in Bureau of Structural Design and Evaluation are greatly appreciated. The help and support from graduate students He Zhang and Dongjian Gao are also appreciated.

### References

1. Jorgensen, R. *Oversize-overweight Permit Operation on State Highways. NCHRP Rep. 80*, National Cooperative Highway Research Program, 1969.
2. Transportation Research Board. *Motor Vehicle Size and Weight Regulations, Enforcement, and Permit Operations. NCHRP Synthesis of Highway Practice 68*, Transportation Research Board, Washington, D.C., 1980.
3. Federal Highway Administration (FHWA). Overweight vehicles - Penalties & permits: An inventory of state practices for fiscal year 1989. *Rep. to Congress from the Secretary of Transportation, FHWAMC-91-003*, Washington, D.C., 1991.
4. Farrar, M., S. Becker, R. Braden, L. Gao, J. Honefanger, K. Keady, J. Mallard, and H. Nassif. *Advances in State DOT Superload Permit Processes and Practices* (No. NCHRP Project 20-68A), 2014.
5. Nassif, H., K. Ozbay, H. Wang, R. Noland, P. Lou, S. Demiroluk, D. Su, C. Na, J. Zhao, and M. Beltran. *Impact of Freight on Highway Infrastructure in New Jersey*. FHWA-NJ-2016-04, New Jersey Department of Transportation, 2016.
6. Lou, P., H. Nassif, D. Su, and P. Truban. Effect of Overweight Trucks on Bridge Deck Deterioration Based on Weigh-in-Motion Data. *Transportation Research Record: Journal of the Transportation Research Board*, 2016. 2592: 86–97.
7. Lou, P., H. Nassif, D. Su, and P. Truban. Impact of Overweight Trucks on the Service Life of Bridge Girders. *Transportation Research Record: Journal of the Transportation Research Board*. 2017. 2642: 103–117.
8. *AASHTO LRFD Bridge Design Specifications for Highway Bridges*, 6th ed. AASHTO, Washington, D.C., 2012.
9. Division of Engineering Services, Caltrans. *Bridge Design Practice (BDP)*, 4th ed. State of California Department of Transportation, 2015.

10. NYSDOT. *NYSDOT LRFD Bridge Design Specifications*. 2016.
11. NJDOT. *Design manual for Bridge & Structures*, 6th ed. NJDOT, Trenton, 2016.
12. WisDOT. *WisDOT Bridge Manual*. Wisconsin Department of Transportation, 2017.
13. Wassef, W. G., J. M. Kulicki, H. Nassif, D. Mertz, and A. S. Nowak. Calibration of AASHTO LRFD Concrete Bridge Design Specifications for Serviceability. *NCHRP Web Document*, 2014.
14. Ghosn, M., and F. Moses. Calibration of a bridge rating formula for overload permits. In *Bridges and Transmission Line Structures*, ASCE, 1987, pp. 15–25.
15. Fu, G., and F. Moses. Overload permit checking based on structural reliability. *Transportation Research Record*: *Journal of the Transportation Research Board*. 1991. 1290: 279.
16. Fu, G., and O. Hag-Elsafi. *Safety-based Bridge-overstress Criteria for Nondivisible Loads* (No. FHWA/NY/RR-97-168), 1997.
17. Fu, G., and O. Hag-Elsafi. Vehicular overloads: Load model, bridge safety, and permit checking. *Journal of Bridge Engineering*, Vol. 5, No. 1, 2000, pp. 49–57.
18. Nowak, A. S. Live load model for highway bridges. *Structural Safety* Vol. 13, No. 1–2, 1993, pp. 53–66.
19. Nowak, A. S. Load model for bridge design code. *Canadian Journal of Civil Engineering*, Vol. 21, No. 1, 1994, pp. 36–49.
20. Pennsylvania DOT. *Part 4 Structure Design Manual, April 2015 Edition*. Pennsylvania Department of Transportation, 2015.
21. WisDOT. *WisDOT Bridge Manual, January 2017*. Wisconsin Department of Transportation, 2017.
22. Connecticut DOT. *Bridge Design Manual, Revisions 2/11*. Connecticut Department of Transportation, 2003.
23. Virginia DOT. *Load Rating and Posing of Structures (Bridge and Culverts), December 2016*. Virginia Department of Transportation, 2016.
24. Nowak, A. S., H. Nassif, and L. DeFrain. Effect of truck loads on bridges. *Journal of Transportation Engineering*, Vol. 119, No. 6, 1993, pp. 853–867.

*The Standing Committee on General Structures peer-reviewed this paper.*

Article

TRR
JOURNAL OF THE TRANSPORTATION RESEARCH BOARD

Transportation Research Record
2018, Vol. 2672(41) 24–33

DOI: 10.1177/0361198118798738
journals.sagepub.com/home/trr
SAGE

# Design and Full-Scale Testing of M50/P1 Bolt-Down Removeable Bollard System

**William F. Williams**[1]

**Abstract**

The purpose of this project was to design and test a new bolt-down bollard system that meets the requirements of American Standards for Testing Materials (ASTM) Designation F2656-15 M50/P1 impact conditions. The test installation consisted of three vertical 10-in. diameter (nominal) bollards with welded base plates bolted to a shallow reinforced concrete foundation. The foundation for this system was sized to reduce the foundation embedment. Shallow foundations are often necessary for use in cities and urban areas where utilities can conflict with deeper foundations. Standard common members and materials were used in the installation to accommodate fabrication and installation in locations all over the world. The bollards can be removed to provide access if necessary. Full-scale testing was performed on the bolt-down bollard system. The bollard system design for this project successfully met the requirements of M50/P1 with a total payload penetration of less than 1 m. The new bollard design successfully met all the performance requirements for ASTM F2656-15 M50/P1. Details of the design and testing of the bolt-down bollard system are provided in this paper. Crash-testing videos and additional information on the design and full-scale testing will be provided in the presentation.

In an effort to assess the performance of anti-terrorist protection barriers, the United States Department of State, Bureau of Diplomatic Security, Physical Security Division, Office of Physical Security Programs developed guidelines to evaluate the performance of perimeter security devices. According to this standard, performance of an anti-terrorist protection security device (barrier/gate) is evaluated and assessed according to its effectiveness in arresting attacking vehicles, and not necessarily for economics, aesthetics, operational cycle time, special maintenance needs, or climate and environment effects. The bolt-down bollard system evaluated here was designed by the author at Texas A&M Transportation Institute. The intended function of this design is to provide perimeter security capable of arresting an attacking vehicle. The design impact condition for this project was M50/P1.

## Background

In August 2007, the American Standards for Testing Materials International (ASTM) developed and published *ASTM Designation: F2656, Standard Test Method for Vehicle Crash Testing of Perimeter Barriers* (*1*). The procedures set out in ASTM F2656 are intended to ensure that perimeter security devices provide a specified level of vehicle impact resistance. The ASTM F2656 test method provides a range of vehicle impact conditions, test designations, and penetration levels that allow agencies to select perimeter security devices that satisfy their specific facility needs. This test method was formally adopted by the U.S. Department of State, Bureau of Diplomatic Security, Physical Security Division, Office of Physical Security Programs, in February 2009 as their official standard for the testing of perimeter security devices. The test reported here was performed and evaluated in accordance with the aforementioned ASTM Designation F2656-15.

## Objective/Scope of Research

The objective of the test reported here was to determine whether the bolt-down bollards are capable of arresting a 15,000 lb truck traveling at a minimum speed of 47 mi/h according to Condition Designation M50 of ASTM F2656-15. This condition designation requires the bolt-down bollards to withstand nominal kinetic energy of 1,250 ft-kips.

[1]Safety & Structural Systems Division, Texas Transportation Institute, College Station, TX

**Corresponding Author:**
Address correspondence to William F. Williams: w-williams@tamu.edu

This paper presents the construction details of the bolt-down bollards, details of the impact vehicle used in the test, particulars of the test performed, and the assessment of the test results.

## Design and Strength Analyses of New Bolt-Down Bollard

A new braced pipe bollard was designed and tested for this project. For M50/P1 loading conditions, a maximum deceleration of 20 g was used to calculate the force for the 15,000-lb truck. The force used for the design was calculated to be 20 g × 15,000 lbs, equaling 300,000-lb force. This force was assumed to act over a time duration of 50 ms. The force applied to the braced 10-in. pipe is shown in Figure 1.

Based on the loading conditions in Figure 1, a free body diagram was developed for the forces applied to the bollard to determine the resisting forces applied to the anchoring system. These forces were also used to design the foundation for the M50/P1 bollard. The foundation was sized to the smallest width, thickness, and length necessary to achieve M50/P1. Strength analyses were performed on several features of the bollard design. An embedded anchor frame to support and anchor the bollard into the 18-in. concrete foundation was designed as part of this project. Strength analyses and sizing of all members, welds, baseplate, and anchor bolts was performed. Based on the limited embedment for the anchor frame, detailed strength analyses were performed to size the required embedment. Based on the design loads, the embedded anchor frame failure mechanism in pullout tension was evaluated. This failure mechanism is shown in Figure 2. All strength analyses were performed using concrete strength values provided by the American Concrete Institute design specifications. All strength calculations were performed using MathCad. Independent variables were established as shown in figures 2 and 3 below. These variables were adjusted as necessary to achieve the desired pullout resistance of the anchor bolts and frame based on the loading conditions shown in Figure 1.

The final design for the bolt-down bollard system consisted of a 42-in. tall 10-in. Schedule 120 ASTM A53 Grade B Pipe welded to a 20-in. × 40-in. × 1½-in. thick baseplate. The bollard was braced with a HSS6x6x1/2 ASTM A500 Grade B tube. This tube was necessary to minimize deflection of the bollard and prevent the impacting truck from sliding up and over the bollard. Any penetration of the truck in this manner would compromise the necessary M50/P1 performance. The bollard was anchored to the 18-in. thick concrete foundation using ten 1½-in. diameter × 8-in. long Grade 5 anchor bolts. These bolts anchored into an embedded anchor frame cast within the concrete foundation. This anchor frame consisted of two W10x68 ASTM A572 Grade 50 W-Shape sections 54 in. in length. The

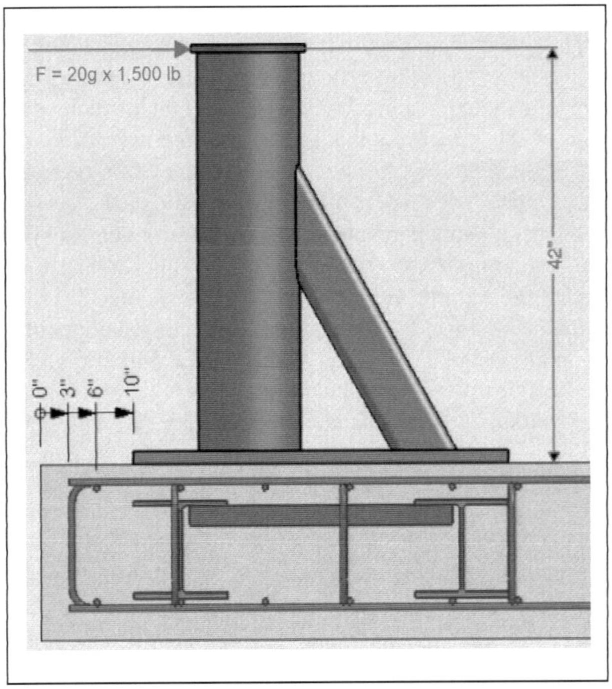


**Figure 1.** Design force for 12-in. pipe bollard.

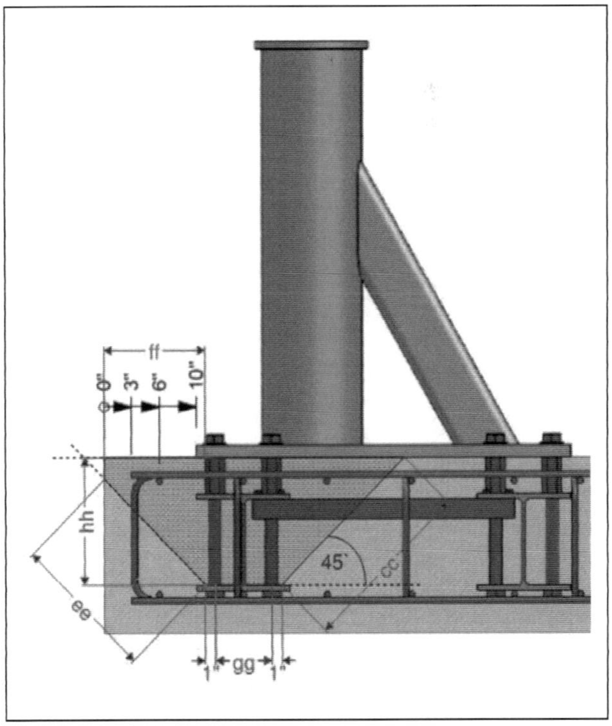


**Figure 2.** Anchor frame failure mechanism end view.

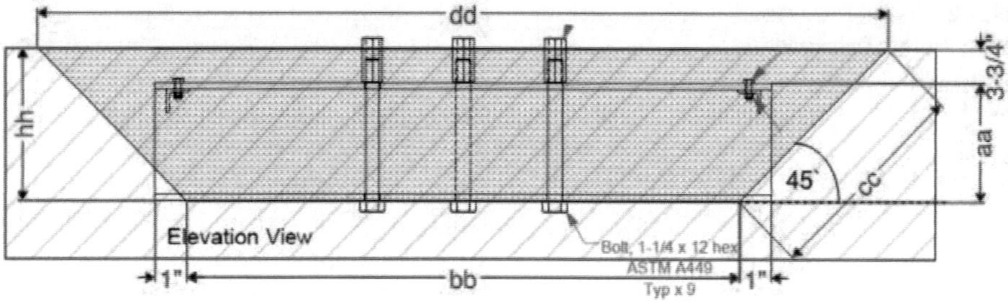


**Figure 3.** Anchor frame failure mechanism front view.

concrete foundation was 8-ft wide × 24 ft in length × 18 in. in thickness. Reinforcement in the foundation consisted of #5 bars at 6 in. on centers each way top and bottom. Three bollards were constructed on the concrete foundation for the test installation. The bollards were spaced 57 in. on centers. These bollards were located in an array perpendicular to the attacking vehicle. The bollards were located (centerline) 22 3/8 in. from the leading edge (attack side) of the concrete foundation.

## Full-Scale Test Installation Details

A full-scale test installation was constructed for the bolt-down bollard system. Details of the test installation constructed for this project are shown in figures 4 to 9. Photographs of the completed crash-test installation are shown in Figure 10.

## Test Conditions

According to ASTM F2656-15, the bolt-down bollards can be rated according to one of three designated condition levels when tested with a medium-duty truck, as shown in Table 1. The nominal levels of kinetic energy that a security device can withstand at each condition level are also indicated in ASTM F2656-15. The test conditions are intended to ensure that perimeter security devices provide a specified level of vehicle impact resistance. Actual vehicle weight and speed must be within a permissible range to receive the specific condition designation. The condition designations are defined by test vehicle type and impact speed.

The test vehicle specified for the test was a medium-duty truck with diesel engine, tested at a vehicle gross weight of 15,000 lb ± 309 lb (6,800 kg ± 140 kg). According to Condition Designation M50 of ASTM F2656-15, which involves the medium-duty truck impacting at 47 mi/h (75 km/h) or above, the security device is required to withstand nominal kinetic energy of 1,250 ft-kips (1,680 KJ).

## Evaluation Criteria

The amount of vehicle penetration of the security device at the required impact velocity determines the dynamic penetration rating for each condition designation. Test vehicle dynamic penetration according to ASTM F2656 is referenced to the following: "the pre-impact location of the intersection of the leading lower vertical edge of the cargo bed and the frame rail on the standard test truck (M)." Penetration ratings according to ASTM F2656-15 are shown in Table 2.

## Full-Scale Crash Testing

A full-scale crash test was performed on the bolt-down bollard installation design on October 21, 2016. The crash test was performed in accordance with ASTM Specification F2656-15. A 2000 Ford F-750 flatbed single-unit truck with diesel engine, shown in Figure 11, was used for the crash test. The test inertia weight of the vehicle was 15,000 lb (6,804 kg). The height to the lower edge of the vehicle front bumper was 22.5 in. (572 mm), and the height to the upper edge of the front bumper was 35.0 in. (889 mm). The vehicle was directed into the installation using a reverse cable tow and guidance system, and was released to be free-wheeling and unrestrained just prior to impact.

The 2000 Ford F-750 flatbed single-unit truck, traveling at an impact speed of 50.2 mi/h (80.8 km/h), impacted the bolt-down bollards at an impact angle of 89.7°. The centerline of the vehicle was aligned with the centerline of the middle bollard. At 0.003 s after impact, the center bollard began to deflect toward the protected side, and at 0.021 s, the hood began to separate from the vehicle. The right front wheel began to toe inward at 0.023 s, and the impact side of the foundation began to lift out of the ground at 0.035 s. At 0.070 s, the cargo bed began to yaw clockwise, and at 0.087 s, the trailing side of the right front tire contacted the right bollard. The cab of the vehicle began to yaw counterclockwise at 0.108 s, and the

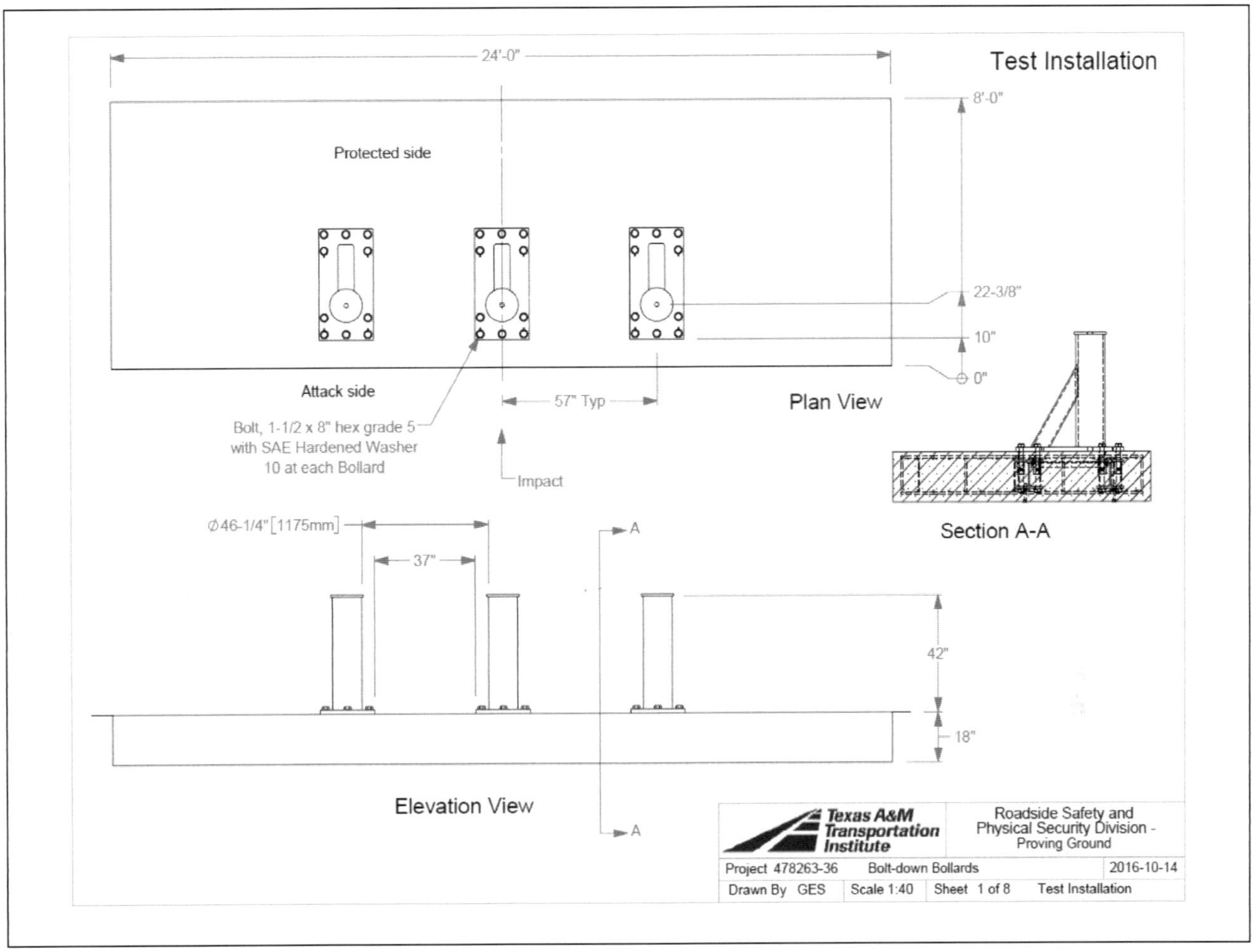


**Figure 4.** Bolt-down bollard layout and section details.

foundation reached maximum height above the existing grade of approximately 1 in. (25 mm) at 0.145 s. At 0.230 s, the cargo bed reached maximum clockwise yaw of 80° and began to yaw counterclockwise. The left leading edge of the cargo bed penetrated into the protected side at 0.250 s. At 0.420 s after impact, the left leading edge of the cargo bed penetrated 17.2 in. (438 mm) into the protected side, and the vehicle began to rebound. Brakes on the vehicle were not applied, and the left front edge of the cargo bed of the vehicle came to rest 9.4 in. (239 mm) into the protected side. The hood of the vehicle detached and was thrown 70 ft (21 m) into the protected side.

Damage to the bolt-down bollards is shown in Figure 12. The foundation was cracked slightly on both sides of the center bollard, radiating from the second row of bolts at the front outward and toward the protected side at 45°. At the front of the foundation, there was a small crack 18 in. (457 mm) to the left of centerline, angling toward the left for 13 in. (330 mm). Multiple hairline cracks were noted to the right of centerline radiating from the impact edge to near the midpoint of the foundation. The concrete foundation was displaced 0.4 in. (10 mm) toward the protected side. Maximum dynamic deflection of the center bollard was not attainable due to being obscured by the vehicle. Maximum permanent deflection of the center bollard was 2.25 in. (57 mm) at the top of the bollard. Photographs of the installation and vehicle after the crash test are shown in Figure 12.

## Summary and Conclusion

The 2000 Ford F-750 flatbed single-unit truck with a diesel engine impacted the bolt-down bollards at 89.7°, with the centerline of the vehicle aligned with the centerline of the center bollard. The minimum nominal impact speed for this M50 test was 47.0 mi/h (75.1 km/h), and the actual impact speed was 50.2 mi/h (80.8 km/h). The bolt-down bollards brought the vehicle to a complete stop, and the vehicle was disabled. The cargo remained on board the vehicle, however the hood and cab detached

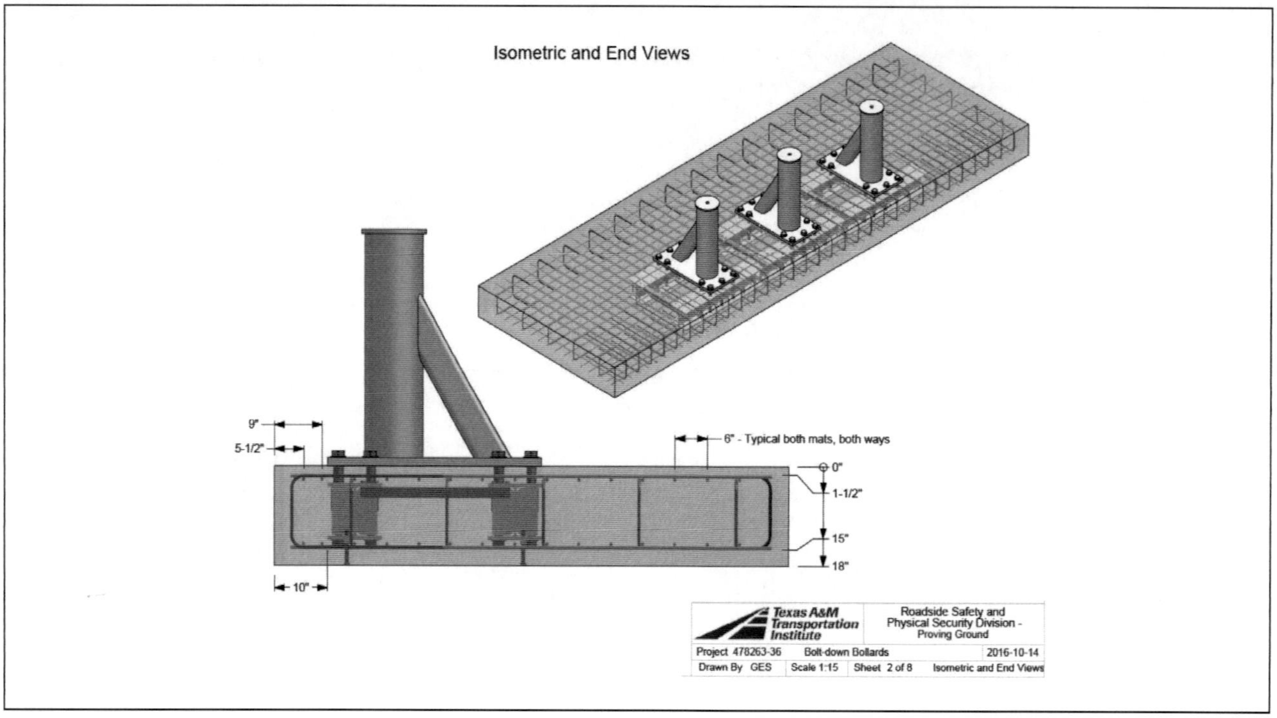


**Figure 5.** Isometric view.

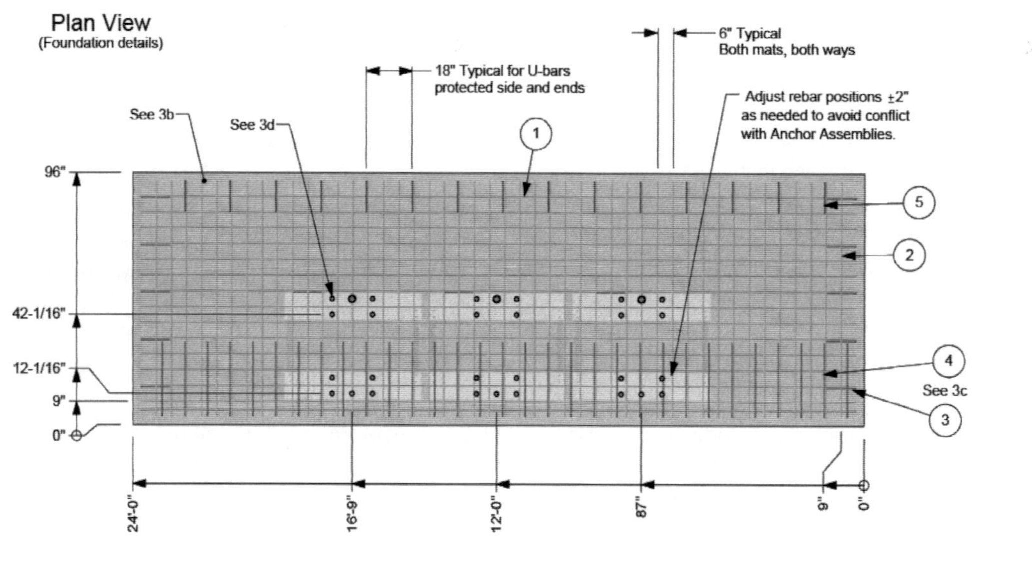


**3a.** All Rebar is grade 60. Coverage is 3" at bottom and sides, 1-1/2" at top. Minimum laps on Ø5/8" (#5) bars is 30".
**3b.** 3000 psi concrete for Foundation.
**3c.** Place Long U-bars at 9" on impact side of the rebar cage, and Large U-bars at 18" on the protected side. Place Small U-bars at 18" on the ends, and otherwise as needed to support the top mat. Adjust rebar positions ±2" as needed to avoid conflict with Anchor Assemblies.
**3d.** Cover Sleeves to protect prevent concrete intrusion.

| # | Part Name | Qty. |
|---|---|---|
| 1 | Longitudinal Bar | 26 |
| 2 | Transverse Bar | 92 |
| 3 | Small U-bar | 13 |
| 4 | Long U-bar | 31 |
| 5 | Large U-bar | 15 |

| Texas A&M Transportation Institute | | Roadside Safety and Physical Security Division - Proving Ground |
|---|---|---|
| Project 478263-36 | Bolt-down Bollards | 2016-10-14 |
| Drawn By GES | Scale 1:40 | Sheet 3 of 8 Plan View |

**Figure 6.** Reinforcing details and plan view.

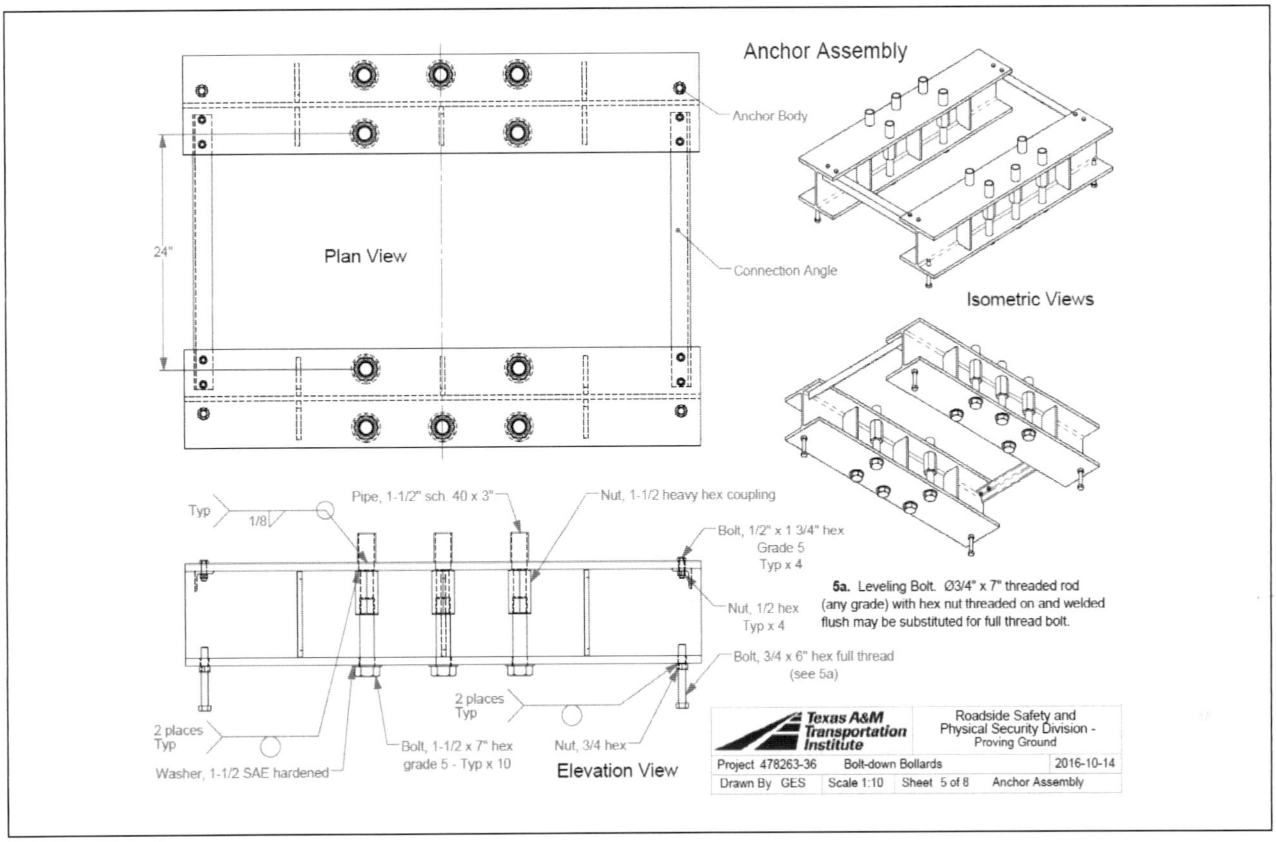


**Figure 7.** Anchor frame details.

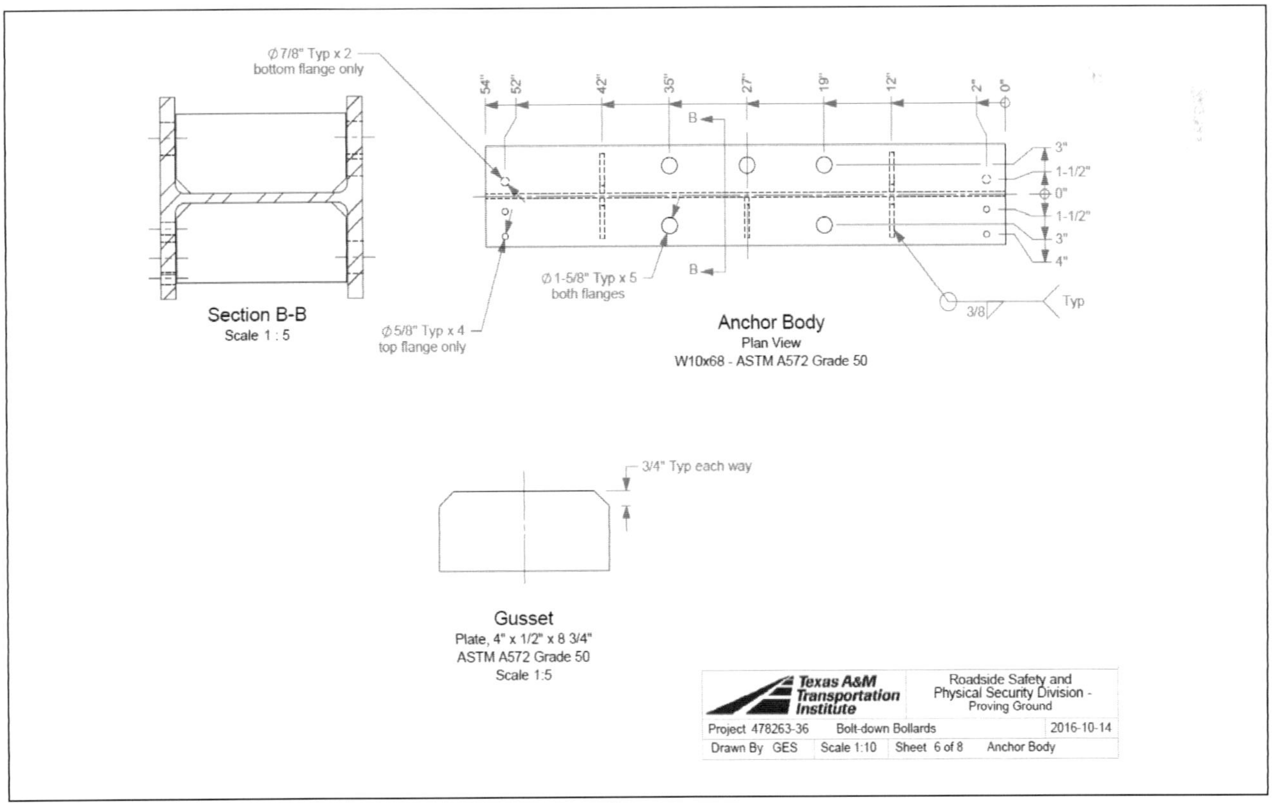


**Figure 8.** Anchor frame details continued.

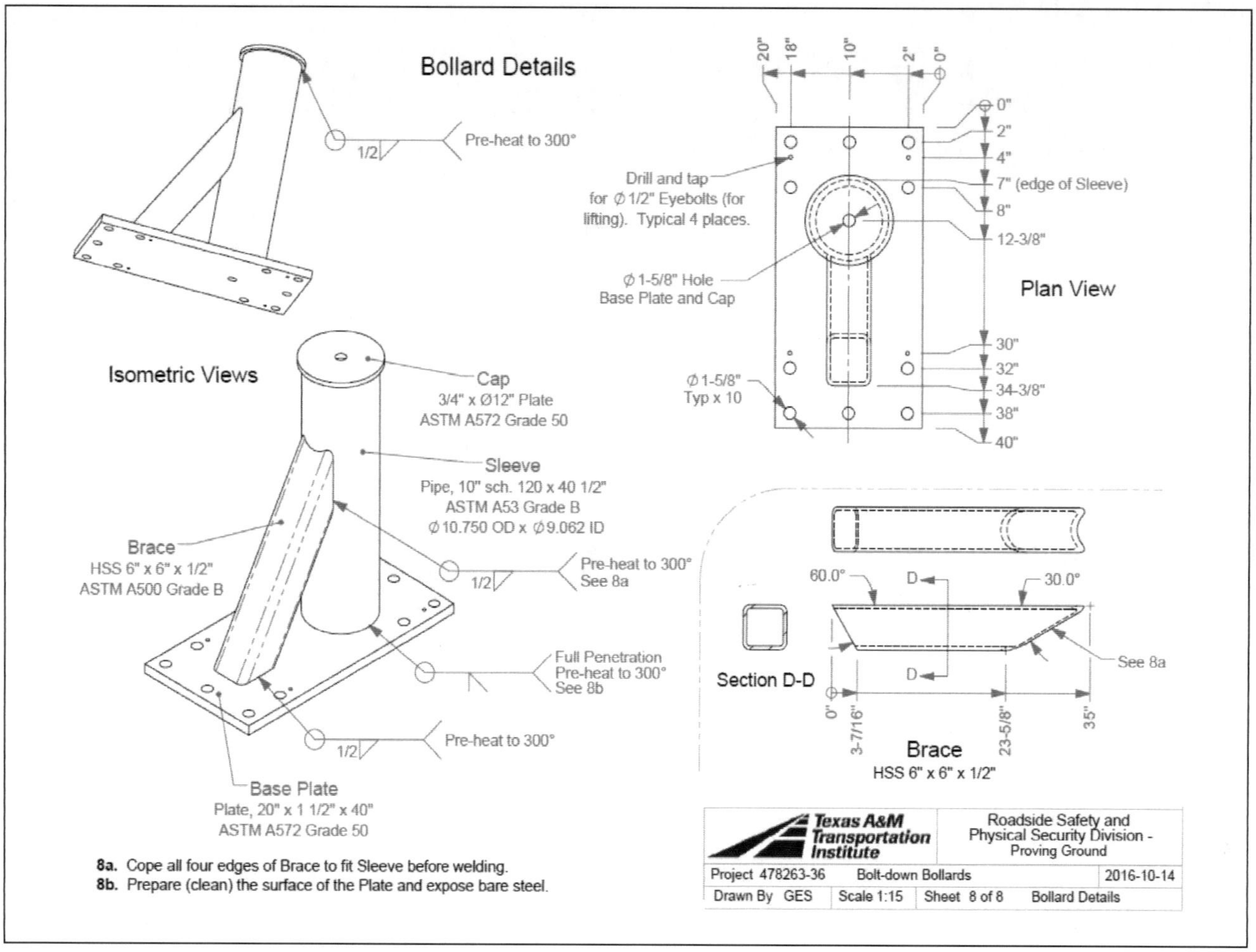


**Figure 9.** Bollard details.

**Table 1.** Impact Condition Designations According to ASTM F2656

| Test vehicle/minimum test inertial vehicle, lb (kg) | Nominal test velocity, mi/h (km/h) | Permissible speed range, mi/h (km/h) | Nominal kinetic energy, ft-kips (KJ) | Condition designation |
|---|---|---|---|---|
| Standard test truck (M) 15,000 (6,800) | 30 (50) | 28.0–37.9 (45.0–60.0) | 451 (656) | M30 |
| | 40 (64) | 38.0–46.9 (60.1–75.0) | 802 (1,110) | M40 |
| | 50 (80) | 47.0–above (75.1–above) | 1,250 (1,680) | M50 |

**Table 2.** Penetration Ratings According to ASTM F2656

| Penetration designation | Dynamic penetration rating |
|---|---|
| P1 | ≤3.3 ft (≤1 m) |
| P2 | 3.31 to 23.0 ft (1.01 m to 7 m) |
| P3 | 23.1 to 98.4 ft (7.01 m to 30 m) |

from the vehicle and was thrown 70 ft (21 m) into the protected area. The leading edge of the cargo bed of the vehicle penetrated 17.2 in. (438 mm) beyond the inside edge of the bolt-down bollards.

ASTM F2656-15 provides a range of vehicle test designations and penetration levels that allow agencies to select perimeter security devices that satisfy their specific facility needs. The amount of vehicle penetration of the

**Figure 10.** Photographs of completed test installation.

**Figure 11.** Photographs of crash vehicle and test installation.

**Figure 12.** Photographs of crash-test installation after the test.

security device at the required impact velocity determines the dynamic penetration rating for each condition designation.

The leading edge of the cargo bed penetrated 17.2 in. (438 mm) beyond the inside edge of the bolt-down bollards. According to ASTM F2656, the bolt-down bollards meet Condition Designation/Penetration Rating M50/P1, which allows penetration of $\leq$ 3.3 ft (1 m) when impacted by a standard truck with a diesel engine at a nominal speed of 50 mi/h (80 km/h).

## Acknowledgments

Thanks to Russ Norris and Phillip Benshoof of the Department of State Diplomatic Security Section. Their assistance and valuable input made this a truly great research project.

## Author Contributions

WW is the sole author of this paper, responsible for the study conception, design, and creation of the engineering details. WW is responsible for the interpretation of results, draft manuscript, and preparation. The author has reviewed the results and approved the final version of the manuscript.

## References

1. ASTM International. *Standard Test Method for Vehicle Crash Testing of Perimeter Barriers*. ASTM Designation: F2656-15. ASTM International, West Conshohocken, Pa., 2015.

*The Standing Committee on General Structures (AFF10) peer-reviewed this paper (18-06237).*

Article

TRR
JOURNAL OF THE TRANSPORTATION RESEARCH BOARD

Transportation Research Record
2018, Vol. 2672(41) 34–43

DOI: 10.1177/0361198118755725
journals.sagepub.com/home/trr
SAGE

# Developing Guidelines for Epoxy Grout Pourback Systems for Controlling Thermal/Shrinkage Cracking at Post-Tensioning Anchorages: Full-Scale Testing and Numerical Analysis

**Irtishad Ahmad[1], Nakin Suksawang[2], Khaled Sobhan[3], John A. Corven[4], Rick Vallier[5], Ehssan Amir Sayyafi[1], and Sharmila Pant[3]**

**Abstract**

Epoxy grout pourbacks at end anchorages of post-tensioning tendons provide an essential level of corrosion protection. The tendon may fail because of strand corrosion, within a few years of installation, if it is not properly protected with epoxy grout pourbacks. Recently, the Florida Department of Transportation (FDOT) found cracking on larger epoxy grout pourbacks, possibly as a result of the exothermic nature of the epoxy grout curing, causing thermal stresses that increase the potential for pourbacks cracking. Coordinated laboratory and numerical investigations were undertaken to develop best practice guidance for eliminating cracking in epoxy grout pourbacks. Potential factors were determined to be the pourback size, shapes, and temperature of the concrete substrate. An epoxy grout pourback material was used to construct full-scale pourback systems consisting of rectangular (R-type) and irregular (S-type) pourbacks with volume to surface, *V/S* ratios of 80, 98, and 113 mm. The full-scale specimens were instrumented with multiple thermocouples and a vibrating wire strain gauge, to capture the temperature profile and localized strain respectively. An ANSYS finite element model was developed to perform a parametric study to better understand the cracking mechanism and the influence of the *V/S* ratio. The model was calibrated and validated using the data obtained from the full-scale test. Based on the research findings, it is recommended that the maximum value of the *V/S* ratio of the epoxy grout pourbacks be limited to 91 for S-type and 107 mm for R-type shapes.

Post-tensioning end anchorages are protected by a four-level system comprising grout, heavy-duty permanent grout caps with sealing O-ring, an applied coating, and enclosure with epoxy grout pourback as illustrated in Figure 1. This was developed because of tendon failure due to strand corrosion in the anchorage (*1*). Figure 2 shows tendon failure at the anchorage in an expansion joint span of the Mid-Bay Bridge in Florida, where no pourback was provided. It was reported that water passed through the mastic coating and into the post-tensioning anchorage, leading to tendon failure because of strand corrosion in the anchorage, within 7 years of installation (*2*). Early segmental bridges incorporated cementitious pourbacks to protect end anchorages. However, these early cementitious pourbacks were found to be ineffective, as observed by significant cracking. Figure 3 shows a single pourback system encasing three anchorages of the span-by-span approaches on the Sunshine Skyway Bridge in Florida, where significant cracking of the cementitious pourbacks was observed (*2*).

Hamilton and Alvarez (*3*) evaluated the performance of epoxy grout pourbacks by constructing full-scale mockups of anchorage ends to determine whether thermal expansion and drying shrinkage caused cracking of the pourback. No discernible cracking was observed or detected on the exterior of the pourback specimens, neither was any sign of change detected in the bond at the interface between the epoxy and concrete. Hamilton and Alvarez recommended that further study of epoxy grout pourback should be conducted.

[1]Florida International University, Miami, FL
[2]Florida Institute of Technology, Melbourne, FL
[3]Florida Atlantic University, Boca Raton, FL
[4]Corven Engineering, Inc., Tallahassee, FL
[5]Florida Department of Transportation, Tallahassee, FL

**Corresponding Author:**
Address correspondence to Nakin Suksawang: nsuksawang@fit.edu

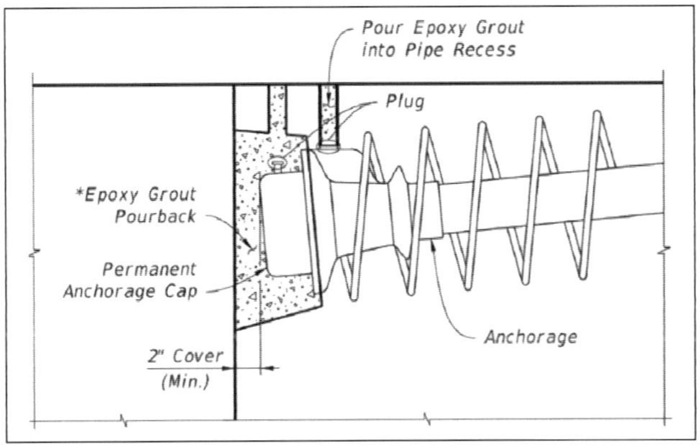


**Figure 1.** Level of protection against corrosion (FDOT Design Standard Index No. 21803).

There has been improvement in tendon protection technology using epoxy grout pourbacks, but not all epoxy grout pourbacks yielded positive results. Larger epoxy grout pourbacks tend to suffer from thermal/shrinkage cracking due to excessive exothermic heat generated during the curing period. The broad objectives of this research were to develop guidelines for epoxy grout pourback system for minimizing thermal/shrinkage cracking at post-tensioning anchorages. The specific objectives of the current research were:

- To design and construct full-scale epoxy grout pourback systems with complex and regular geometric shapes;
- To monitor the time–temperature and deformation behaviors of the pourbacks using embedded thermocouples and strain sensors, respectively;
- To conduct finite element thermal and stress analysis on the epoxy grout pourback systems;
- To validate the finite element results from the observed behavior in the full-scale field experiments; and
- To provide design guidelines in terms of critical volume to surface (*V*/*S*) ratios to minimize the probability of cracking.

To develop a comprehensive experimental program to evaluate the performance and to better understand the epoxy grout pourback failure mechanism, field inspection of epoxy grout pourbacks was conducted on two bridges: the SR 826/836 Interchange Bridge #11 in Miami, FL, and the LeRoy Selmon Crosstown Expressway in Tampa, FL. Both bridges were newly constructed, post-tensioned, concrete segmental bridges. The SR 826/836 Interchange Bridge #11 had no epoxy grout pourback problems, whereas some cracking of the pourbacks was detected on the LeRoy Selmon Crosstown Expressway Bridge. There are two reasons for this: first, the diurnal temperature variation is lower in Miami than in Tampa; and, second, the pourbacks on the SR 826/836 Interchange Bridge #11 were much smaller. Unlike the SR 826/836 Interchange Bridge #11, the dead-ends on the LeRoy Selmon Crosstown Expressway Bridge comprise multiple anchorage points that are spread throughout the entire depth of the bridge segment. To simplify the pourback construction, one large formwork encapsulating multiple anchorages was used, and only a few epoxy grout pourbacks experienced cracking, as illustrated in Figure 4. Cracks occurred across the entire width, extended across the widest or narrowest regions of the pourback or both, ranged in size and locations, and appeared to originate near the corners of the mass. From field observation, cracking is expected to decrease if the exotherm experienced by the pourback is reduced by individual anchor protection, implying a decreased need for epoxy grout, but an increased need in the amount of formwork and labor.

## Methods

The research was carried out primarily in two phases: Phase I, full-scale testing of epoxy pourback systems; and Phase II, Finite Element Analysis and Validation Studies. All necessary physical and mechanical properties of the selected epoxy grout are provided in the manufacturer's product data sheet (*4*), the compressive strength and the peak exothermic temperature were determined in the laboratory for verification purposes.

Compressive strength of the epoxy grout was determined in accordance with ASTM C-579 (method B) at 13°C, 24°C, and 43°C (*5*). At each temperature, three 51-mm cube specimens were tested at 8, 16, and 24 hours and 3, 7, 14, and 28 days. The peak exothermic temperature of the epoxy grout was measured in the laboratory following ASTM D 2471 specifications (*6*). Epoxy specimens were prepared at four different *V*/*S* ratios of 25, 11, 8, and 5 mm with corresponding dimensions of 305 mm × 305 mm × 76 mm (standard), 114 mm × 114 mm × 38 mm, 76 mm × 76 mm × 25 mm, and 76 mm × 76 mm × 13 mm, respectively.

### *Phase I: Full-Scale Testing of Epoxy Pourback Systems*

*Construction of Pourback Specimens.* The full-scale epoxy grout testing consisted of two sets of epoxy pourbacks: (1) irregular shaped pourbacks (S-type) as illustrated in Figure 5; and (2) rectangular shaped pourbacks (R-type) as illustrated in Figure 6. For each set, three different *V*/*S* ratios were investigated: 79, 98, and 113 mm. These *V*/*S* ratios were based on actual pourback size commonly found in the field. All pourbacks had equal grout thickness of 165 mm. The S-type dimensions were 610 mm × 965 mm, 991 mm × 1568 mm, and 1168 mm × 1854 mm for *V*/*S* ratios of 79, 98, and 113 mm, respectively. The R-type dimensions were 406 mm × 962 mm, 749 mm × 1143 mm, and 889 mm × 1321 mm for the same *V*/*S* ratios of 79, 98, and 113 mm, respectively.

**Figure 2.** Tendon failure on the Mid-Bay Bridge.

**Figure 3.** Pourback cracking on the Sunshine Skyway Bridge.

The pourback specimens were cast against reinforced concrete walls, with wooden forms with plastic lining were used for molds. The forms were made such that a 305-mm gap was left open to allow pouring of the epoxy grout. The grout was mixed in a 19 l bucket, in accordance with the manufacturers specification, using a hand mixer. It was then hand-poured into the form. Grout caps, injected with cementitious grout, were also used to simulate field conditions.

*Instrumentation.* The pourback specimens were instrumented, using 12 thermocouples and a vibrating wire (VW) strain gauge for each specimen, to monitor the temperature gradients in the pourback specimens and the strain at the center of the pourback specimens. The data were collected every 10 min for a period of 48 hours after casting. More details of instrumentation plans are available in (*2*).

## *Phase II: Finite Element-Modeling with ANSYS*

Finite element thermal analysis was performed with ANSYS (*7*) to conduct a parametric study for better understanding of the cracking mechanism of epoxy grout pourback system. The three-dimensional element type SOLID70 3D was used for the transient thermal analysis, while SOLID185 3D was used for the stress analysis. Figure 7 presents a typical finite element modeling algorithm, used in this study for transient thermal distribution and thermally induced stresses on the structural behavior (*8*). For the thermal and stress analyses, the thermal conductivity, specific heat, coefficient of thermal expansion, modulus of elasticity, Poisson's Ratio, and the density of the material were the needed inputs. Table 1 shows the essential material properties for the concrete slab and the epoxy grout system that were used for the thermal and stress analysis.

# Results

## *Peak Exothermic Temperature Test*

Figure 8 illustrates the variations of the peak exothermic temperature with the *V*/*S* ratios, increasing *V*/*S* ratios, consistent with the field behavior of full-scale specimens. It should be noted that the peak exothermic temperature of the manufacturer's product data sheet corresponds to a value of 60°C as illustrated by the horizontal dash line in Figure 8. As the *V*/*S* ratio approaches 25 mm, the measured peak exothermic temperature was found to be 104°C. Considering that the *V*/*S* ratio of field specimens is significantly higher, this difference may have some implications in the thermal/shrinkage cracking behavior.

## *Behavior of Full-Scale Specimens*

*Strain Data.* The strain data was obtained from the VW strain gauges. Figure 8 illustrates the strain in the middle of the specimens for all six full-scale specimens. Based on the product data sheet for the epoxy grout specimens (*1*), the cracking strain is approximately $2000 \times 10^{-6}$ mm/mm, as indicated by the dotted horizontal line in Figure 9. It is observed that the cracking in epoxy pourbacks begins at early-age approximately 20 h (1200 min) after casting, particularly for specimens S2.5 and S3 (S-type pourback with *V*/*S* ratios of 101 and 113 mm, respectively). From the strain data, specifically for S3, the cracks observed on the specimens were not surface

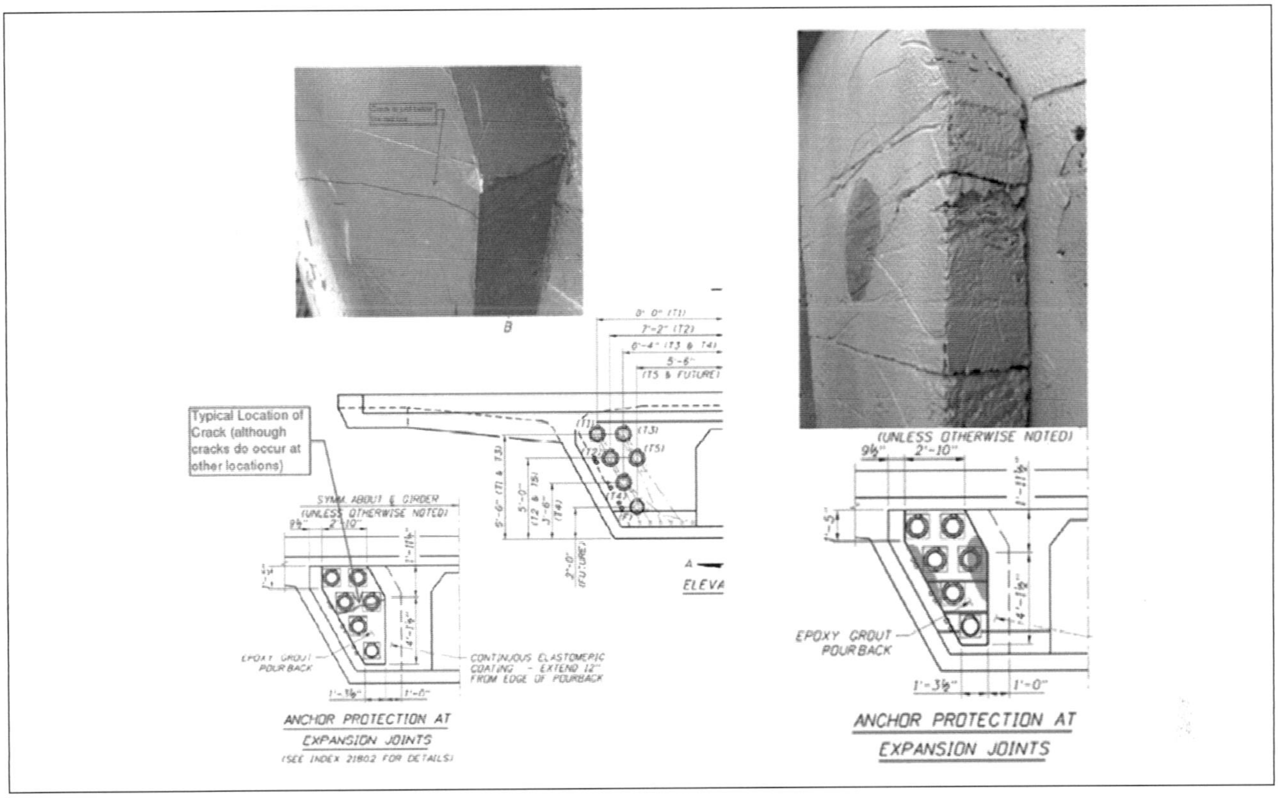


**Figure 4.** Typical location of cracks.

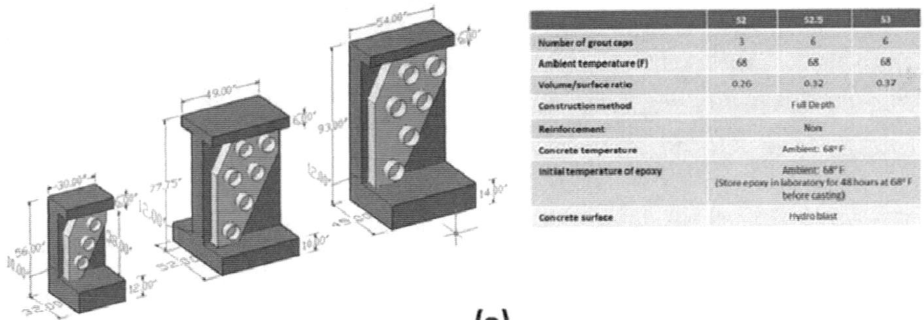

| | S2 | S2.5 | S3 |
|---|---|---|---|
| Number of grout caps | 3 | 6 | 6 |
| Ambient temperature (F) | 68 | 68 | 68 |
| Volume/surface ratio | 0.26 | 0.32 | 0.37 |
| Construction method | Full Depth | | |
| Reinforcement | Non | | |
| Concrete temperature | Ambient: 68° F | | |
| Initial temperature of epoxy | Ambient: 68° F (Store epoxy in laboratory for 48 hours at 68° F before casting) | | |
| Concrete surface | Hydro blast | | |

(a)

(b)

**Figure 5.** S-type pourback systems: (*a*) complex geometric shapes and properties; and (*b*) fully constructed field specimen.

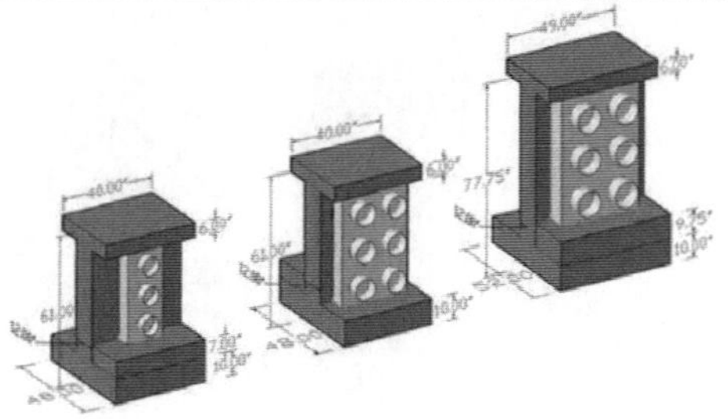

| | R2 | R2.5 | R3 |
|---|---|---|---|
| Number of grout caps | 3 | 6 | 6 |
| Ambient temperature (F) | 68 | 68 | 68 |
| Volume/surface ratio | 0.26 | 0.32 | 0.37 |
| Construction method | Full Depth | | |
| Reinforcement | Non | | |
| Concrete temperature | Ambient: 68° F | | |
| Initial temperature of epoxy | Ambient: 68° F (Store epoxy in laboratory for 48 hours at 68° F before casting) | | |
| Concrete surface | Hydro blast | | |

(a)

(b)

**Figure 6.** R-type pourback systems: (*a*) regular geometric shapes and properties; and (*b*) fully constructed field specimens.

cracks but penetrated all the way through the epoxy pourbacks. It is highly recommended that any cracks found in the pourbacks should be injected and sealed as soon as possible, to avoid chloride-laden water to leak through.

*Visual Inspection.* The specimens were visually inspected daily. Three specimens, S2.5, S3, and R3, had cracks. The S-type pourbacks had diagonal cracks in the center of the specimen, whereas the rectangular-shaped pourback had a horizontal crack forming toward the top (Figure 10).

## *Finite Element Analysis of Full-Scale Specimens*

*Comparison of Time–Temperature Behavior.* For the calculation of thermal stresses by the finite element method, it was necessary to obtain the temperature of each node with respect to time as body forces. The time–temperature curves were developed and compared with the observed behavior of the full-scale specimens. The heat generation values were obtained through a trial and error procedure (*2*). The time–temperature behavior measured in the full-scale specimens, and calculated by ANSYS thermal analysis, was compared for all thermocouple locations. Time–temperature curve calculated by ANSYS at a sensor location in specimen R3 ($V/S$ = 113 mm) and the corresponding experimental values are shown in Figure 11. It is found that there is a good agreement between the ANSYS results and the experimental observation. More details are available in (*2*).

*Maximum Temperature.* The calculated and experimental values of the maximum temperatures for different $V/S$ ratios are compared in Figure 12. In general, it is found that the maximum temperature increases as the $V/S$ ratio increases, with good agreement between the analytical and experimental values.

*Crack Locations.* Figure 13 shows the experimentally observed crack locations and the locations of the tensile cracks determined by the maximum tensile stresses computed by finite element analysis utilizing the strength criteria provided in manufacturer's product data sheet (*1*). The maximum tensile strength of the epoxy grout ranges from 24 to 28 MPa as reported in the product data sheet. Figure 13 indicates agreement between the actual and predicted crack locations for specimens S3, R3, and S2.5.

*Variation of Maximum Tensile Stress with V/S Ratios.* Figure 14 shows the variations maximum tensile stresses with $V/S$ ratios computed by finite element analysis. The dotted line in Figure 13 indicates the tensile strength (24 MPa) of the E-Bond epoxy reported in the manufacturer's product data sheet. A significant finding from this plot is that the S-type specimens reach cracking stress when $V/S$ ratio exceeds 98 mm. For the R-type specimens, the cracking stress is reached at a higher $V/S$ ratio of 113 mm.

## Conclusions

The full-scale testing of epoxy grout pourbacks has been conducted using the R-type and S-type pourbacks with $V/S$ ratios of 79, 98, and 113 mm, typically found on Florida segmental

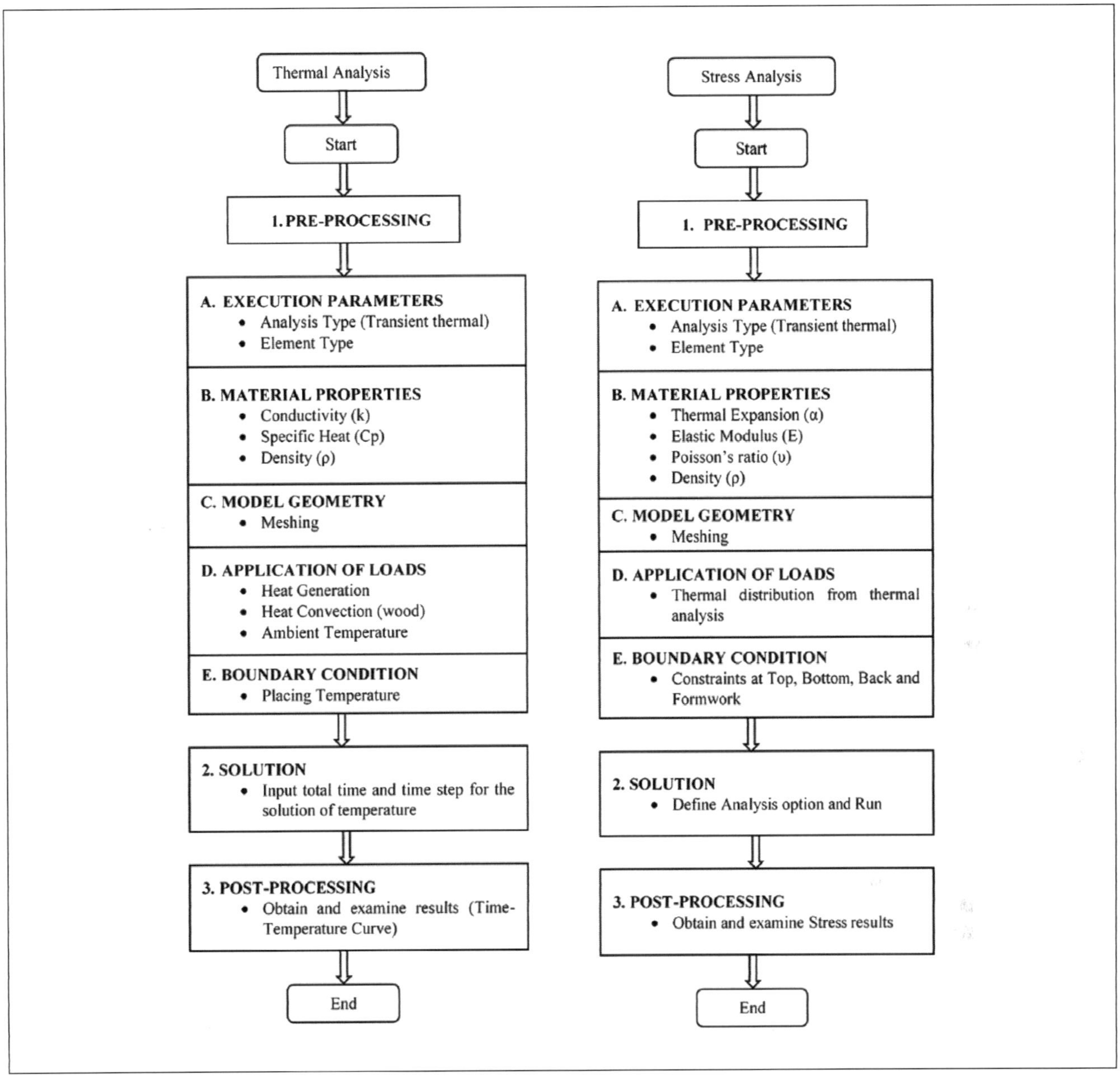


**Figure 7.** Process of thermal and stress analysis (7).

bridges. The full-scale specimens have been instrumented with multiple thermocouples and a VW strain gauge to capture the temperature profile and localized strain, respectively. A finite element analysis has also been performed to validate the experimental data and to determine the *V/S* ratio limits for epoxy grout pourbacks. The following conclusion and recommendations could be made from this research:

1. Based on the full-scale data and observation, specimens with *V/S* ratio of 113 mm had full-depth cracks. In addition, full-depth cracks were also observed on the S-type specimen with *V/S* ratio of 98 mm. The cracks appeared within the first 24 h and concentrated in regions with high exothermic temperature. Therefore, it was concluded that these cracks were caused by exothermic temperature rather than shrinkage.
2. The peak exothermic temperature as measured in the laboratory and in the full-scale specimens was found to be significantly higher than the data provided by the manufacturer's product data sheet.
3. Cracks observed on the specimens were not surface cracks but penetrated all the way through the epoxy pourbacks. Hence, it is highly recommended that any cracks found in the pourbacks should be injected and sealed as soon as possible to avoid chloride-laden water to leak through.

**Table 1.** Material Properties Used in Finite Element Analysis of Full-Scale Specimens

| Concrete slab | Standard test | Units | Value |
|---|---|---|---|
| Thermal conductivity | | kJ/(m·h·°C) | 2.3 |
| Specific heat | ASTM E1269 -11 | kJ/(kg·°C) | 0.23 |
| Heat-transfer film coefficient (air exposure) | | kcal/(m$^2$·h·°C) | 4.3 |
| Density | | kg/m$^3$ | 2400 |
| Young's modulus | ASTM C580 | MPa | 30,000 |
| Poisson's ratio | | – | 0.18 |
| Thermal expansion coefficient | ASTM C531 | m/m/°C | $1.0 \times 10^{-6}$ |
| Epoxy grout system | Standard test | Units | Value |
| Thermal conductivity | | kJ/(m·h·°C) | 2.3 |
| Specific heat | ASTM E1269 -11 | kJ/(kg·°C) | 0.23 |
| Heat-transfer film coefficient (wood forms) | | kcal/(m$^2$·h·°C) | 8.5 |
| Heat-transfer film coefficient (air exposure) | | kcal/(m$^2$·h·°C) | 4.3 |
| Density | | kg/m$^3$ | 1986 |
| Young's modulus | ASTM C580 | MPa | 12,000 |
| Poisson's ratio | | – | 0.4 |
| Thermal expansion coefficient | ASTM C531 | m/m/°C | $5.4 \times 10^{-6}$ |
| Plastic end caps | Standard test | Units | Value |
| Thermal conductivity | | kJ/(m·h·°C) | 0.68351 |
| Specific heat | ASTM E1269 -11 | kJ/(kg·°C) | 1.05 |
| Density | | kg/m$^3$ | 1400 |
| Young's modulus | ASTM C580 | MPa | 3000 |
| Poisson's ratio | | – | 0.4 |
| Thermal expansion coefficient | ASTM C531 | m/m/°C | $3.6 \times 10^{-5}$ |

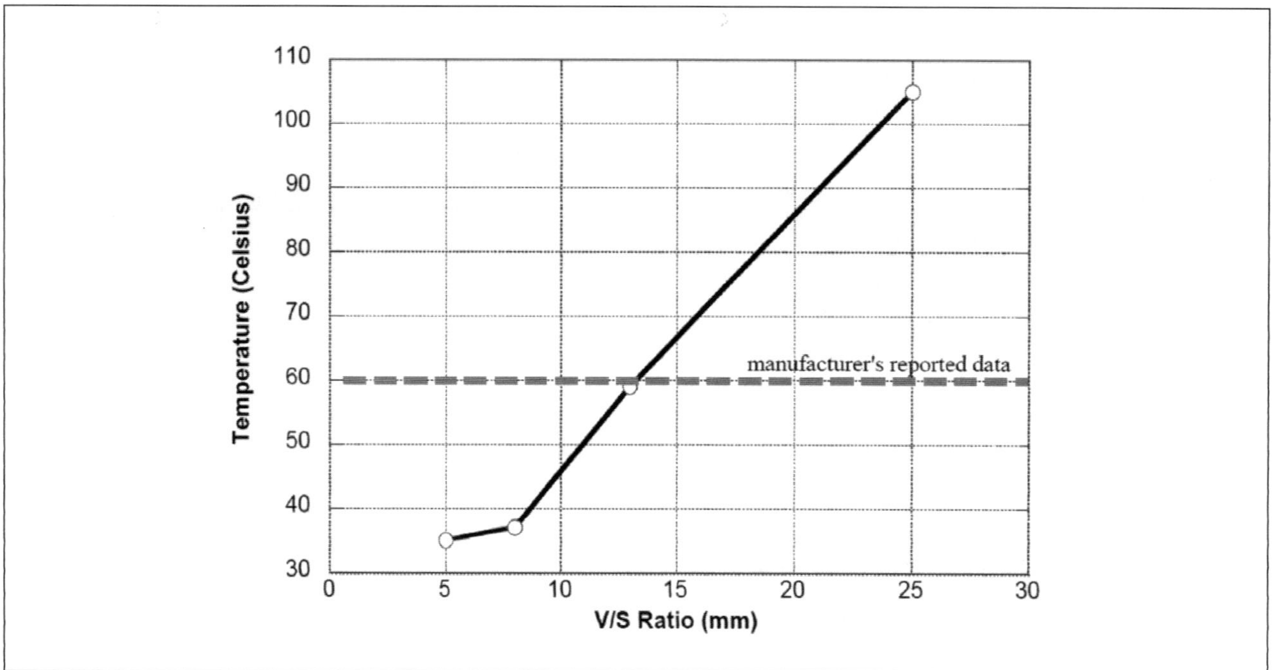


**Figure 8.** Variations in peak exothermic temperature with *V/S*.

4. The time-temperature curves predicted by the ANSYS finite element model closely matched the data obtained from field experiments. Thermal stresses predicted by finite element analysis around the vicinity of the actual physical crack observed in the field showed close agreement with the limiting tensile strength.

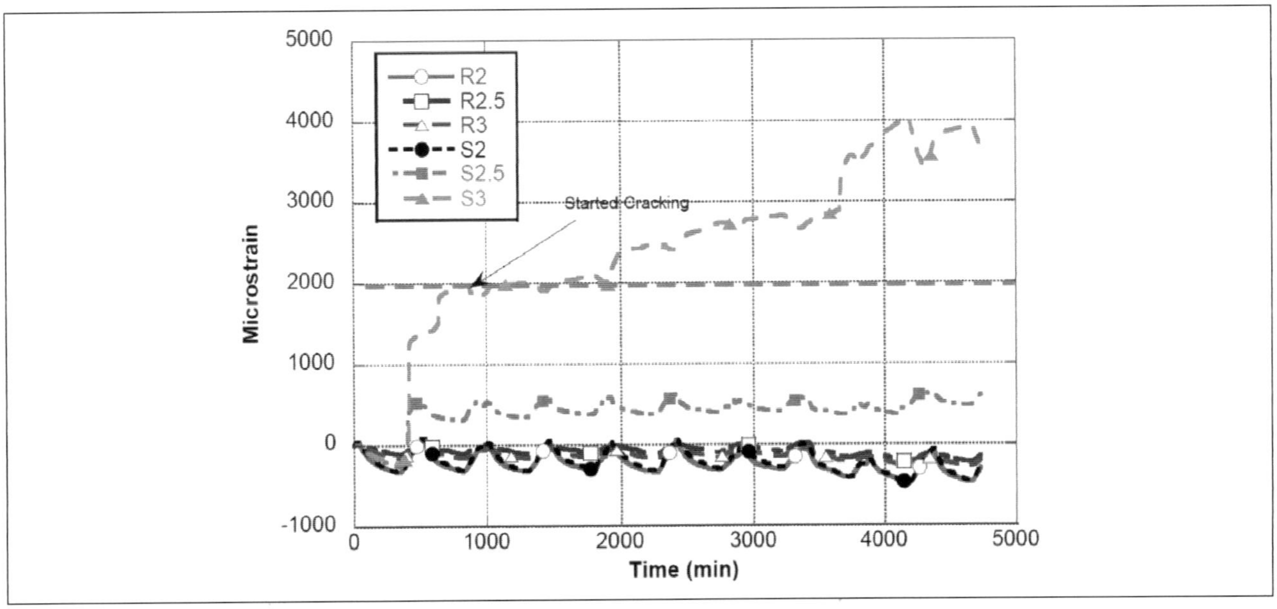


**Figure 9.** VW strain versus time for full-scale specimens.

**Figure 10.** Observed cracks in pourback specimens.

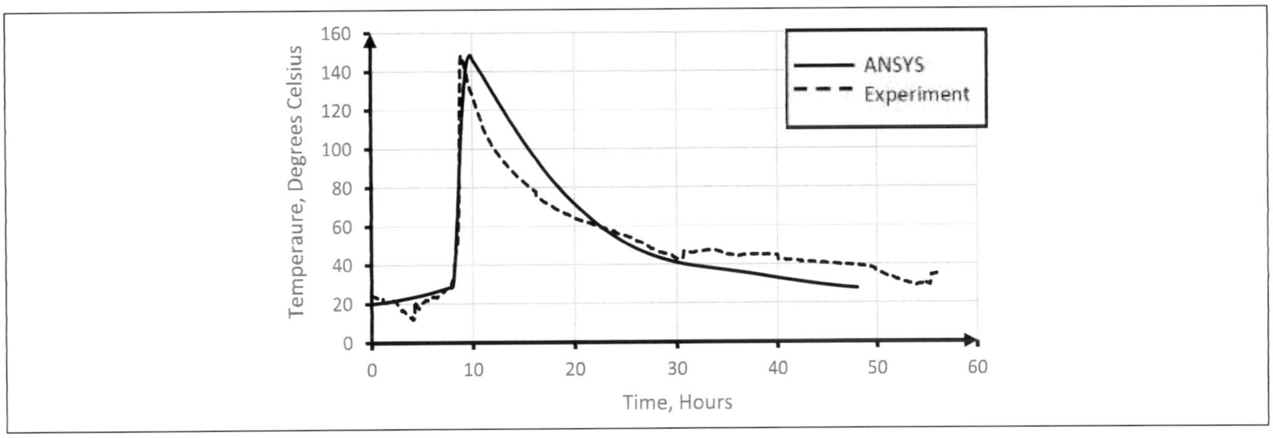


**Figure 11.** Comparison of time–temperature plots in specimen R3.

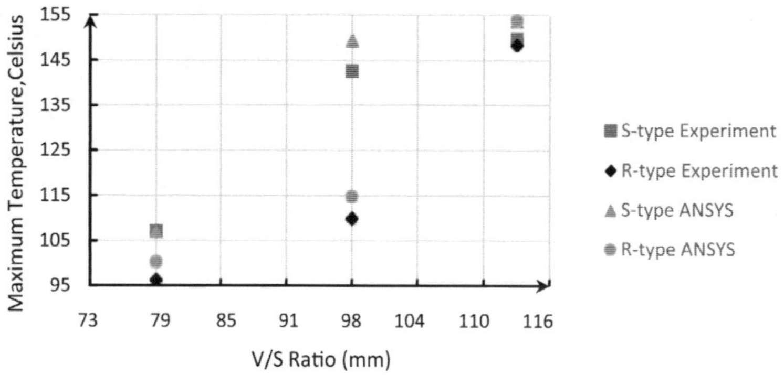


**Figure 12.** Variation in maximum temperature with *V/S* ratio.

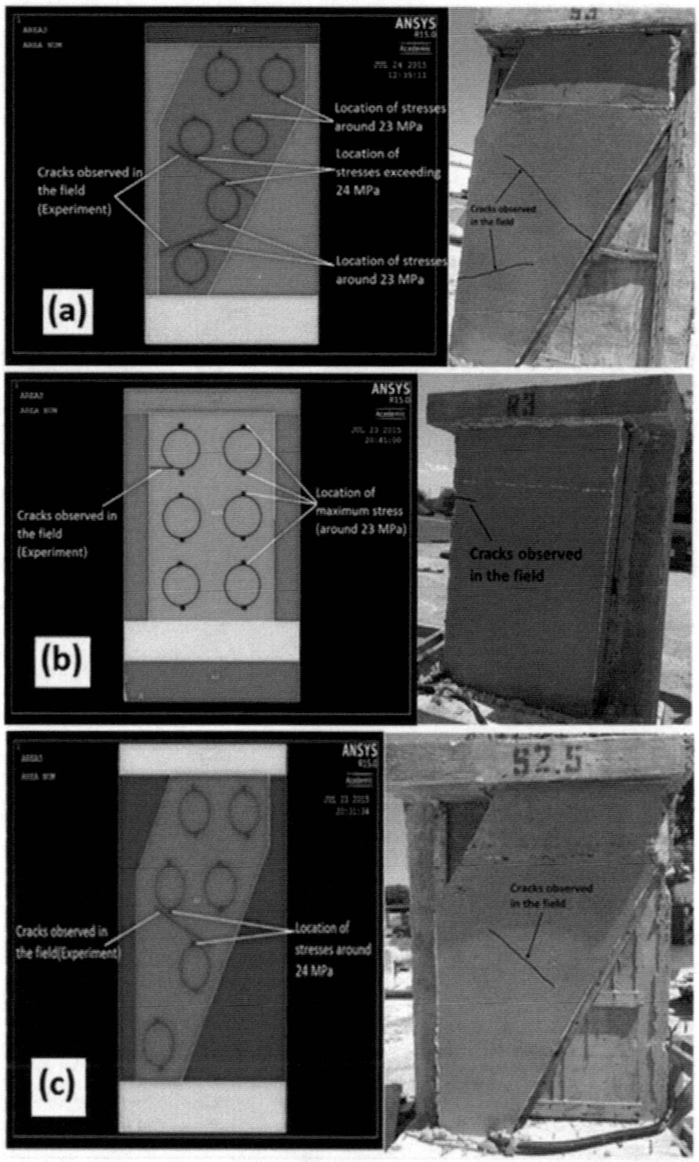


**Figure 13.** Comparison of crack locations from finite element analysis and observed behavior: (*a*) specimen S3; (*b*) specimen R3; and (*c*) Specimen S2.5.

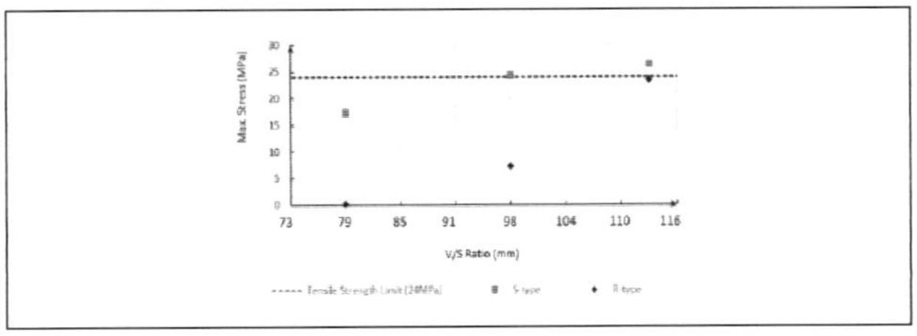


**Figure 14.** Variation in maximum tensile stress with *V/S* ratio based on ANSYS.

5. As higher peak exothermic temperature is associated with higher thermal stresses and cracking potential, it is recommended that epoxy materials with lower peak exothermic temperature should be preferred.
6. It is recommended that the maximum value of V/S ratio of the epoxy grout pourbacks be limited to 91 and 107 mm for S-type and R-type shapes, respectively. This could be achieved by reducing the size of the pourback, however, if not possible individual anchorage or a small group of anchorages could be protected with smaller pourbacks that do not exceed the recommended *V/S* ratio.

## Acknowledgment

The authors thank the Florida Department of Transportation for funding this project.

## References

1. Corven Engineering, Inc. *New Directions for Florida Post-Tensioned Bridges*, Report prepared for Florida Department of Transportation, Volume 1–10, Tallahassee, FL, 2002.
2. Ahmad, I., Suksawang, N., Sobhan, K., Corven, J.A., Sayyafi, E. A., Pant, S., and Martinez, F. "Develop Epoxy Grout Pourback Guidance and Test Methods to Eliminate Thermal/Shrinkage Cracking at Post-Tensioning Anchorages: Phase II," *Final Report*, Contract No. BDV29-977-13, Florida Department of Transportation, 2016.
3. Hamilton, H.R., and Alvarez, G. A. *Post-Tensioning Grout Bleed, Duct, and Anchorage Protection Test*, Final Report submitted to FDOT, Contract No. BC-354 RPWO #73, 2002, p. 54.
4. Florida Department of Transportation. *Approved Product List*, 2015. https://fdotwp1.dot.state.fl.us/ApprovedProductList/ProductTypes/Index/45.
5. ASTM C579-01. "Standard Test Methods for Compressive Strength of Chemical-Resistant Mortars, Grouts, Monolithic Surfacings, and Polymer Concretes," American Society for Testing and Materials, West Conshohocken, PA, 2012.
6. ASTM D2471-99. "Standard Test Method for Gel Time and Peak Exothermic Temperature of Reacting Thermosetting Resins," American Society for Testing and Materials, West Conshohocken, PA, 2008.
7. Stolarski, T., Y. Nakasone, and S. Yoshimoto. *Engineering Analysis with ANSYS Software*, Elsevier, Burlington, MA, 2011.
8. Kim, S. G. "Effect of Heat Generation from Cement Hydration on Mass Concrete Placement," *MS Thesis*, Iowa State University, Ames, IA, 2010.

*The Standing Committee on Concrete Bridges (AFF30) peer-reviewed this paper (18-01740).*

Article

TRR
JOURNAL OF THE TRANSPORTATION RESEARCH BOARD

Transportation Research Record
2018, Vol. 2672(41) 44–55

DOI: 10.1177/0361198118768527
journals.sagepub.com/home/trr
SAGE

# Influence of Thermal Sweep on Girder Stability during Construction

**Jeffrey M. Honig[1], Zachary S. Harper[1,2], and Gary R. Consolazio[1]**

**Abstract**

During construction, girder stability of precast, prestressed concrete bridge girders is adversely affected by fabrication imperfections. Consequently, limits on lateral sweep imperfection caused by fabrication tolerances are imposed by design standards, thus reducing the possibility of girder instability and rollover. However, thermal sweep, induced by solar heating during early stages of construction, can add to pre-existing fabrication tolerances thereby amplifying girder imperfections and reducing stability. In the present study, lateral thermal gradients available in the literature were adopted and enhanced for purposes of computing thermal girder sweep. A variety of girder types—PCI BT-63, Florida-I Beams, and AASHTO Type-V—were then investigated to quantify the influence that lateral thermal sweep has on the stability of individual precast concrete bridge girders under lateral wind load. Previously validated finite element analysis modeling and analysis techniques were used to conduct a parametric study that included 10 girder types, varying span lengths, and five geographic locations. Results revealed that thermal sweep may cause wind carrying capacity reductions of the order of 30 to 60% for typical span lengths, and even greater reductions at span lengths that approach maximum design limits. Consequently, it is crucial that thermal sweep, caused by environmental solar-heating conditions, be considered in construction-stage girder stability analyses.

Precast, prestressed concrete girders are often chosen as a means of achieving economical highway bridge construction. However, as achievable span lengths for such girders continue to increase (due to improvements in materials), consideration of girder stability—specifically during early stages of construction—remains a primary concern from a structural safety and design perspective. During construction, individual girders are initially lifted by crane (Figure 1) and placed into position atop flexible bearing pads located on the bridge supports (e.g., abutments or piers). A critical phase of construction, with regard to stability, occurs after girder placement (but prior to the casting of the deck), when girders are supported only by bearing pads and can be subject to lateral wind loads (Figure 2*a*).

In particular, the stage at which wind loading is often most critical occurs when the first girder is erected. At this stage of construction, other girders are not present to brace against, hence the initial girder can only be anchored to the pier at the ends of the span (i.e., intermediate bracing is not feasible). For long-span bridge designs, in which girder stability is a primary concern, girder erection can potentially be scheduled to minimize the wind exposure period for the initial girder, so that—statistically—it is less likely that peak wind forces will occur. However, meeting such a schedule is not always feasible, and adverse weather conditions cannot necessarily be anticipated. For example, in Florida, strong afternoon thunderstorms with significant wind speeds can form rapidly during summer months. In such situations, it is important to be able to assess, in advance, whether it will be safe to simply set a girder on bearing pads or whether end-span anchors will be needed to prevent girder collapse under the effects of wind loading.

**Figure 1.** Placement of a precast concrete bridge girder during construction. (Photo credit: The Lane Construction Corporation.)

[1]Engineering School of Sustainable Infrastructure and Environment, Department of Civil and Coastal Engineering, University of Florida Gainesville, FL
[2]Project Engineer, Quest Structures, Orinda, CA

**Corresponding Author:**
Address correspondence to Gary R. Consolazio: grc@ce.ufl.edu

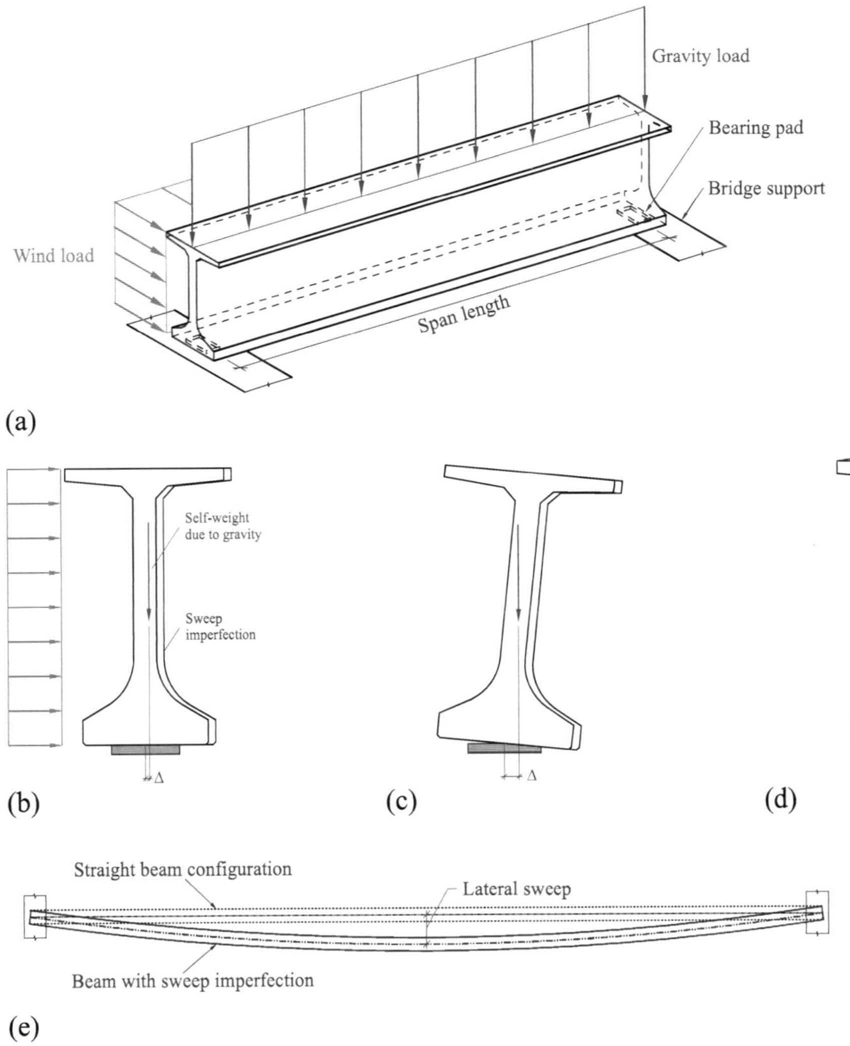


**Figure 2.** Bridge girder loading conditions and susceptibility to buckling: (*a*) overview, (*b*) initial state with applied loads, (*c*) rollover instability, (*d*) lateral-torsional instability, and (*e*) plan view showing lateral sweep.

Girder instability arises when structural deformations caused by load application act to increase the moment arm of the load to such an extent that static equilibrium cannot be achieved. As deformations grow, additional moment (often called secondary moment) causes the structure to deform further, which increases the moment arm even more. In a stable system, this process continues until the structure converges to a deformed state in which static equilibrium is achieved. However, if the load exceeds some critical value (i.e., the buckling load), the system becomes unstable, in which case the process diverges and the structural deformations increase without bound (i.e., the structure collapses). Long-span bridge girders are susceptible to two primary modes of instability: girder rollover and lateral-torsional buckling (Figure 2, *b, c*, and *d*).

Deviations from ideal straightness—such as lateral sweep—tend to increase the potential for girder instability. Sweep is defined as lateral bowing, and is quantified as the maximum deviation from a perfectly straight line (Figure 2*e*) connecting one end of the girder to the other. Lateral sweep arises partially due to fabrication imperfections created after release of prestressing strands, where unintended eccentricity may be produced, resulting in lateral deflection. In an attempt to limit such deflections, quality-control measures require that fabrication-related sweep be restricted to less than ⅛ in. (3.175 mm) for every 10 ft (3.048 m) of girder length (*1*).

Following fabrication, and specifically during transportation and early stages of construction before placement of the bridge deck, precast concrete girders are often subject to

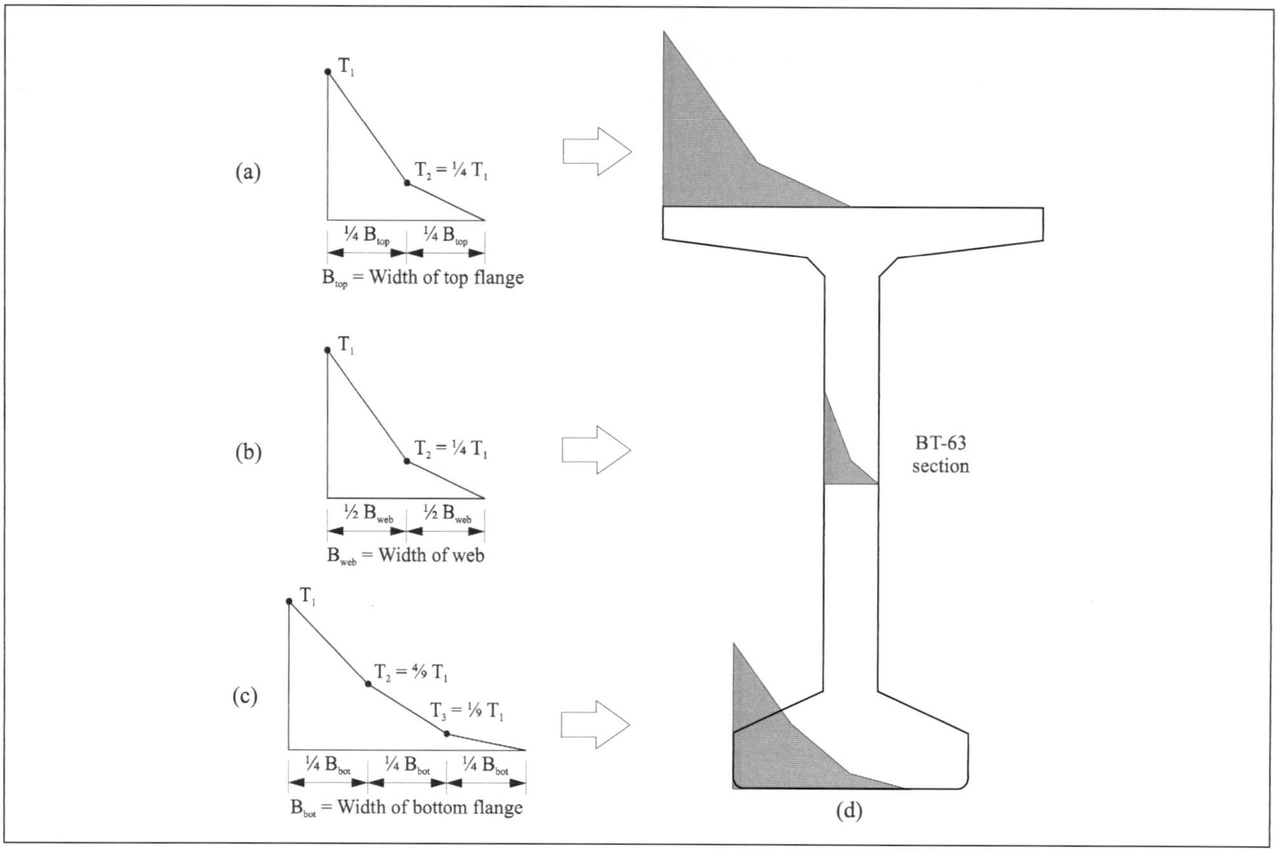


**Figure 3.** Lateral thermal gradients for prestressed concrete girders, per Lee (*4*): (*a*) top flange, (*b*) web, (*c*) bottom flange, and (*d*) applied to BT-63 girder cross-section.

environmental conditions that may increase sweep imperfection. A major source of additional sweep is uneven temperature change (i.e., thermal gradients across the width of the girder cross-section). If one side of the girder is warmer than the other (e.g., due to uneven exposure from the sun), the warmer side will expand more than the cooler side, thus producing curvature and thermal sweep. This thermal component of sweep can add to existing fabrication sweep, thereby increasing girder instability and potential for collapse during construction.

In 2007, a collapse of nine precast concrete bridge girders occurred during construction of a highway bridge in Arizona. An investigation of the incident (*2*) determined that a combination of pre-existing fabrication sweep and superimposed thermal sweep likely contributed to the collapse. It is therefore important that consideration be given to the effects of thermal sweep when designing for girder stability during bridge construction (e.g., when assessing the need for girder end-anchorages). For the research presented in this paper, girder stability analyses were conducted to quantify the extent to which thermal sweep reduces the lateral wind load carrying capacity of typical bridge girders. Varying cross-sectional girder shapes, span lengths, and environmental (heating) conditions were considered.

## Computation of Thermal Sweep

In the AASHTO Load and Resistance Factor Design (LRFD) Bridge Design Specifications (*3*), vertical gradients of temperature are specified for purposes of computing internal forces (and consequential stresses and vertical deflections) caused by non-uniform thermal effects in both concrete and steel superstructures. However, corresponding specifications are not provided for non-uniform gradients of temperature occurring in the lateral direction. To address this lack of guidance pertaining to lateral gradients, Lee (*4*) used experimentally measured temperature data for prestressed concrete girders subjected to solar (environmental) heating, and complementary thermal finite element analyses (FEA), to develop simplified approximations of lateral temperature gradients. Piecewise linear gradients were proposed for the top flange, web, and bottom flange (Figure 3, *a, b*, and *c*). By merging the simplified lateral gradients together with procedures such as moment-curvature beam analysis, maximum lateral deflections (i.e., sweep) induced by thermal gradients can be computed for girders of arbitrary cross-sectional shape and span length. Lee also determined that winter and summer seasonal solar-heating conditions produced the maximum and minimum lateral thermal sweep levels,

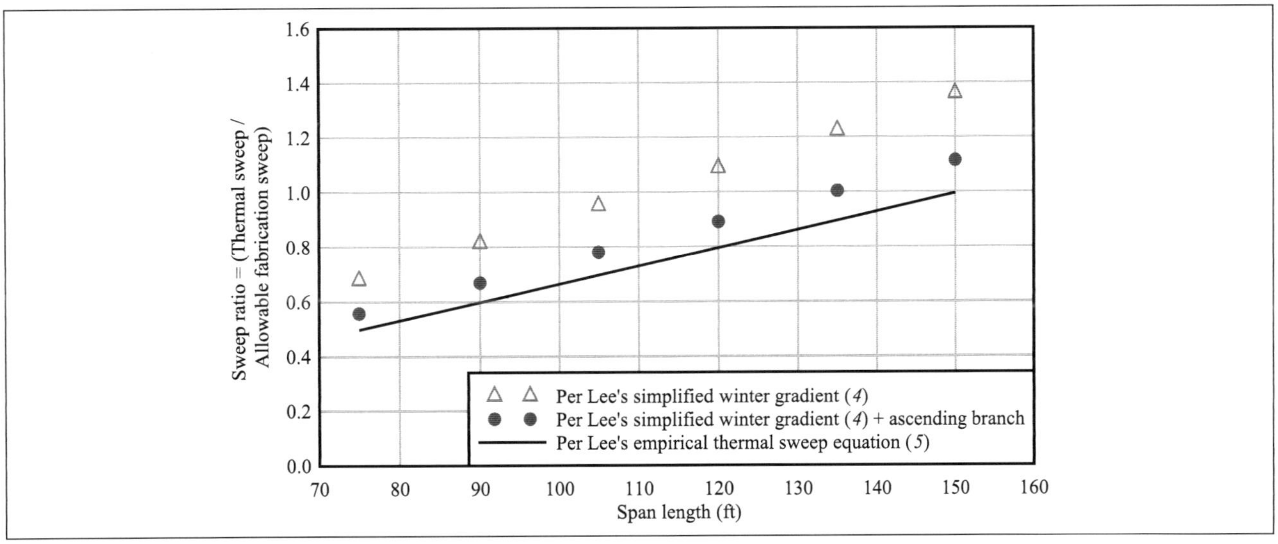


**Figure 4.** Sweep ratios for a BT-63 per simplified gradients proposed by Lee (*4*) and per direct empirical sweep prediction (*5*).

respectively. Given that winter conditions were identified as yielding the largest lateral thermal movements, for conservativism Lee (*4*) recommended that winter temperature gradients be used when calculating thermal sweep for prestressed concrete bridge girders.

In a related study, Lee (*5*) also developed an empirical methodology by which lateral sweep can be computed directly from girder span length, thus avoiding the need to form (and apply) simplified lateral thermal gradients. Empirical expressions for thermal sweep were developed for four typical AASHTO-PCI bridge girder cross-sectional shapes: Type-I, Type-IV, Type-V, and BT-63. In developing this latter approach, Lee used two-dimensional (2D) FEA—validated through the use of experimentally collected temperature data—to calculate both nonlinear lateral temperature gradients and maximum lateral deflections caused by those gradients (*5*). Environmental data—e.g., solar radiation level (based on geographic location), ambient air temperature, wind speed—were incorporated into the model and were chosen to represent "peak severity" conditions, as determined from a review of data recorded over a 30-year period.

In each of Lee's studies (*4, 5*), the environmental parameters considered were primarily based on conditions in Atlanta, Georgia as the experimental measurements used in model validation were obtained at this same location. However, to facilitate application of the proposed methodologies to a broader set of geographic locations, Lee also proposed peak severity environmental conditions for seven additional cities distributed across the United States.

In the present study, the simplified gradients proposed by Lee were: modified to enhance their accuracy, applied to a wide variety of girder cross-sectional sizes and shapes, and ultimately employed in an investigation of the influence of thermal sweep on girder buckling capacity. To assess the accuracy of Lee's simplified gradients (*4*) prior to applying them to new girder shapes and span lengths, they were first applied to the BT-63 girder cross-sectional shape (Figure 3*d*) studied by Lee. Combining the gradients, the BT-63 shape, and moment-curvature analysis, lateral sweep deflections were computed for various span lengths, and compared with the results obtained with the more accurate empirical lateral deflection equation reported by Lee (*5*). For purposes of comparing results from the two approaches, a "sweep ratio" parameter was defined as maximum thermal sweep divided by maximum allowable fabrication sweep, where maximum fabrication sweep was computed in accordance with PCI (*1*). Calculations were performed for a BT-63 girder at varying span lengths for winter conditions in Atlanta using both methods. A comparison of the data revealed that application of simplified gradients (Figure 4, triangles) produced thermal sweep levels significantly larger (i.e., more conservative) than those produced by use of the direct (and more accurate) sweep deflection equation (Figure 4, solid line).

Lee's simplified gradients were derived from more refined nonlinear gradients computed from FEA. Each FEA gradient (Figure 5*a*) consisted of a continuously descending "left side" branch—with the maximum left side temperature denoted as $T_1$ ("T-one") by Lee, corresponding to direct solar heating—and an ascending "right side" branch—corresponding to secondary solar heating. Lee approximated the left side descending branch FEA gradient with a piecewise linear approximation, and intentionally—and conservatively—omitted the ascending right side branch altogether (*4*) (Figure 5*b*). Secondary ("right side") heating tends to offset curvature produced by primary (left side) heating. Consequently, omitting any approximation of the FEA right side ascending branch from the simplified gradients leads to computed thermal sweep levels that are significantly conservative (Figure 4, triangles) relative to more accurate predictions. To reduce the level of conservatism, in the present

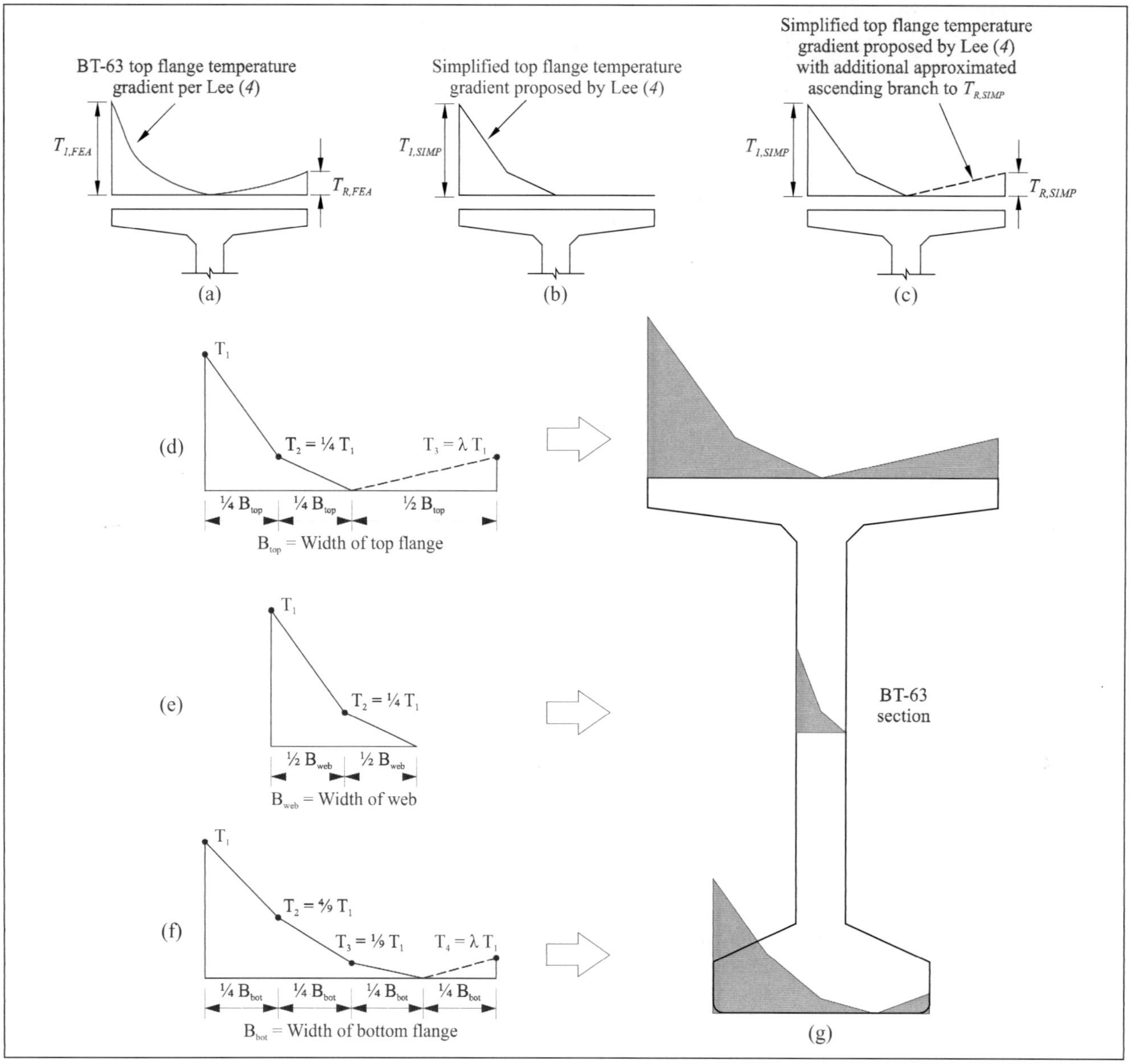


**Figure 5.** Modified transverse thermal gradients for prestressed concrete girders: (*a*) from finite element analysis of a typical BT-63 girder (4), (*b*) simplified gradient proposed by Lee (4), (*c*) approximated right side ascending branch, (*d*) top flange thermal gradient, (*e*) web thermal gradient, (*f*) bottom flange thermal gradient, and (*g*) modified thermal gradients applied to a BT-63 girder cross-section.

study an approximate ascending branch was added to the simplified ("left side only") gradients proposed by Lee (*4*). Using the refined nonlinear FEA temperature gradients documented by Lee (*4*), the following temperature ratio ($\lambda$) was defined:

$$\lambda = \frac{T_{R,FEA}}{T_{1,FEA}} \quad (1)$$

where $T_{R,FEA}$ and $T_{1,FEA}$ are the maximum right side and left side temperatures, respectively, both obtained from Lee's refined thermal FEA results. A simplified linear right side ascending branch (Figure 5*c*) was constructed by defining the maximum simplified right side temperature as:

$$T_{R,SIMP} = \lambda(T_{1,SIMP}) \quad (2)$$

where $T_{1,SIMP}$ was the maximum left side temperature used in Lee's simplified gradients (*4*). An analogous approach was used to add a linear ascending branch to the simplified bottom flange gradient as well. Since Lee's FEA gradients in the web did not contain an ascending right side branch, due to the narrow width of the web, no change was made to the web temperature gradient. To test the effectiveness of the ascending

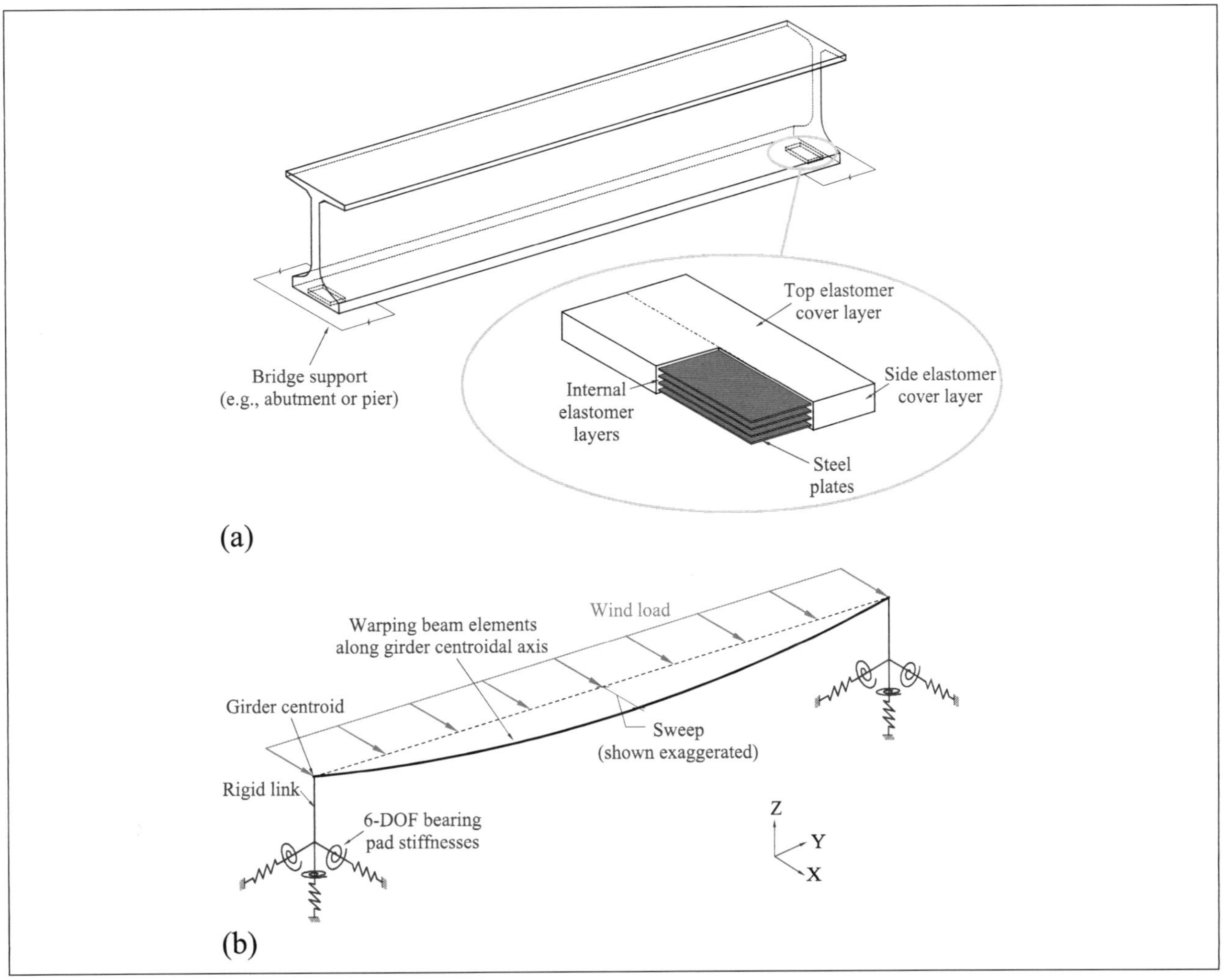


**Figure 6.** Concrete girder on an elastomeric bearing pad: (*a*) location and structure of the neoprene pads (isometric view), and (*b*) finite element model of a single unanchored concrete girder (isometric view).

linear branches, the "enhanced simplified" gradients (Figure 5, *d, e*, and *f*) were applied to the BT-63 girder cross-section (Figure 5*g*) to compute thermal sweep levels at various span lengths. As shown in Figure 4 (solid circles), the enhanced "Lee plus ascending branch" approach produced sweep levels that were reasonably close to the more accurate direct empirical sweep equation (solid line), while also maintaining a reasonable margin of conservatism (*4*). Based on these results, the enhanced simplified gradients (Figure 5, *d, e*, and *f*) were used to compute thermal sweep data in the remainder of the study presented in this paper.

## Modeling and Analysis Procedures

To assess the stability of unanchored concrete girders during construction, specifically, after being lifted and placed into final position (Figure 6*a*), individual girders were modeled (Figure 6*b*) and structurally analyzed using the ADINA FEA code (*6*). In each analysis, the primary objective was to quantify the magnitude of lateral wind load that would destabilize the girder, thus causing lateral torsion buckling and eventual rollover. Given that girder stability is reduced by the presence of lateral imperfection, all models in this study incorporated lateral sweep that represented the additive effects of both thermal sweep and construction tolerance. Maximum thermal sweep was computed using the enhanced simplified lateral gradients approach, and construction sweep was set at ⅛ in. (3.175 mm) for every 10 ft (3.048 m) of girder span length, per PCI (*1*). Geometrically, sweep was implemented using a half-sine wave function with maximum sweep at midspan.

In addition to lateral sweep, vertical camber was also included in each girder model. Introduced due to strand eccentricity, camber has the effect of elevating the center of gravity of a girder above the supports (typically maximizing at midspan). Importantly, when a girder is initially placed on the supports during construction, the weight of the bridge deck is not yet present. Consequently, individual girder camber levels during construction are considerably larger than those that will be present in the final completed bridge. Since stability is influenced by the elevation of the girder center of gravity, camber was included in all models analyzed in this study. As documented in Consolazio et al. (*7*), maximum girder camber levels implemented in the stability analysis models were computed based on trial beam designs that incorporated typical span lengths and placement of all prestressing tendons as low in the bottom flange as feasible.

Girder capacities were quantified using geometrically nonlinear large-displacement analyses in which static loads were applied in incremental steps, taking into account the deformed state of the girder at each load step. By analyzing the resultant load-displacement data—using a modified-Southwell analysis, as reported in Consolazio et al. (*7*)—it was possible to determine when lateral girder displacements began growing asymptotically, indicating a buckling collapse. To enable parametric studies involving many hundreds of analyses to be performed, a computationally efficient analysis approach (*8*) was employed that 1) was capable of accurately representing lateral-torsional buckling and rollover behavior, and 2) had been previously validated through physical laboratory testing (*9*). Among other analytical features, each bridge girder was modeled using warping beam elements, an advanced beam element formulation available in ADINA (*6*) that incorporates degrees of freedom associated with torsionally induced out-of-plane warping of the cross-section. Importantly, the kinematic formulation of the warping beam element includes coupling between bending (flexural) and torsional deformation modes. Use of warping beam elements requires calculation of a comprehensive set of cross-sectional properties, several of which require that the warping function be known for the cross-sectional shape under consideration. For complex shapes such as bridge girders, the warping function cannot be calculated in a mathematically closed-form manner, but must instead be computed numerically. In the present study, each girder cross-section of interest (BT-63, etc.) was meshed and analyzed (using scalar-field FEA) to compute the warping function as well as the necessary cross-sectional properties (torsional constant, warping constant, etc.). Additional details of this process are provided in Consolazio et al. (*7*).

Each girder was assumed to be supported on steel-reinforced elastomeric bearing pads (Figure 6*a*) that were capable of undergoing four primary modes of deformation (roll, torsion, shear, axial). Bearing pad stiffnesses corresponding to six separate degrees of freedom (X-, Y-, Z-translations, and X-, Y-, Z-rotations) were modeled using six independent springs, for which the relevant properties were computed using methods discussed in Consolazio et al. (*7*) and Harper and Consolazio (*10*).

Each model was subjected to two types of structural loads: wind and gravity. Lateral wind loads were applied to the model as tributary nodal loads and were applied in the direction of increasing girder sweep (Figure 6*b*). Concurrently, gravity load was applied as a vertical "acceleration load" (i.e., a mass-proportional body force) in units of *g*, the acceleration due to gravity, so that a load of 1 g represented the self-weight of the girder. In field conditions, girders are always subjected to a constant gravity load of 1 g. However, in the structural models analyzed in this study, gravity loading was used as the mechanism of initiating instability. For each girder model, a uniform wind load was applied (and held constant), as gravity load was linearly increased—beyond 1 g if possible—until girder instability occurred. Linearly increasing the vertical gravity load after applying lateral wind load ensured robust detection of girder buckling by the modified-Southwell data processing algorithm that was used (*7*). Vertical load carrying capacity was then evaluated several times at different lateral wind pressures, iterating until the vertical capacity was within 1% of 1 g (i.e., girder self-weight). The resulting wind pressure for each case was denoted as the lateral wind capacity of that girder, representing the maximum wind load that could be sustained by the girder without collapsing.

## Influence of Thermal Sweep on Girder Rollover

### *Girder Types Investigated*

To investigate the influence that thermal sweep has on the stability of individual precast concrete bridge girders under lateral wind load, a parametric study was performed using FEA models of 10 different girder cross-sectional shapes (Figure 7) and suitable span length ranges for each. For each combination of girder type and span length, two separate conditions were analyzed: 1) without thermal sweep, and 2) with thermal sweep. In this manner, the extent to which thermal sweep reduced lateral wind carrying capacities was isolated and numerically quantified for each case. Multiple heating environments corresponding to geographic locations from across the continental United States were also considered for each girder and span length combination.

Primary girder types investigated in this study were Florida-I Beams (FIBs), which consist of a group of eight different cross-sectional shapes ranging in depth from 36 to 96 in. Additionally, BT-63 and Type-V girder cross-sections—both with a depth of 63 in.—were also included. As such, comparisons could be made among three different commonly used 63-in. deep sections: FIB63, BT-63, and Type-V.

Span length ranges investigated for each FIB section (Table 1) were based on Florida Department of Transportation (FDOT) design guidance (*11*). Span length ranges for the BT-63 and Type-V sections were chosen to coincide with those employed for the FIB63, so that parameter variations among the 63-in. deep sections were limited strictly to cross-sectional properties (Table 1). Individual span lengths investigated were incremented at 5-ft intervals within each range shown in Table 1 (e.g., FIB36: 75 ft, 80 ft, 85 ft, etc. to 120 ft). Material

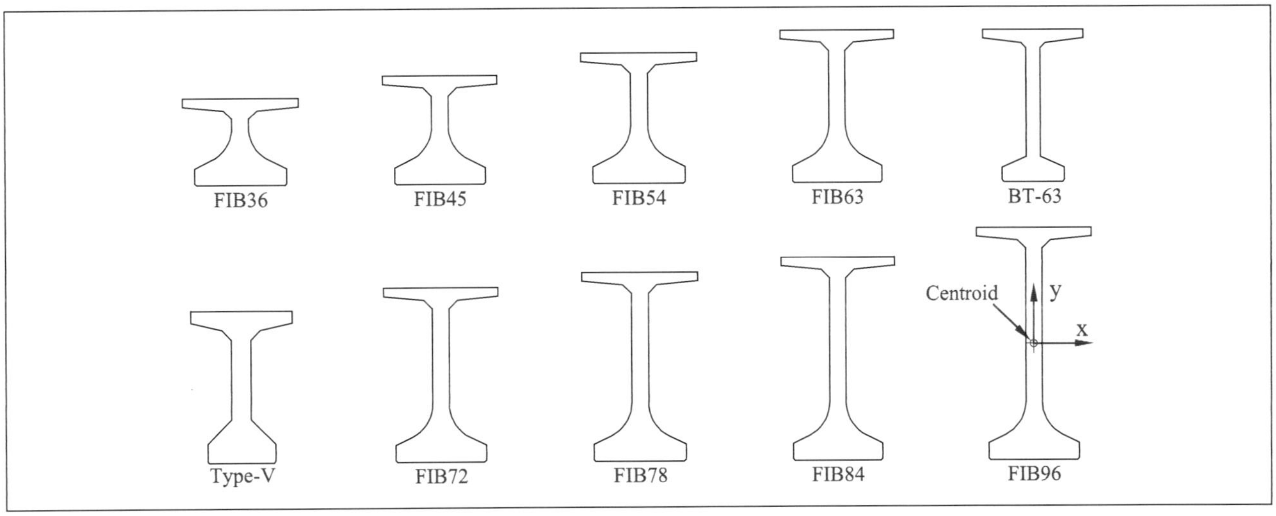


**Figure 7.** Girder cross-sections considered.

**Table 1.** Girder Cross-Sectional Properties and Span Lengths Investigated

| Section | Depth (in.) | A (in.$^2$) | $I_{xx}$ (in.$^4$) | $I_{yy}$ (in.$^4$) | J (in.$^4$) | Span length range (ft) |
|---|---|---|---|---|---|---|
| FIB36 | 36 | 807 | 127,700 | 81,283 | 30,864 | 75–120 |
| FIB45 | 45 | 870 | 226,810 | 81,540 | 31,885 | 95–140 |
| FIB54 | 54 | 933 | 360,270 | 81,798 | 32,939 | 110–155 |
| FIB63 | 63 | 996 | 530,790 | 82,055 | 33,973 | 120–170 |
| BT-63 | 63 | 713 | 392,520 | 37,472 | 11,771 | 120–170 |
| Type-V | 63 | 1,013 | 521,160 | 61,235 | 38,849 | 120–170 |
| FIB72 | 72 | 1,059 | 741,060 | 82,314 | 35,041 | 135–190 |
| FIB78 | 78 | 1,101 | 904,610 | 82,484 | 35,693 | 145–195 |
| FIB84 | 84 | 1,143 | 1,087,800 | 82,657 | 36,421 | 155–205 |
| FIB96 | 96 | 1,227 | 1,516,200 | 83,002 | 37,859 | 170–225 |

*Note:* A = Cross-sectional area; Ixx = Strong-axis moment of inertia; Iyy = Weak-axis moment of inertia; J = St. Venant torsional constant.

properties assumed for the prestressed concrete girders were $fc'$ = 8.5 ksi (which are minimum permissible values for Florida bridge construction), unit weight = 150 pcf, and Poisson's ratio = 0.2. Based on these values, and PCI guidance (*12*), the concrete elastic modulus was computed as E = 5,589 ksi.

## *Atlanta, Georgia, (United States): Illustrative Example*

Procedures used to quantify the influence of thermal sweep on lateral wind capacity are illustrated in this section for Atlanta and subsequently expanded to additional distributed geographic locations in the following section. For each girder cross-sectional shape, thermal sweep levels were computed at selected span lengths using the enhanced simplified gradient formation approach (described earlier) and moment-curvature analysis. All thermal sweep levels were calculated (Figure 8*a*) using the winter temperature data provided by Lee (*4*). Next, the thermal sweep deflections were divided by the maximum allowable fabrication sweep deflections to form thermal sweep ratios. Based on the clearly observable linear correlation between sweep ratio and span length (Figure 8*b*), a linear best-fit relationship was formed to represent all FIB sections. For the BT-63 and Type-V sections, slightly differing linear fits were independently constructed in the same fashion. The best-fit linear expressions (for FIB, BT-63, Type-V) were then used as mathematically convenient means of computing thermal sweep levels for each span length considered in the parametric study. For each girder analyzed, the total sweep deflection was modeled as the superposition of thermal sweep (from the best-fit expression) and maximum allowable fabrication sweep.

Results from girder stability analyses conducted for the Atlanta geographic location are summarized in Figure 9. Individual girder wind load capacities, plotted as a function of span length, are presented for all 10 girder types in Figure 9, *a* and *b*, for cases without and with thermal sweep, respectively. At longer span lengths, capacities computed

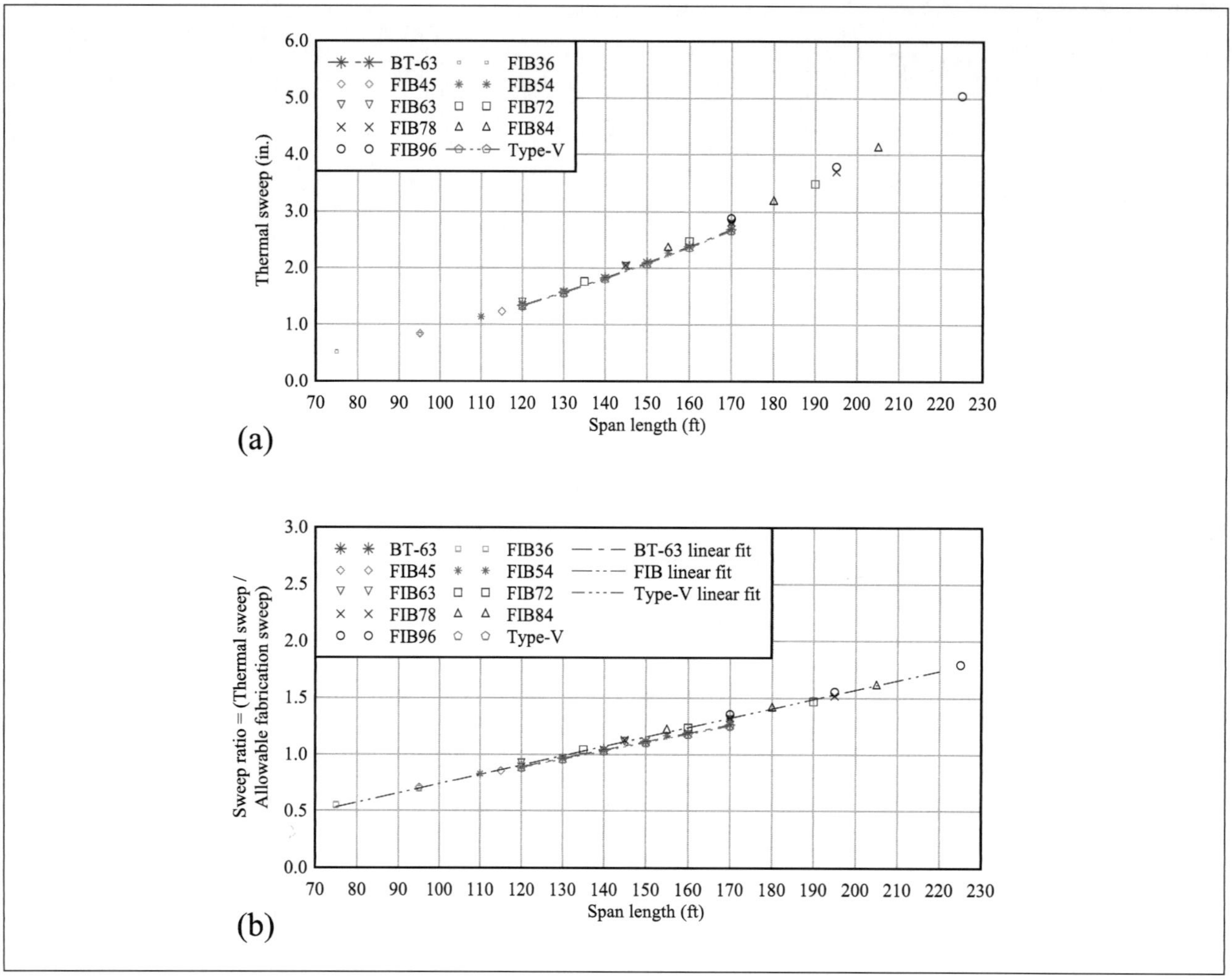


**Figure 8.** Thermal sweep data (Atlanta, Georgia, USA, winter conditions): (*a*) thermal sweep data, and (*b*) thermal sweep ratios with linear fits.

with consideration of thermal sweep (Figure 9*b*) are noticeably reduced relative to those that ignore thermal sweep (Figure 9*a*). These capacity reductions are even more apparent in Figure 9*c* in which, for example, at a span length of approximately 180 ft, inclusion of thermal sweep reduces the lateral wind load carrying capacity by approximately 70%. More generally, Figure 9*c* illustrates that lateral wind capacities decline sharply with increasing span length, and that capacity reductions of the order of 30 to 60% are possible for typical span lengths and typical girder shapes. Information of this type can help inform decision-making that relates to requiring temporary bracing or potentially delaying girder placement when unfavorable wind conditions exist.

## Distributed Geographic Locations

To broaden the scope of this study with regard to variations of environmental (heating) conditions, four additional geographic locations—distributed across the continental United States, and studied by Lee (*4*)—were also included: Alamosa, Colorado; Brownsville, Texas; Caribou, Maine; and Medford, Oregon. In the same fashion as discussed for Atlanta using winter data provided by Lee (*4*) for each additional city, thermal sweep data and subsequent sweep ratios were computed, and then used to form linear fits (analogous to Figure 8*b*). The additional city-specific linear fits were then used to introduce thermal sweep into an expanded set of girder stability analyses. Reductions in lateral wind load carrying capacity that occurred due to inclusion of thermal sweep are presented in Figure 10. Data for Atlanta taken from Figure 9*c* are repeated in Figure 10*a* for convenience of comparison. The four additional geographic locations are presented in Figure 10, *b, c, d,* and *e*, and the collection of data for all five cities is shown in Figure 10*f*. Similar to the results shown for Atlanta (Figure 10*a*), computed reductions of girder wind capacities for the additional cities (Figure 10, *b, c, d,* and *e*) indicate that at moderate to long span lengths, thermal sweep significantly

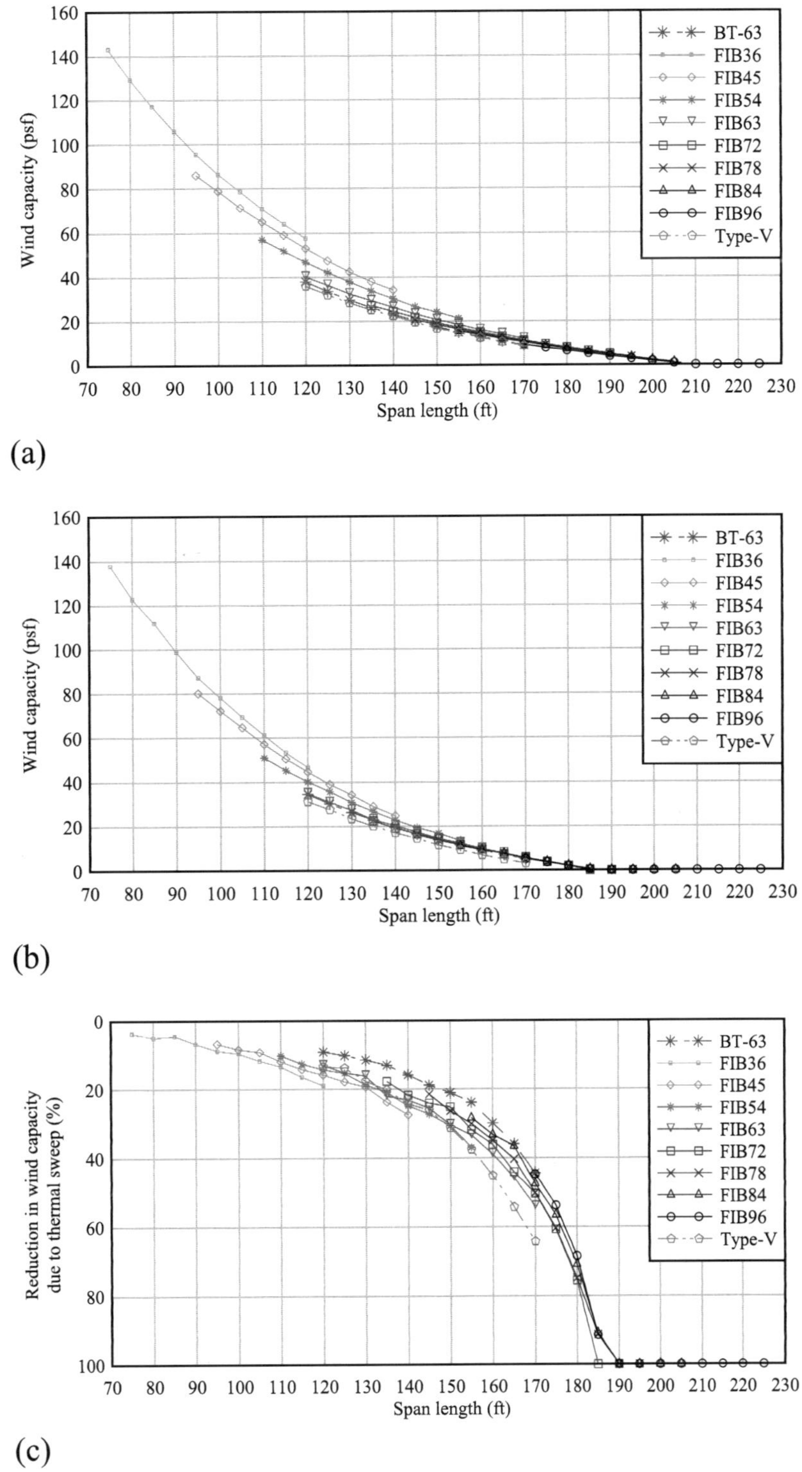


**Figure 9.** Wind capacities for unanchored girders: (*a*) without inclusion of thermal sweep, (*b*) with inclusion of thermal sweep for Atlanta winter data, and (*c*) reduction in wind capacities due to thermal sweep (for Atlanta winter data).

reduces the lateral wind load carrying capacity of typical bridge girders. Consequently, thermal sweep should be considered in construction-stage girder stability analyses, independent of geographic location or girder type (Figure 10*f*).

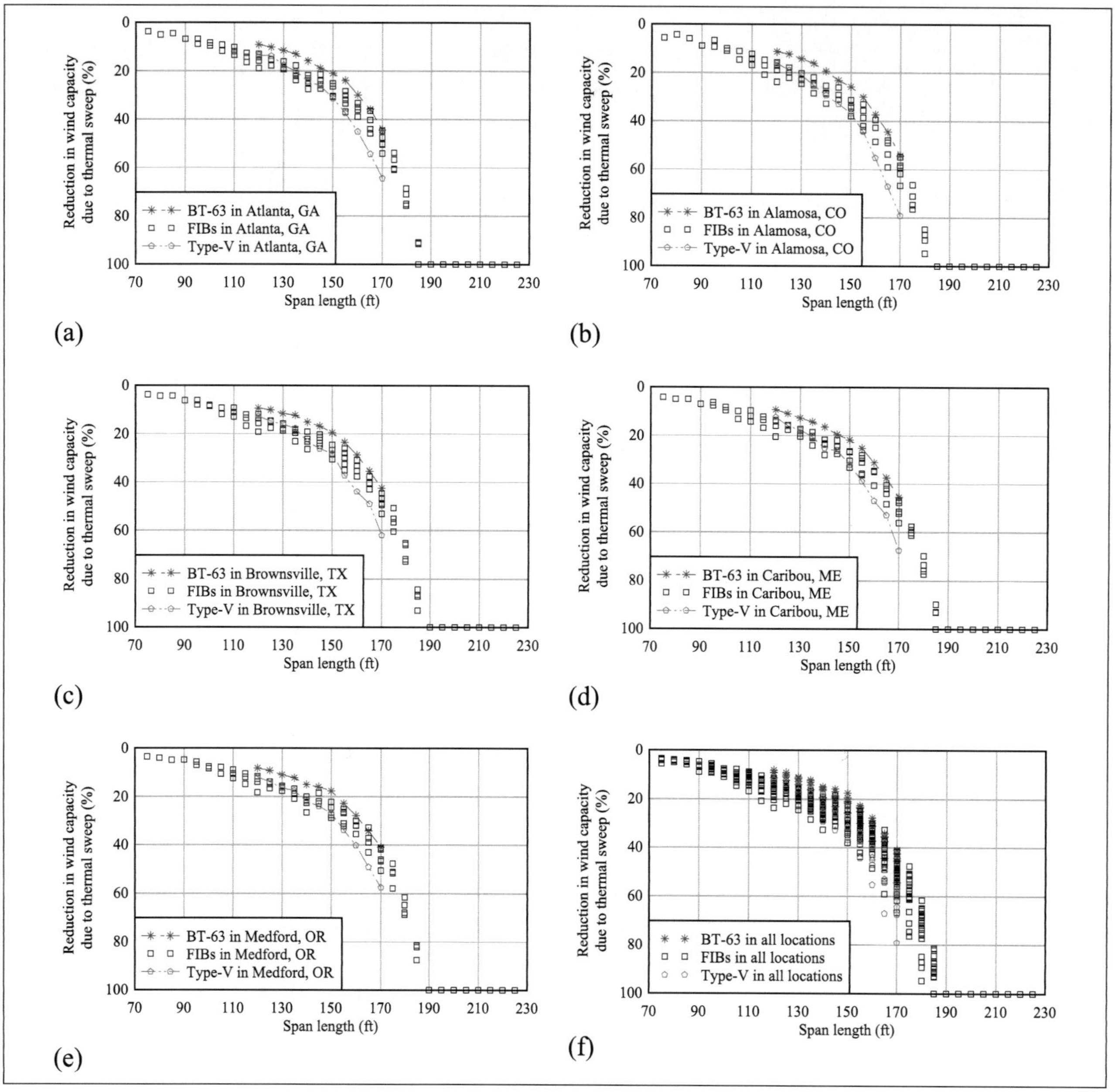


**Figure 10.** Reduction in girder wind capacities due to thermal sweep computed for: (*a*) Atlanta, Georgia, (*b*) Alamosa, Colorado, (*c*) Brownsville, Texas, (*d*) Caribou, Maine, (*e*) Medford, Oregon, and (*f*) all geographic locations compiled.

## Conclusions

In this paper, lateral temperature gradients were used to quantify the influence that thermal sweep has on the stability of unanchored concrete bridge girders under wind loading conditions during construction. Since AASHTO (*3*) only provides specifications for non-uniform vertical temperature gradients, without corresponding specifications for lateral gradients, published research (*4*) was reviewed, enhanced, and used to compute lateral thermal sweep deflections at various span lengths for the following girder types: FIBs, PCI BT-63, and AASHTO Type-V. Girder stability analyses were then conducted, using FEA models of differing girder shapes and span lengths, under two conditions: 1) with inclusion of thermal sweep, and 2) without thermal sweep. (In both cases, sweep due to fabrication tolerances was also included). To cover a broad range of environmental heating conditions found across the United States, five different geographically distributed cities were investigated. For each city, girder type, and span length, the reduction of girder wind carrying capacity that was caused by thermal sweep was quantified. A key finding from the parametric study was that capacity reductions of the order of 30 to 60% may be expected—due to the influence of thermal sweep—for typical girder types and typical span length. It should be noted that in this study, for both cases (with and without thermal effects), girder wind carrying capacities were computed without any factors of safety. If the results presented in this paper are to be used in developing design guidance, appropriate factors of safety should be introduced. Considering the severe consequences of stability failure, factors of safety associated with stability would typically be chosen to be larger than those associated with ultimate design strength. Thermal sweep should thus be considered in construction-stage girder stability analyses, particularly in situations where maximum feasible span lengths are being approached in bridge design.

## Acknowledgment

Majority funding for the work presented in this paper was provided by the Florida Department of Transportation (FDOT) under contracts BDK75 977-33 and BDV31-977-46.

## References

1. Precast/Prestressed Concrete Institute (PCI). *Bridge Design Manual*, 3rd ed. PCI, Chicago, Ill., 2011.
2. Oesterle, R. G., M. J. Sheehan, H. R. Lotfi, W. G. Corley, and J. J. Roller. *Investigation of Red Mountain Freeway Bridge Girder Collapse*. Final Report, CTL Group Project No. 262291. CTL Group, Skokie, Ill., 2007.
3. American Association of State Highway and Transportation Officials (AASHTO). *AASHTO LRFD Bridge Design Specifications*, 7th ed. AASHTO, Washington, D.C., 2014.
4. Lee, J. H. Investigation of Extreme Environmental Conditions and Design Thermal Gradients during Construction for Prestressed Concrete Bridge Girders. *Journal of Bridge Engineering*, Vol. 17, No. 3, 2011, pp. 547–556. https://doi.org/10.1061/(ASCE)BE.1943-5592.0000277.
5. Lee, J. H. Behavior of Precast Prestressed Concrete Bridge Girders Involving Thermal Effects and Initial Imperfections during Construction. *Engineering Structures*, Vol. 42, 2012, pp. 1–8. https://doi.org/10.1016/j.engstruct.2012.04.003.
6. ADINA. *Theory and Modeling Guide, Volume 1: ADINA Solids & Structures*. ADINA R&D, Inc, Watertown, Mass., 2016.
7. Consolazio, G. R., K. R. Gurley, and Z. S. Harper. *Bridge Girder Drag Coefficients and Wind-Related Bracing Recommendations*. Structures Research Report No. 2013/87322. University of Florida, Gainesville, Fla., 2013.
8. Consolazio, G. R., and J. M. Honig. *Distribution Factors for Construction Loads and Girder Capacity Equations*. Structures Research Report No. 2017/124011. University of Florida, Gainesville, Fla., 2017.
9. Consolazio, G. R., H. R. Hamilton, and M. S. Berry. *Experimental Validation of Bracing Recommendations for Long-Span Concrete Girders*. Structures Research Report No. 2012/72909–74040. University of Florida, Gainesville, Fla., 2012.
10. Harper, Z. S., and G. R. Consolazio. Calculation Method for Quantifying Axial and Roll Stiffnesses of Rectangular Steel-Reinforced Elastomeric Bridge Bearing Pads. *Transportation Research Record: Journal of the Transportation Research Board*, 2013. 2331: 3–13.
11. Florida Department of Transportation (FDOT). *Instructions for Design Standard No. 20010: Prestressed Florida-I Beams*. FDOT, Tallahassee, Fla., 2012.
12. Precast/Prestressed Concrete Institute (PCI). *PCI Design Handbook*, 7th ed. PCI, Chicago, Ill., 2010.

*The Standing Committee on Concrete Bridges (AFF30) peer-reviewed this paper (18-01771).*

Article

TRR
JOURNAL OF THE TRANSPORTATION RESEARCH BOARD

Transportation Research Record
2018, Vol. 2672(41) 56–64

DOI: 10.1177/0361198118768529
journals.sagepub.com/home/trr
SAGE

# Thermal Field Characteristics of Reinforced Concrete Box Girder during High-Temperature Asphalt Pavement Paving

**Dong Zheng[1], Zhen-dong Qian[1], De-xuan Liu[1], Xiang-fei Zhang[1], and Yang Liu[1]**

## Abstract

During high-temperature asphalt pavement paving, the bridge is subjected to significant temperature variations, and thereby the nonlinear thermal field distribution that occurs in the box girder might threaten the safety and serviceability of the bridge structures. Therefore, this study aims to investigate the thermal field characteristics of reinforced concrete box girder with steel fiber-reinforced concrete leveling layer during asphalt pavement paving. Firstly, a three-dimensional thermal field model of a reinforced concrete segmental box girder was developed and validated using the transient thermal field theory, and the dynamic paving process was simulated by the element deletion method. Secondly, the thermal field model was applied to analyze the vertical, transversal, and longitudinal temperature distribution during asphalt pavement paving. Thirdly, the effects of various sensitive factors on the thermal field of bridge components during asphalt pavement paving were evaluated. Finally, the extreme temperature load model was established through the statistical analysis method, in which the sensitive factors were introduced. Results demonstrate that the proposed numerical simulation method is feasible for thermal field characteristic analysis of reinforced concrete box girder during asphalt pavement paving. In addition, the research results can provide the theoretical foundation for temperature load selection and structural safety design of the concrete box girder during asphalt pavement paving.

In concrete bridge structures, a complex thermal field induced by temperature variations has frequently caused large stress and deformation which may be comparable to the load-induced response (*1*). This stress and deformation, if not taken into account accurately, can lead to unexpected thermal cracking and structural displacements which have an adverse impact on the structural safety and serviceability of concrete bridges (*2,3*). Therefore, it is essential to investigate the thermal field characteristics of the concrete bridge under varying conditions.

Basically, the thermal field distribution in the bridge structure is affected by three groups of factors (*4,5*). The first is the periodic ambient variation, including air temperature, solar radiation, wind speed, and so forth. Starting from the 1960s, many studies have been devoted to the thermal behaviors of bridges under different environmental conditions. For example, Tong et al. developed a two-dimensional finite element model to analyze the combined influence of various environmental parameters on the temperature distribution in a steel bridge (*6*). The second is the property of bridge components, including cross-section geometry, thermal parameters of materials, and so forth. With the increasing requirements on the performance of bridges, some concrete box girders with unconventional cross-sections and high-performance structural materials have been developed and studied in recent years (*7*). Kong et al. integrated the field measurement and numerical modeling method to investigate the thermal field distributions of bridges with glass fiber-reinforced polymer composite panels (*8*). The third group of factors is sudden temperature variation, including hydration heat of cement, fire loading caused by traffic accidents, and so on (*9–11*). Subramaniam et al. studied the effect of hydration heat release on the thermal movements of concrete deck and steel girders (*10*). Xi et al. utilized a numerical approach to analyze the static behavior of a steel beam under the effect of fire (*11*).

To date, most of the previous studies have focused on the thermal behaviors of different bridges under the effect of periodic ambient variations, but relatively little attention has been given to the role of asphalt pavement paving in the temperature distribution of bridge structures. Hot-mixed asphalt (HMA) concrete, as a common material

[1]Intelligent Transport System Research Center, Southeast University, Nanjing, P.R. China

**Corresponding Author:**
Address correspondence to Zhen-dong Qian: qianzd@seu.edu.cn

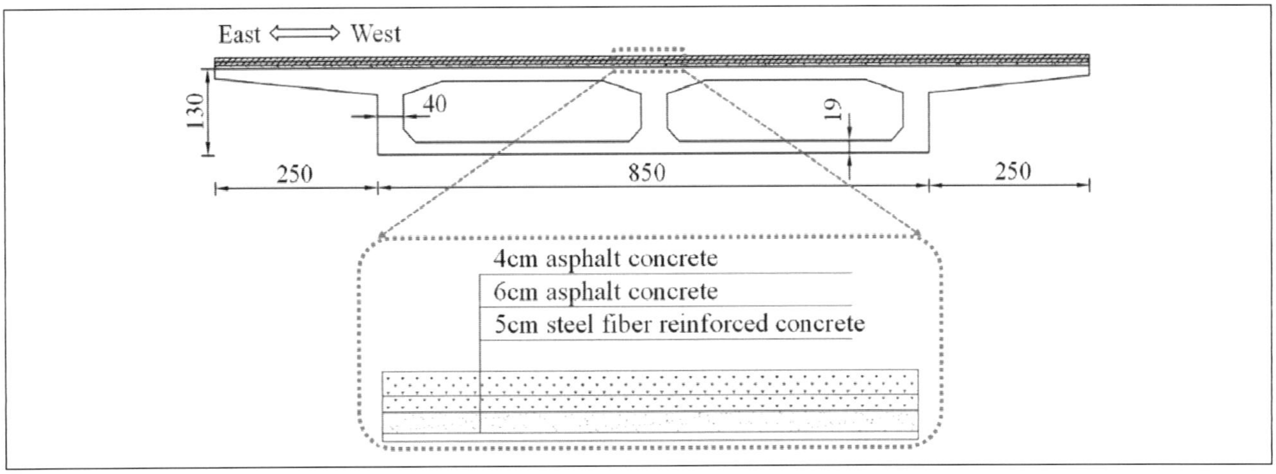


**Figure 1.** Cross-section of the Dantu Main Line Bridge (unit: cm).

used for bridge deck pavement, can be considered as a moving heat source because of its high paving temperature. Once HMA concrete is paved on the bridge deck at a certain speed, the bridge will be inevitably subjected to significant non-uniform temperature changes, which may affect the structural performance (*12,13*). Furthermore, many concrete bridges apply steel fiber-reinforced concrete (SFRC) as the leveling-layer material. In addition to excellent impact strength and toughness, SFRC also has high thermal conductivity, which can affect the temperature gradients in the bridge (*14,15*).

The purpose of this study is to analyze the thermal field characteristics of a reinforced concrete box girder bridge with SFRC leveling layer during high-temperature asphalt pavement paving. Firstly, based on the transient temperature field theory, a validated thermal field model of concrete box girder was established, and the dynamic process of asphalt pavement paving was simulated using the element deletion algorithm. Secondly, the vertical, transversal, and longitudinal temperature distributions of the concrete box girder during asphalt pavement paving were investigated. Thirdly, a sensitivity study was conducted to evaluate the effects of various factors on the thermal field distribution of bridge components during asphalt pavement paving. Finally, the extreme temperature load during asphalt pavement paving was modeled using the statistical analysis method, in which the sensitive factors were introduced.

## Thermal Field Model

### Bridge Description and Thermal Parameters of Materials

The Dantu Main Line Bridge, selected for modeling the thermal field in this study, is a reinforced concrete box girder bridge with 1.3 m-high cross-sections (*13,16*). The bridge is located in Zhenjiang, China, and the orientation of the bridge axis is along the north–south direction. The box girder, as shown in Figure 1, is 13.5 m in width. On the top of the box girder section, the paving scheme is 5 cm-thick SFRC concrete as the leveling layer, 6 cm-thick asphalt concrete as the sublayer pavement, and 4 cm-thick asphalt concrete as the upper pavement. In addition, the vertical temperature changes of a mid-span cross-section were measured on site by resistance thermometer (RT-1) and vibrating wire readout (VW-102A) in real time during asphalt pavement paving (*16*).

**Table 1.** Thermal Parameters of Different Materials

| Materials | $k$ (W/mK) | $c$ (J/kgK) | ρ (kgm$^{-3}$) | α | ε |
|---|---|---|---|---|---|
| Reinforced concrete | 1.7 | 850 | 2500 | 0.65 | 0.9 |
| SFRC | 2.2 | 900 | 2600 | 0.65 | 0.9 |
| Asphalt concrete | 1.0 | 810 | 2100 | 0.8 | 0.81 |

The thermal parameters of materials are required for numerical simulation, including thermal conductivity ($k$), specific heat capacity ($c$), density (ρ), absorptivity coefficient ($\alpha$), and emissivity coefficient (ε). The thermal parameters of different materials used in this study are listed in Table 1, as indicated by other investigations (*15–17*).

### Temperature Boundary Conditions

For a bridge subjected to natural environmental conditions, heat is mainly transferred by means of thermal radiation, heat convection, and heat conduction (*8*). In this study, the continuous temperature variation of the pavement and box girder bridge deck is induced by the combined actions of solar radiation, natural convective heat transfer, and radiation heat transfer. Moreover, the bottom and sides of the box girder are subjected to natural convective heat transfer and radiation heat transfer, while only radiation heat transfer is assumed to occur within the concrete boxes. Therefore, based on the transient thermal field theory, the temperature boundary conditions in the research can be expressed as:

$$\begin{cases} -k\dfrac{\partial T_s}{\partial y} = h_c f[T_s(t) - T_a] + q(t) \\ \qquad + \varepsilon\sigma\left[(T_s(t) - T_z)^4 - (T_a - T_z)^4\right] \\ -k\dfrac{\partial T_b}{\partial \varphi} = h_c f[T_b(t) - T_a] + \varepsilon\sigma\left[(T_b(t) - T_z)^4 - (T_a - T_z)^4\right] \\ -k\dfrac{\partial T_{in}}{\partial y} = \varepsilon\sigma\left[(T_{in}(t) - T_z)^4 - (T_a - T_z)^4\right] \end{cases} \quad (1)$$

where $k$ is the material thermal conductivity coefficient; $h_c$ is the natural convective heat transfer coefficient, where $h_c = 4.0v + 5.8$, in which $v$ is the wind speed (*16*); $T_s$ is the surface temperature of the bridge deck pavement; $T_b$ is the temperature of the bottom and sides of the box girder; $T_{in}$ is the inner-surface temperature of the box; $T_a$ is the ambient air temperature; $q(t)$ is the solar radiation intensity, where $q(t)$ is calculated by Equation 2; $\varepsilon$ is the material emissivity coefficient; $\sigma$ is the Stefan–Boltzmann constant, where $\sigma$ = 5.6697E-8 W/(m$^2$K$^4$); $T_z$ is the absolute zero.

$$q(t) = \begin{cases} 0 & 0 \le t \le 12 - c/2 \\ q_0 \cos m\pi(t-12)/12 & 12 - c/2 \le t \le 12 + c/2 \\ 0 & 12 + c/2 \le t \le 24 \end{cases} \quad (2)$$

where $q_0$ is the maximum solar radiation intensity at noon, $q_0 = 0.131\,mQ$; $m = 12/c$; $Q$ is the total solar radiation of one day; $c$ is the effective sunshine duration.

## *Thermal Field Model and Initial Thermal Field*

According to the geometrical dimension in Figure 1, a 21 m-long reinforced concrete segmental box girder bridge was modeled using thermal-type solid elements (DC3D20) and the finite element software ABAQUS 6.11, as shown in Figure 2. As for the mesh generation, the approximate globe size of mesh was 0.5 m, and the partial mesh within 0.3 m on both sides of the pavement edge was denser.

Before the asphalt pavement paving, the temperature distribution of the concrete box girder with SFRC leveling layer under meteorological conditions was chosen as the initial thermal field of the model. Moreover, in order to validate the model effectiveness, the boundary parameters of thermal field model were determined based on the in-situ monitoring data of the Dantu Main Line Bridge in Zhu (*16*).

After that, the element deletion method was adopted to simulate the dynamic process of asphalt pavement paving. The detailed simulation process is summarized as follow:

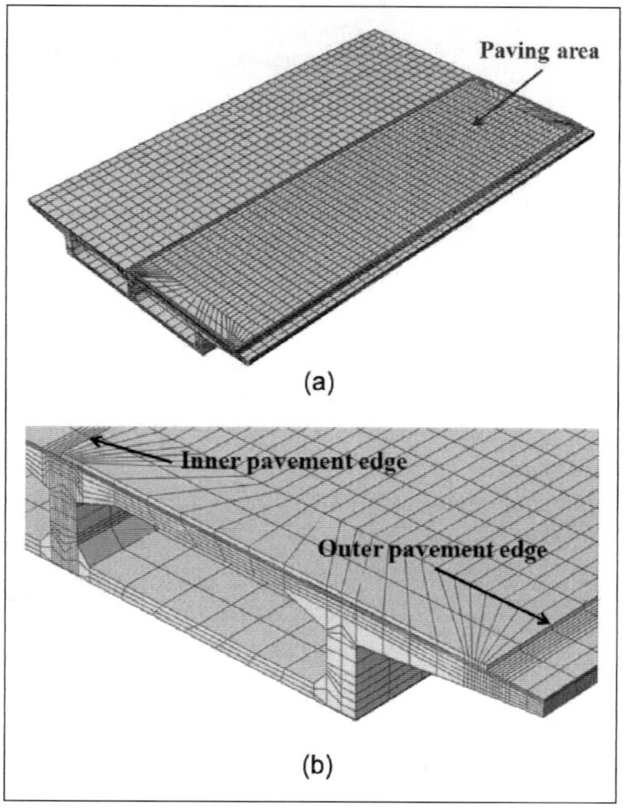


**Figure 2.** Thermal field model: (*a*) Meshing diagram; (*b*) Partial meshing diagram.

1. The pavement thickness was set as 6 cm, and the initial temperature of asphalt pavement was set as 142°C in accordance with the measured field data (*16*);
2. The 21 m-long pavement model was evenly divided into 21 parts, and all the solid elements of pavement were deactivated;
3. Next, 22 heat transfer analysis steps were set. Specifically, the first 21 steps were used to simulate the dynamic paving process by reactivating the pavement elements progressively at 30-s interval (equivalent to the paving speed of 2 m/min). Moreover, the 22nd step was used to simulate the temperature distribution of the box girder within 7200 s after paving.

## *Data Acquisition and Validation of the Model*

In order to investigate the 3D temperature distribution characteristics of the box girder during the paving process, the scheme of data acquisition in the numerical simulation was first formulated, as shown in Figure 3. In this scheme, seven cross-sections were selected first, between which the interval

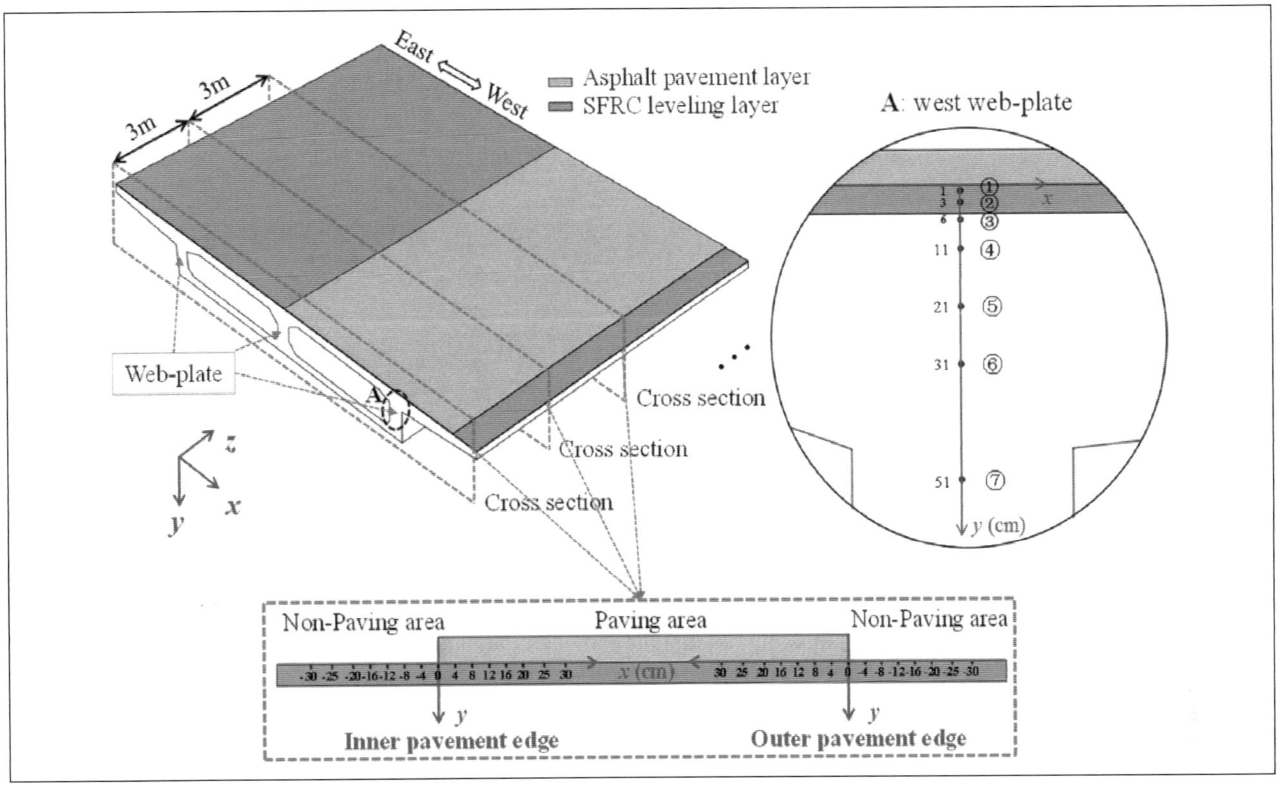


**Figure 3.** Scheme of data acquisition.

was 3 m. Furthermore, in each cross-section, the west web-plate (see Figure 3) was chosen to lay out seven vertical measuring locations within 51 cm below the top of SFRC leveling layer, and 30 transversal measuring locations were distributed within 0.3 m on both sides of the pavement edge.

The validation of thermal field model was conducted based on the vertical temperature results measured by Zhu (*16*). As a result of significant temperature variations after the asphalt pavement paving, some vertical measuring locations near the top (i.e., Point①, Point③ and Point④) were selected for the model validation, at which the comparison results between the simulated and measured temperature were illustrated in Figure 4. Considering the errors in the in-situ measurement and simulation simplification, a temperature deviation of within 5°C in Figure 4 can support the effectiveness of thermal field model. Therefore, the thermal field model is reasonable for the thermal field characteristic analysis of reinforced concrete box girder during asphalt pavement paving.

## Numerical Analysis and Discussion of Results

### 3D Temperature Distribution Analysis

In this section, according to the aforementioned scheme of data acquisition (see Figure 3), the 3D temperature distribution characteristics of the bridge components during asphalt pavement paving were investigated using the validated thermal field model.

*Vertical Temperature Distribution.* The vertical temperature distribution characteristics are summarized based on the simulated results in the mid-span cross-section, as shown in Figure 5. It can be concluded that in the paving process, the closer the location is to the top of the leveling layer, the faster the speed of rise in temperature is. In addition, the highest temperature-rising speed approximates 3°C/min, and the highest vertical temperature is about 54°C at 36 min (2160 s) at Point①, whereas the temperature of Point⑦ is basically not affected by the asphalt pavement paving. At 36 min (2160 s), the vertical temperature difference in the web-plate reaches a peak, which might threaten the durability of bridge components. In conclusion, with the increase of vertical depth ($y$) of the location, the impact of asphalt pavement paving on the vertical thermal field is fading, and the vertical temperature distribution presents distinct hysteresis. Moreover, for the mid-span cross-section, 36 min (2160 s) after starting to pave is suggested as the critical time.

*Transversal Temperature Distribution.* In order to investigate the transversal temperature distribution of bridge components during asphalt pavement paving, the temperatures of the predetermined transversal measuring locations in the mid-span cross-section were acquired at the critical time, as given in Figure 6. It can be seen that the transversal

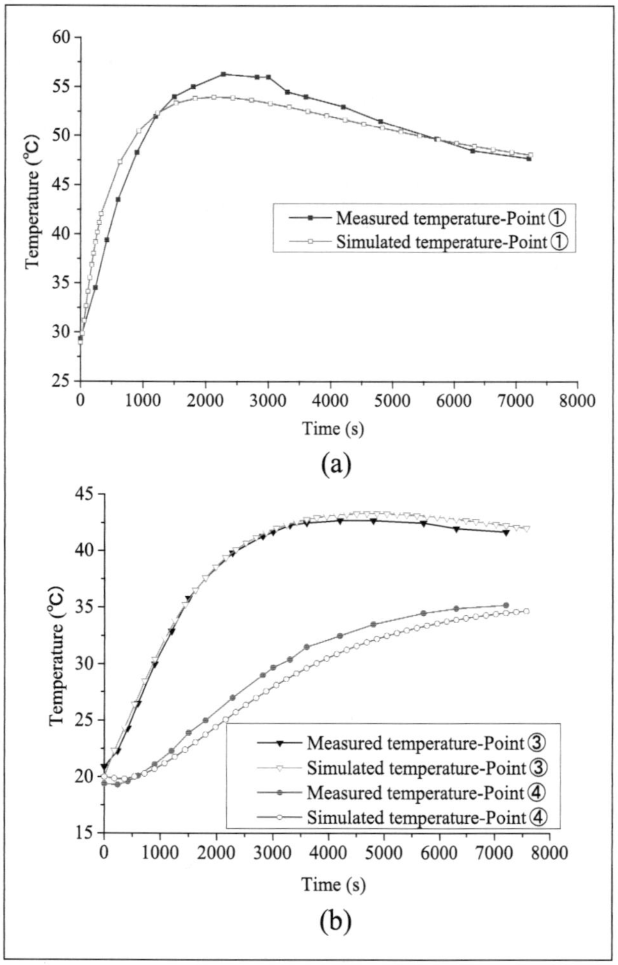


**Figure 4.** Comparison between the simulated and measured results: (*a*) Point①, (*b*) Point③ and Point④.

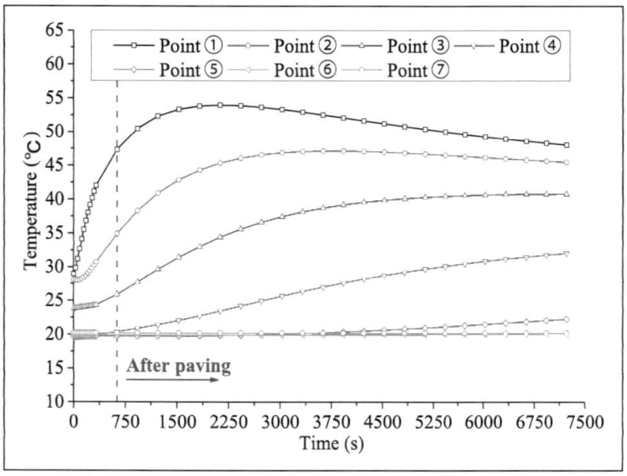


**Figure 5.** Vertical temperature distribution of the mid-span cross-section.

temperature distributions around the inner pavement edge and outer pavement edge are similar, and the transversal

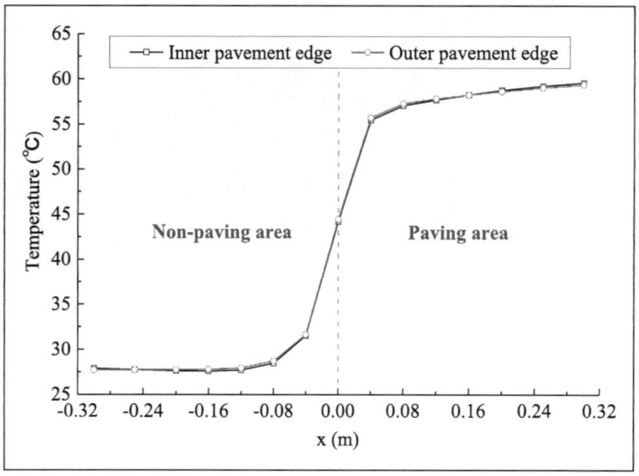


**Figure 6.** Temperature distributions at the transversal measuring locations at the critical time.

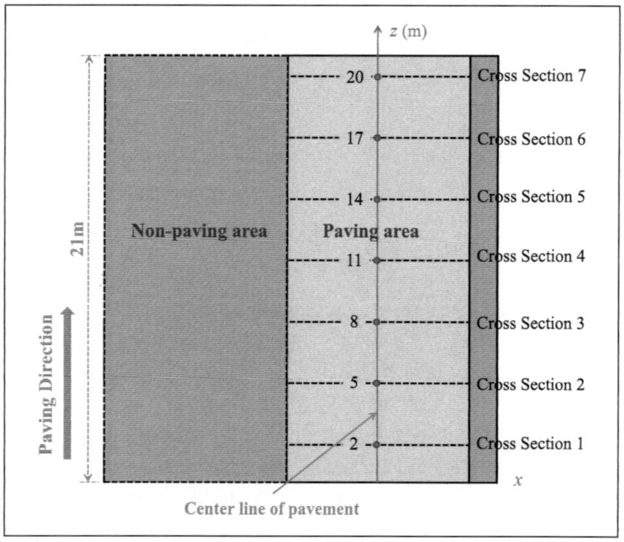


**Figure 7.** Cross-section position and longitudinal measuring locations.

temperature area of change is mainly distributed within a 0.12-m region around the pavement edge. Furthermore, the transversal temperature curves in this region approximately present an antisymmetric distribution.

*Longitudinal Temperature Distribution.* In the selected cross-section, seven longitudinal measuring locations were arranged along the center line of pavement at the top of the leveling layer, as shown in Figure 7. Furthermore, the longitudinal temperature distribution of bridge components during asphalt pavement paving was simulated and obtained in this section.

Figure 8 shows the temperature–time curves at the longitudinal measuring locations of various cross-sections. From Figure 8, it can be seen that all cross-sections have a similar temperature changing law in which the temperature rises

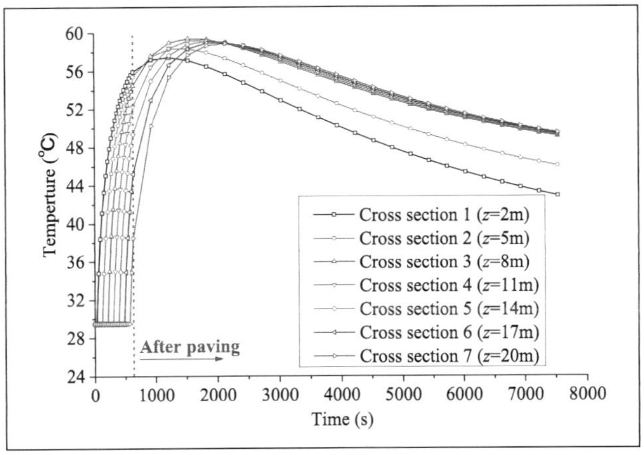


**Figure 8.** Temperature–time curves at the longitudinal measuring locations.

dramatically in the paving process, reaches a peak after paving, and then drops gradually. In detail, the temperature reached the peak of about 59°C within 30 min after paving at the average rising speed of approximately 2.4°C/min.

## Sensitivity Study

In practice, environmental conditions such as construction season, wind speed, and solar radiation density evidently affect the temperature distribution in bridge structures (*18,19*). Besides, with the increasing requirements for pavement performance, many modified asphalts have been developed and used into HMA concretes. Compared with conventional asphalts, modified asphalts require higher mixing temperature, thus leading to higher paving temperature which can expand the heat quantity of pavements. For instance, gussasphalt concrete has the paving temperature of 230°C. Therefore, it is worthwhile to further investigate the effects of various sensitive factors on temperature distribution in the box girder structure.

In this section, four sensitive factors were studied, including construction season, wind speed, solar radiation density, and paving temperature. Because of the similar regularity of the temperature distribution in different cross-sections (Figure 7), the temperature distribution in the mid-span cross-section was chosen for the sensitivity study. Sequentially, the temperature of partial vertical and transversal measuring locations in the mid-span cross-section was calculated under different operating conditions. In detail, for each operating condition, only the sensitive factor concerned was changed while other sensitive factors were kept constant. The detailed settings of various operating conditions are listed in Table 2.

*Effect of Sensitive Factors on Vertical Temperature Distribution.* Figure 9 illustrates the vertical temperature differences between Point① and Point⑦ under various operating conditions. It can be seen in Figure 9 that in the paving process, only the construction season and paving temperature

**Table 2.** Settings of Various Operating Conditions

| Operating condition | | $T_a$ (°C) | $v$ (m/s) | $q$ (W/m$^2$) | $T_p$ (°C) |
|---|---|---|---|---|---|
| 1 | H-t S | 40 | 1 | 800 | 140 |
| | L-t S | 20 | 1 | 600 | 140 |
| 2 | | 20 | 1, 4, 7 | 600 | 140 |
| 3 | | 20 | 1 | 200, 400, 600 | 140 |
| 4 | | 20 | 1 | 600 | 140, 170, 230 |

*Note:* H-t S = the high-temperature construction season; L-t S = the low-temperature construction season; $T_a$ = the air temperature; $v$ = the wind speed; $q$ = the solar radiation density; $T_p$ = the paving temperature.

significantly affect the distribution of vertical temperature difference. In addition, as can be found in Figure 9*a*, compared with the low-temperature construction season, the high-temperature construction season leads to an increase of the highest vertical temperature difference by more than 30%. Thus, asphalt pavement is suggested to be paved during the low-temperature construction season if feasible. Moreover, the effect of wind speed is opposite to that of solar radiation density, as shown in Figure 9, *b* and *c*. In detail, an increase in wind speed promotes heat convection to lower the temperature difference, whereas a greater amount of solar radiation density leads to an increase in the temperature difference. From Figure 9*d*, it is obvious that with the increasing paving temperature, the highest vertical temperature difference is larger, which means the temperature effects in the box girder can be reduced through developing some pavement materials with low paving temperature.

*Effect of Sensitive Factors on Transversal Temperature Distribution.* The critical temperature distributions at partial transversal measuring locations around the inner pavement edge under different operating conditions are given in Figure 10. It can be concluded from Figure 10*a* that the transversal temperature is reduced by 27.5°C on average at the low-temperature season compared with the high-temperature season, indicating that the transversal temperature is very sensitive to construction season. In addition, as can be seen from Figure 10, *b–d*, wind speed and solar radiation density have more significant impacts on the transversal temperature distribution in the non-paving area. On the contrary, the transversal temperature in the paving area is more susceptible to the paving temperature. Furthermore, an increase in solar radiation density and paving temperature leads to an increase in the transversal temperature, while an increase in wind speed drops the transversal temperature.

## Extreme Temperature Load Model

As the premise for the thermodynamic response analysis of bridges, the temperature load model can be intuitively used to predict the thermal field distributions of bridge components during asphalt pavement paving. In this research, an extreme temperature load (ETL) model was established, involving

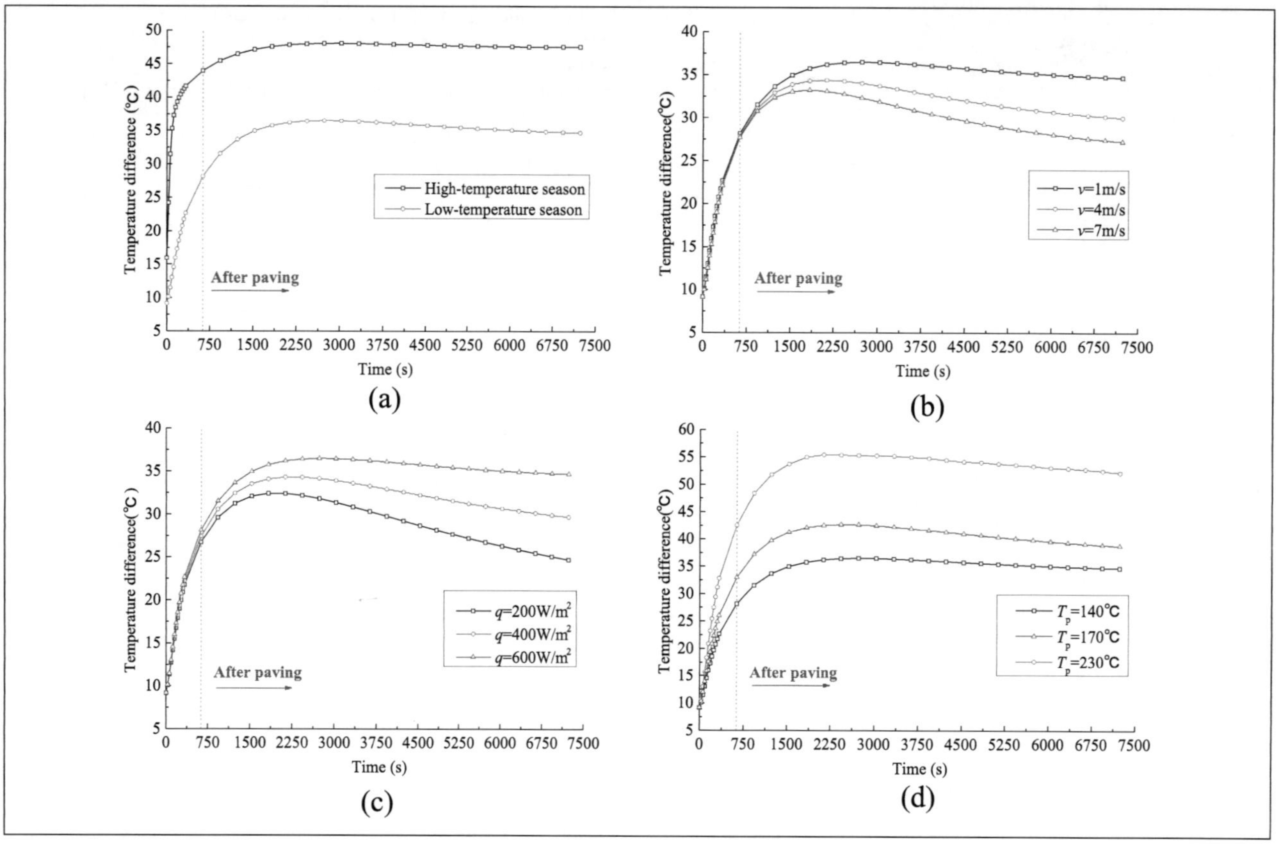


**Figure 9.** Vertical temperature differences under various operating conditions: (*a*) Construction seasons; (*b*) Wind speeds, *v*; (*c*) Solar radiation intensity, *q*; (*d*) Paving temperatures, $T_p$.

multiple sensitive factors. The statistical equations of the ETL model as indicated by other investigations (*12,20,21*) are given in Equation 3.

$$\begin{cases} \Delta T_{zx} = \mu_0\mu_1\mu_2\mu_3\mu_4 T_1 sin\left(\lambda\lambda_1\lambda_2\lambda_3\lambda_4 b\pi x\right) + \gamma_0\gamma_1\gamma_2\gamma_3\gamma_4 T_2 \\ \Delta T_{zy} = \varepsilon_0\varepsilon_1\varepsilon_2\varepsilon_3\varepsilon_4 T_3 e^{-\beta_0\beta_1\beta_2\beta_3\beta_4 ay} \end{cases} \quad (3)$$

where $\Delta T$ is the maximum temperature difference; $x$ is the transversal location in the bridge, $-0.12 \text{ m} \le x \le 0.12 \text{ m}$; $y$ is the vertical location in the bridge, $0 \text{ m} \le x \le 0.51 \text{ m}$; $z$ is the longitudinal location in the bridge, $0 \text{ m} \le z \le 21 \text{ m}$; $\varepsilon_0$, $\beta_0$, $\mu_0$, $\lambda_0$, and $\gamma_0$ are variable parameters related to the longitudinal location, represented as the function of $z$; $\varepsilon_i$, $\beta_i$, $\mu_i$, $\lambda_i$ and $\gamma_i$ are variable parameters related to sensitive factors, $i$ =1, 2, 3, 4; $a$, $b$, $T_1$, $T_2$ and $T_3$ are constant parameters, where $a$ = 15.539, $b$ = 5.291, $T_1$ = 15.032, $T_2$ = 14.523 and $T_3$ = 45.219.

Considering the requirements of asphalt pavement paving in Chinese technical specifications JTG F40-2004 (*22*), two paving speeds (i.e., 2 m/min and 6 m/min) were chosen in this study, and the statistical equations under each paving speed were fitted by the Levenberg–Marquardt method. Results of variable parameters in the ETL model are tabulated in Table 3.

## Conclusions

This paper focuses on the thermal field characteristics of a reinforced concrete segmental box girder bridge with SFRC leveling layer during asphalt pavement paving, based on the numerical simulation technique. The following conclusions can be summarized:

1. The effect of asphalt pavement paving on the vertical thermal field is attenuated with the increasing depth in the web-plate, and the vertical temperature distribution presents distinct hysteresis. In addition, the highest vertical temperature is approximately 54°C at 36 min after starting to pave.
2. The transversal temperature changing area is mainly distributed within 0.12 m around the pavement edge, and the transversal temperature curves approximately present antisymmetric distribution.
3. In the longitudinal direction, the temperature–time curves of different cross-sections present a similar pattern, and the highest temperatures are basically reached within 30 min after paving, with an average speed of increase of about 2.4°C/min.
4. The sensitivity study shows that construction season significantly affects the thermal field distribution.

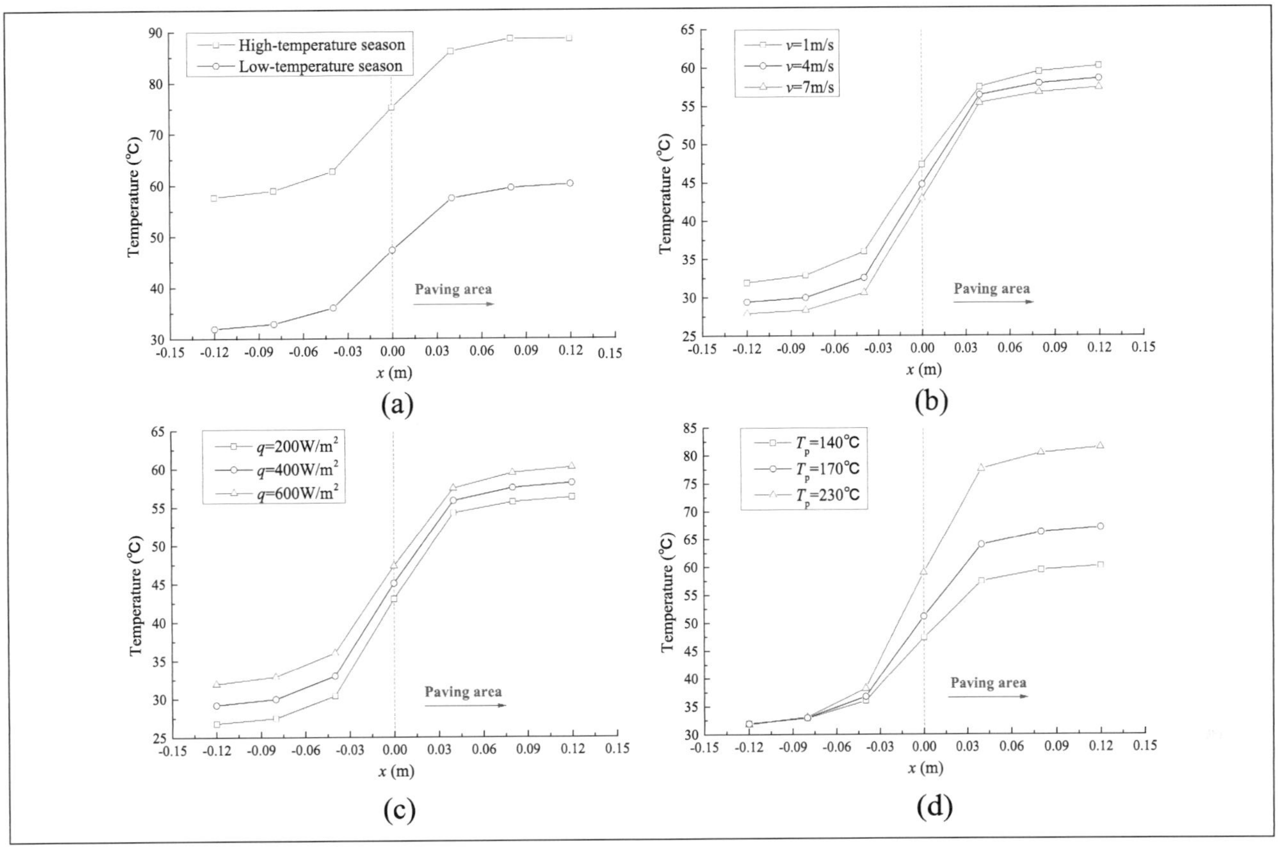


**Figure 10.** Transversal temperature distributions under various operating conditions: (*a*) Construction seasons; (*b*) Wind speeds, *v*; (*c*) Solar radiation intensity, *q*; (*d*) Paving temperatures, $T_p$.

**Table 3.** Results of the Variable Parameters in the ETL Model

| Parameters | | $\varepsilon_i$ | $\beta_i$ | $\mu_i$ | $\lambda_i$ | $\gamma_i$ |
|---|---|---|---|---|---|---|
| $i = 0$ | $V_p = 2$ m/min | $-0.028e^{-0.031z}+1.020$ | $0.320e^{0.016z}+0.623$ | $0.002z+0.974$ | $0.001z+0.988$ | $-0.115e^{-0.018z}+1.095$ |
| | $V_p = 6$ m/min | $-0.022e^{-0.011z}+1.020$ | $0.358e^{0.005z}+0.624$ | $0.0008z+0.992$ | $0.0004z+0.996$ | $-0.098e^{-0.006z}+1.092$ |
| $i = 1$ | H-t S | 1.22 | 0.58 | 1.1 | 1 | 1.12 |
| | L-t S | 1 | 1 | 1 | 1 | 1 |
| $i = 2$ | | $0.164e^{-0.085v}+0.850$ | $-0.330e^{-0.352v}+1.232$ | $-0.131e^{-0.312v}+1.096$ | $-0.075e^{-0.342v}+1.053$ | $-0.044e^{-0.737v}+1.021$ |
| $i = 3$ | | $0.0002q+0.869$ | $-0.099e^{0.0018q}+1.299$ | $-0.049e^{0.0017q}+1.136$ | $-0.013e^{0.0022q}+1.050$ | $-0.030e^{0.0018q}+1.088$ |
| $i = 4$ | | $0.006T_p+0.157$ | $0.352 \ln T_p-0.738$ | $0.009T_p-0.229$ | $0.041 \ln T_p+0.799$ | $0.009T_p-0.188$ |

*Note:* $V_p$ = the paving speed; 1 m/s ≤ $v$ ≤ 7 m/s; 200 W/m² ≤ $q$ ≤ 600 W/m²; 140°C ≤ $T_p$ ≤ 230°C; $R^2$ of all fitting equations in Table 3 is higher than 90%.

Wind speed and solar radiation density mainly affect the temperature distributions in the non-paving area, whereas the paving temperature is the major contributor to temperature variations in the paving area. Moreover, the low-temperature construction season with high wind speed and pavement materials with low paving temperature are conducive to reduce the thermal effects in bridge superstructures.

5. The Extreme Temperature Load model that considers four sensitive factors (i.e., construction seasons, wind speed, solar radiation density, and paving temperature) is established through Levenberg–Marquardt method, which can be used to assist further improvement of thermal design standard values in bridge codes.

## Acknowledgments

This study was supported by the National Natural Science Foundation of China (No. 51378122 and 51678146).

## References

1. Song, Z., J. Xiao, and L. Shen. On Temperature Gradients in High-Performance Concrete Box Girder under Solar Radiation. *Advances in Structural Engineering*, Vol. 15, No. 3, 2012, pp. 399–416.

2. Xia, Y., B. Chen, X. Q. Zhou, and Y. L. Xu. Field Monitoring and Numerical Analysis of Tsing Ma Suspension Bridge Temperature Behavior. *Structural Control and Health Monitoring*, Vol. 20, No. 4, 2013, pp. 560–575.
3. Mondal, P., and J. T. Dewolf. Development of Computer-Based System for the Temperature Monitoring of a Post-Tensioned Segmental Concrete Box-Girder Bridge. *Computer-Aided Civil and Infrastructure Engineering*, Vol. 22, No. 1, 2007, pp. 65–77.
4. Rodriguez, L. E., P. J. Barr, and M. W. Halling. Temperature Effects on a Box-Girder Integral-Abutment Bridge. *Journal of Performance of Constructed Facilities*, Vol. 28, No. 3, 2014, pp. 583–591.
5. Dilger, W. H., A. Ghali, M. Chan, and M. S. Cheung. Temperature Stresses in Composite Box Girder Bridges. *Journal of Structural Engineering*, Vol. 109, No. 6, 1983, pp. 1460–1478.
6. Tong, M., L. G. Tham, F. T. K. Au, and P. K. K. Lee. Numerical Modelling for Temperature Distribution in Steel Bridges. *Computers and Structures*, Vol. 79, No. 6, 2001, pp. 583–593.
7. Krkoška, L., and M. Moravčík. The Analysis of Thermal Effect on Concrete Box Girder Bridge. *Procedia Engineering*, Vol. 111, 2015, pp. 470–477.
8. Kong, B., C. S. Cai, and F. Pan. Thermal Field Distributions of Girder Bridges with GFRP Panel Deck versus Concrete Deck. *Journal of Bridge Engineering*, Vol. 19, No. 11, 2014, pp. 04014046.
9. Reddy, D. V., K. Sobhan, and J. D. Young. Fire Resistance of Structural Concrete Retrofitted with Carbon Fiber–Reinforced Polymer Composites. *Transportation Research Record: Journal of the Transportation Research Board*, 2015. 2522: 151–160.
10. Subramaniam, K. V., J. Kunin, R. Curtis, and D. Streeter. Influence of Early Temperature Rise on Movements and Stress Development in Concrete Decks. *Journal of Bridge Engineering*, Vol. 15, No. 1, 2010, pp. 108–116.
11. Xi, F., Q. M. Li, and Y. H. Tan. Dynamic Response and Critical Temperature of a Steel Beam Subjected to Fire and Subsequent Impulsive Loading. *Computer and Structures*, Vol. 135, No. 8, 2014, pp. 100–108.
12. Liu, Y., Z. Qian, and H. Hu. Thermal Field Characteristic Analysis of Steel Bridge Deck during High-Temperature Asphalt Pavement Paving. *KSCE Journal of Civil Engineering*, Vol. 20, No. 7, 2016, pp. 1–11.
13. Liu, Q. W., J. Zhu, B. H. Tang, and Y. S. Cai. Experiment on Temperature Distribution of Reinforced Concrete Box Girder under Asphalt High Temperature Paving. *China Journal of Highway and Transport*, Vol. 20, No. 4, 2007, pp. 96–100.
14. Yoo, D. Y., Y. S. Yoon, and N. Banthia. Predicting the Post-Cracking Behavior of Normal-and High-Strength Steel-Fiber-Reinforced Concrete Beams. *Construction and Building Materials*, Vol. 93, 2015, pp. 477–485.
15. SabbaghYazdi, S. R., and T. A. SaadatAbadi. Sequential Computations of Two-Dimensional Temperature Profiles and Thermal Stresses on an Unstructured Triangular Mesh by GFVM Method. *International Journal of Civil Engineering*, Vol. 9, No. 3, 2011, pp. 171–182.
16. Zhu, J. *Study on Temperature Fields under the Action of Solar Radiation and Bituminous Deck Pavement of the Reinforced Concrete Box Girder Bridge*. MS thesis. Southeast University, China, 2005.
17. Chen, X. Q., Q. W. Liu, and J. Zhu. Measurement and Theoretical Analysis of Solar Temperature Field in Steel-Concrete Composite Girder. *Journal of Southeast University (English Edition)*, Vol. 25, No. 4, 2009, pp. 513–517.
18. Xiao, J. Z., Z. W. Song, and C. Z. Zhang. On Sensitivity of Thermal Parameters for Concrete Box Girder in the Sunshine. *Key Engineering Materials*, Vol. 385–387, 2008, pp. 825–828.
19. Abid, S. Temperature Distributions and Variations in Concrete Box-Girder Bridges: Experimental and Finite Element Parametric Studies. *Advances in Structural Engineering*, Vol. 18, No. 4, 2015, pp. 469–486.
20. British Standards Association. *Steel, Concrete and Composite Bridges-Part 2: Specification for Loads*. British Standards Association, London, 1978.
21. Gu, B., Z. J. Chen, and X. D. Chen. Temperature Gradients in Concrete Box Girder Bridge under Effect of Cold Wave. *Journal of Center South University*, Vol. 21, 2014, pp. 1227–1241.
22. Ministry of Transport of P. R. China. *Technical Specifications for Construction of Highway Asphalt Pavements*. China Communications Press, Beijing, 2014.


*The Standing Committee on Concrete Bridges (AFF30) peer-reviewed this paper (18-01993).*

Article

TRR
JOURNAL OF THE TRANSPORTATION RESEARCH BOARD

Transportation Research Record
2018, Vol. 2672(41) 65–74

DOI: 10.1177/0361198118756892
journals.sagepub.com/home/trr
SAGE

# Performance of Skew Reinforcing in Inverted-T Bridge Caps

**Satya Sapath Roy[1], Jamshaid Sawab[1], Tianmin Zhou[1], Y. L. Mo[1], and Thomas T. C. Hsu[1]**

**Abstract**

Inverted-T bridge caps (ITBCs) have been widely used in most bridges in Texas in recent years. In some typical cases, the bridge caps are skew when two roads are not aligned in a perpendicular manner. The traditional method of flaring the transverse reinforcement out in skew ITBCs introduces significant complexity in design and during construction. An alternative is to provide skew reinforcing which will substantially reduce the design complexities and construction period. In this paper, three ITBC specimens were subjected to shear action to evaluate and compare the performance of ITBCs with traditional reinforcing with those with skew reinforcing. The three specimens include 0-degree skew, 30-degree skew with traditional reinforcing, and 30-degree skew with skew reinforcing, worthy of verification and quantification as demonstrated in this paper. The physical test results show that the skew reinforcement arrangement does not weaken the bridge cap's capacity, and the structural performance remains almost the same as that of the traditional arrangement. In addition, the serviceability performance of skew reinforcements is better as compared with traditional reinforcements.

Reinforced concrete (RC) inverted-T bridge caps (ITBCs) are used extensively in Texas and throughout the rest of the United States because they are aesthetically pleasing and offer a practical means to increase the vertical clearance of bridges. Many of these inverted-T bridge structures are skew, some in excess of 45°, based on the angle of the bridges crossing roadways, waterways, or railways. Most of these ITBCs in Texas are designed using the traditional empirical procedures outlined in the TxDOT (Texas Department of Transportation) Bridge Design Manual-LRFD (*1*) that conforms to the AASHTO (American Association of Highway and Transportation) LRFD 2014 Bridge Design Specifications (*2*). There are no precise calculation methods or guidelines given in the AASHTO LRFD or TxDOT Bridge Design Manual-LRFD to design skew ITBCs. The TxDOT Bridge Design Manual (*1*) states only that hanger and ledge reinforcement should be placed perpendicular to the centerline of the skew bent. The detailing of the skew ends of the bent should be done with a section of skew stirrups and ledge reinforcements. Typically, the transition of straight bars to the skew bars is carried out over a column support where the transverse reinforcement spacing is less critical. The designer of ITBC flares the bars out to match the skew angle while trying to maintain a minimum and maximum spacing based on the outcome of the design calculations.

Such detailing of transverse reinforcement in skew ITBCs brings complexity in the design and construction process. This transverse reinforcement has a profound influence on the overall shear capacity of the bent cap as well as the performance of the support ledge. Therefore, any kind of improper detailing can cause poor placement of concrete and cracks within the concrete structure which may reduce the load-carrying capacity and increase future maintenance cost. Faster and easier construction can be obtained if skew transverse reinforcing steel is utilized, and it can provide an alternative approach which will significantly reduce the design complexities and construction period. Accordingly, there is a concern about whether using skew transverse reinforcement will provide reasonable structural behavior for the skew ITBC regarding overall cracking occurrence and structural performance. Therefore, it is necessary to analyze the structural performance of the skew ITBC with skew transverse reinforcement and compare its performance with the traditional ITBC.

ACI (American Concrete Institute) 318-14 (*3*) states that nominal torsional moment strength increases if the area enclosed by the centerline of the outermost closed transverse

[1]Civil and Environmental Engineering, University of Houston, Houston, TX

**Corresponding Author:**
Address correspondence to Y. L. Mo: yilungmo@egr.uh.edu

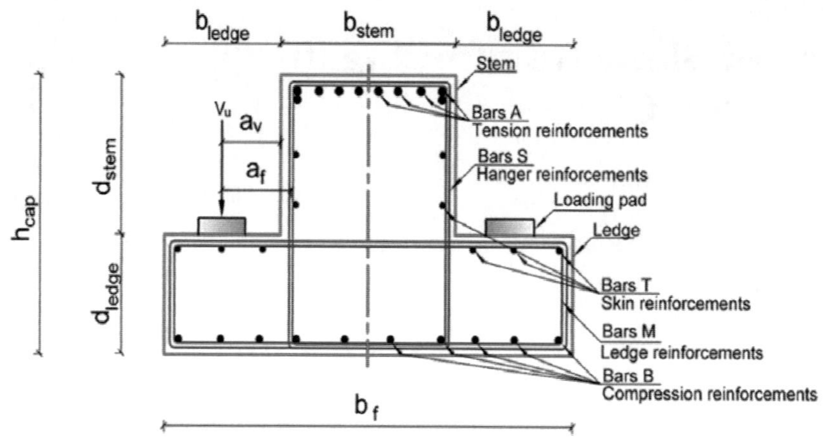


**Figure 1.** Main components of inverted-T bridge cap (ITBC).
$b_{Ledge}$ = width of ledge, $d_{Ledge}$ = depth of ledge, $d_{Stem}$ = depth of stem, $h_{cap}$ = total height of cap, $b_f$ = total width of flange, $a_v$ = distance from the face of the stem to the center of loading pad, $a_f$ = distance from the central line of hanger reinforcement to the center of loading pad.

torsional reinforcement is increased. By providing skew transverse reinforcement in the skew ITBC as a replacement of traditional reinforcement, the area enclosed by the centerline of the outermost closed loop reinforcement is increased. Thus, skew transverse reinforcement has the potential to decrease the effects of torsion coming onto the skew ITBC. But, there has not been adequate research to support this theory. Lack of experimental research has thwarted this approach so far. Thorough experimental and analytical research is needed to verify the potential beneficial effects of using skew reinforcement instead of traditional reinforcement.

Zhu et al. (*4*) studied crack control for ledges in ITBC. With the design criteria based on the ultimate strength concept, the ITBCs are susceptible to high service-load stresses in reinforcement that cause excessive crack width. The design method for crack width prediction, method of crack control, and proposed code provisions for the interior spans and the exterior end faces of the bent cap were demonstrated with the help of the strut and tie model (*5*), and then compared with the results from full-scale specimen tests. The results show that by calculating the force that will produce a critical crack width and comparing it with the load designed for service limit state, the crack development of ITBC could be estimated and controlled.

Larson et al. (*6*) investigated several factors influencing the strength and serviceability behavior of inverted-T beams, including: the length and the depth of the ledge, the web reinforcement ratio, and the loading distributions, etc. It was found that the web reinforcement ratio was the determining factor for the shear strength. A total of 33 full-scale tests were conducted, and serviceability criteria for minimizing diagonal cracking in inverted-T beams under service loads were developed based on the physical test results.

So far, no significant research has been undertaken to study the performance of skew transverse reinforcement in ITBC. The objective of this study is to investigate the structural behavior and failure mode of the skew ITBC utilizing skew transverse reinforcement, emphasizing the following aspects: (i) skew angle, and (ii) detailing of transverse reinforcement. In this paper, three ITBC specimens were subjected to shear loading to evaluate and compare overall cracking occurrence and structural performance.

## Experimental Program

### *Test Program*

The capacity of a RC ITBC is limited to the strength of the weakest components that participate in transferring the applied loads. The various components of ITBCs, with the most typical design terminology, are shown in Figure 1. Mirza and Furlong (*7–9*) defined the six failure modes of the ITBC. The first three modes are local failure, which are nonductile in nature (*10*). They are: (i) failure of hanger stirrups, (ii) ledge failure caused by punching shear, and (iii) ledge failure caused by loss of shear friction. The three global failure modes—flexure, shear, and torsion—are ductile failures. In this study of symmetrical ITBCs, torsion is not a failure mode. So, we focus on the shear behavior of traditional and skew transverse stirrups in ITBCs for two different skew angles (0-degree, 30-degree). Accordingly, during the design stage of the test specimens, it was ensured that the flexure failure and the three local failure modes must be prevented before the yielding capacity of the transverse stirrups is reached. For this study, the shear capacity of the test specimen based on the UH method (*11*) is estimated as 161 kips. This shear capacity is the lowest among flexure, torsion, shear friction, punching shear, and hanger capacity, and was used to capture, evaluate, and compare the performance of traditional and skew reinforcing in the skew ITBC.

**Table 1.** Test Matrix and Test Parameters

| Specimen I.D. | Skew angle | | Detailing of transverse reinforcement | | Amount of transverse reinforcement | f'c (ksi) |
|---|---|---|---|---|---|---|
| | 0° | 30° | Traditional (T) | Skew (S) | 2M | |
| ITBC-0-T-2M | X | | X | na | X | 6.5 |
| ITBC-30-T-2M | na | X | X | na | X | 7.5 |
| ITBC-30-S-2M | na | X | na | X | X | 7.3 |

*Note:* ITBC = Inverted-T Bridge Cap; Skew Angle (0°/30°), Detailing of Transverse Reinforcement [Traditional(T)/Skew(S)]; Amount of Transverse Reinforcement [2M − 2* Minimum Transverse Reinforcements Specified by AASHTO LRFD (2)]; na = not applicable.

The flexure, shear, torsion, hanger, punching shear, and shear friction capacity of a 0-degree ITBC, which is defined as the control specimen in this research, were predicted based on the ASSHTO LRFD (*2*) and ACI (*3*) code provisions. Based on the predicted capacities, the three ITBC test specimens were fabricated. The test matrix with different test parameters for the ITBC specimens is presented in Table 1. There are three critical parameters considered in the test matrix, which are skew angle, detailing of transverse reinforcement, and amount of transverse reinforcements. The two skew angle values considered in this study are 0 and 30 degrees. Two types of transverse reinforcement detailing are considered, namely traditional reinforcing (presently used by TxDOT) and skew reinforcing. The amount of transverse reinforcement considered is twice the minimum amount of transverse reinforcement specified by AASHTO LRFD (*2*). This transverse reinforcement ratio was achieved by providing a spacing of 5.5 in. Another transverse reinforcement ratio considered was the minimum amount of transverse reinforcements specified by AASHTO LRFD (*2*); this will be evaluated in a future study.

The total length of the ITBC specimen considered for this study is 18 ft. Figure 2*a* shows the sectional view of ITBC test specimens with the typical reinforcement details. Figures 2, *b, c*, and *d* show the partial plan view of reinforcement details of the ITBC-0-T-2M, ITBC-30-T-2M and ITBC-30-S-2M specimens, respectively.

Deformed mild steel #3 reinforcing bars were used as transverse reinforcement. The yield strength ($f_{yt}$) of #3, #4, #6 and #7 rebars are 66.0 ksi, 67.0 ksi, 69.57 ksi and 73.30 ksi, respectively. All rebars details are described in Figure 2*a*. All the three test specimens, ITBC-0-T-2M, ITBC-30-T-2M and ITBC-30-S-2M were cast in three different times, and the average concrete compressive strength (f'c) varied from 6.5 to 7.5 ksi.

## *Test Setup*

The test setup of the loading frame with the ITBC test specimen is depicted in Figure 3. There are three vertical loading frames, which rest on top of two longitudinal wide flange beams, which are anchored to the strong floor. Each vertical steel frame is equipped with a corresponding hydraulic actuator. Three hydraulic actuators, two with a capacity of 600 kips each and one with a capacity of 330 kips, are used to apply the loads to the skew ITBC specimens. The actuators are controlled by a versatile Flex-Test GT system that allows both the load control as well as the displacement control procedures. Displacement control procedure was used with calibrated increments to ensure the same amount of load in each of the six pedestals during the test. The spreader beam attached to the actuators transfers the load from actuators to the ledge of the ITBC test specimen through the loading pads, as shown in Figure 3. The loading pads are put on the lead sheet to distribute the load uniformly and to eliminate the stress concentration under the loading pads. Each test lasted 3 to 5 hours.

## *Instrumentation*

Each ITBC specimen is supported at the bottom by three reaction load cells, two with a capacity of 500 kips each and one with a capacity of 1000 kips, which are used to check the force equilibrium of the structural system during the tests. The roller support has a load cell with a capacity of 1000 kips, and the hinge support has two load cells with a capacity of 500 kips each. The linear variable differential transformer (LVDT) and string potentiometers are attached to the specimen to monitor and record the displacement of the test specimen during the test. The arrangement for this is shown in Figure 4, *a* and *b*. LVDTs were placed vertically at the two supports and at both ends of the specimen to measure the settlement of the supports and the total deflection of each specimen, respectively. The net deflection of the specimens was calculated by subtracting the support settlement from the total deflection. In addition to LVDT, string potentiometers were placed under each loading point of the specimen to monitor the deflection during the test. The reaction force obtained from the load cell was plotted against the displacement obtained from the LVDT to establish the desired load-deflection curve of each specimen.

The strain gauges were pasted in the transverse web stirrups, ledge reinforcements, and longitudinal rebars to monitor and record the tensile strain in the test specimens. When

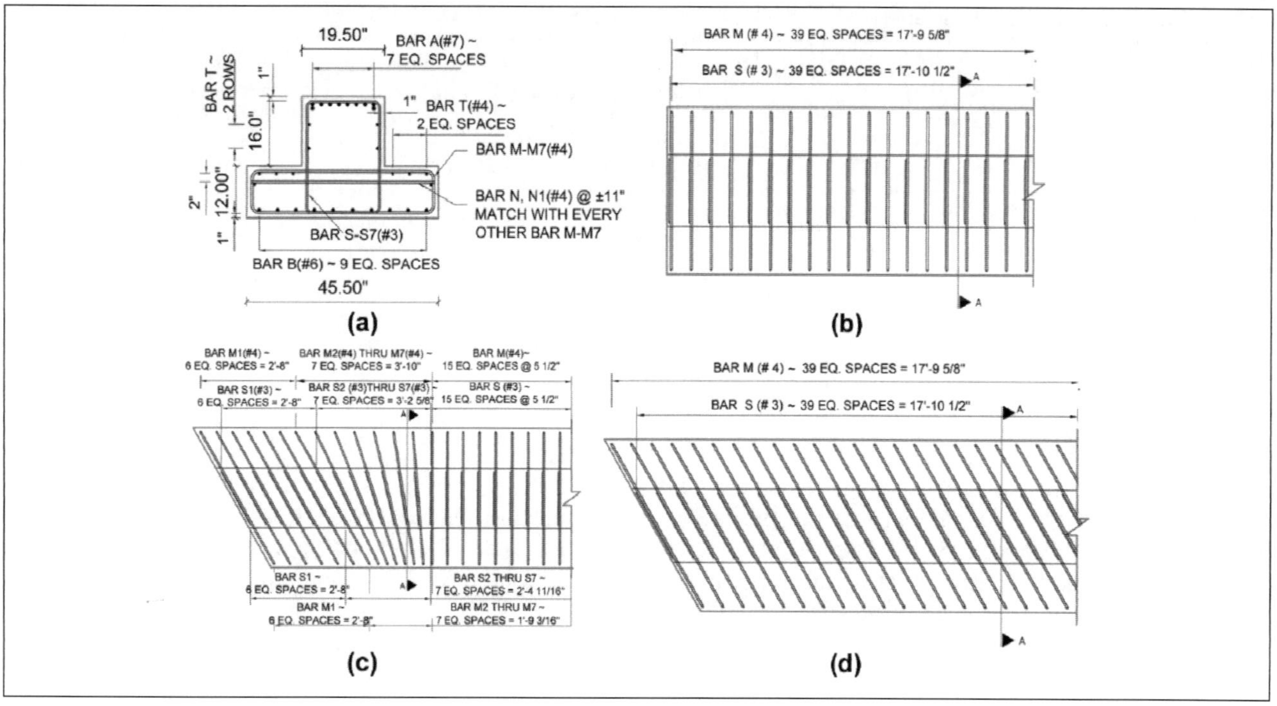


**Figure 2.** Reinforcing details of test specimens.
(*a*) Typical reinforcing at section AA, (*b*) Partial plan reinforcing details of the ITBC-0-T-2M, (*c*) Partial plan reinforcing details of the ITBC-30-T-2M, (*d*) Partial plan reinforcing details of the ITBC-30-S-2M.

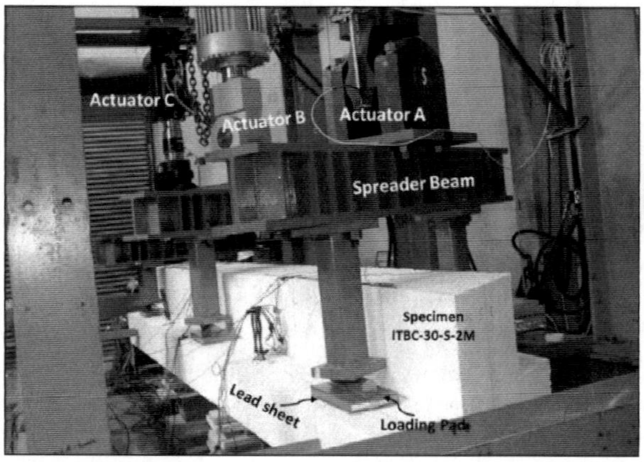


**Figure 3.** Test setup of inverted-T bridge cap test specimen with ITBC-30-S-2M.

the tensile strain in the stirrups reached 0.0023, the rebars were considered as yielded. Figure 4*c* shows the typical arrangement and locations of strain gauges in transverse stirrups for all ITBC specimens. The strain gauges were given a number from 1 to 6 on the north side and 1 to 7 on the south side as shown. Transverse stirrups were represented as NS1–NS20 on the north side and SS1–SS20 on the south side of the ITBC specimens. The first letter (N/S) indicates north or south end of the specimen; the second letter (S) represents the S bar; and the third number represents the stirrup number or location, as shown in Figure 4*c*.

## Experimental Results and Discussion

### *Specimen ITBC-0-T-2M*

The reaction force–displacement curve for specimen ITBC-0-T-2M is shown in Figure 5. As shown in Figure 6*a*, during the test, first flexural crack and first flexural shear cracks were observed at a load of 109 kips and 190 kips, respectively. The loads applied by the actuators were increased and the strains in the transverse stirrups (S bars) were monitored. First yielding of transverse stirrups in the overhang part occurred at a support reaction of 308 kips. At this loading stage, the strains in the shear stirrups (NS7, NS8, SS7, and SS8) were measured as greater than 0.0023, which was the yielding strain of the rebars. The flexural shear crack patterns are shown in Figure 6*b*. The maximum flexural shear crack width at yielding was measured as 0.012 in.

Furthermore, crack 3 occurred between the ledge and the web on the end face of the specimen at a support reaction of 336 kips. The width of crack 3 increased significantly at a peak support reaction of 338 kips, and the load-carrying capacity of the specimen started to drop. The peak support reaction of 338 kips corresponds to an individual actuator load of 226 kips. The failure mode of the specimen was primarily attributed to the shear failure caused by the yielding of

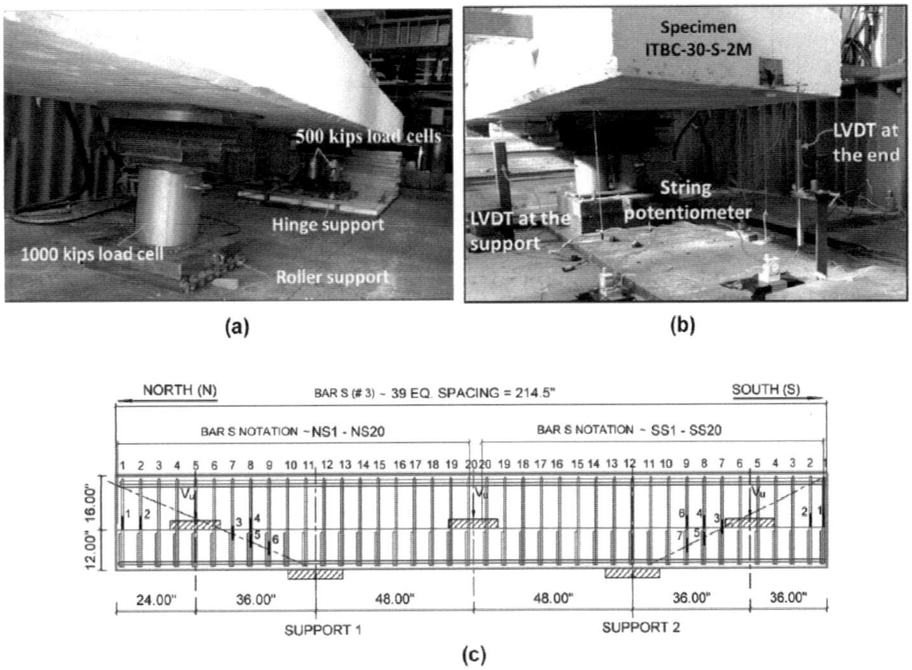


**Figure 4.** Instrumentation.
(*a*) Load cell arrangement at the support locations, (*b*) LVDT and string potentiometer for deflection measurements, (*c*) Typical strain gauges arrangement and location.

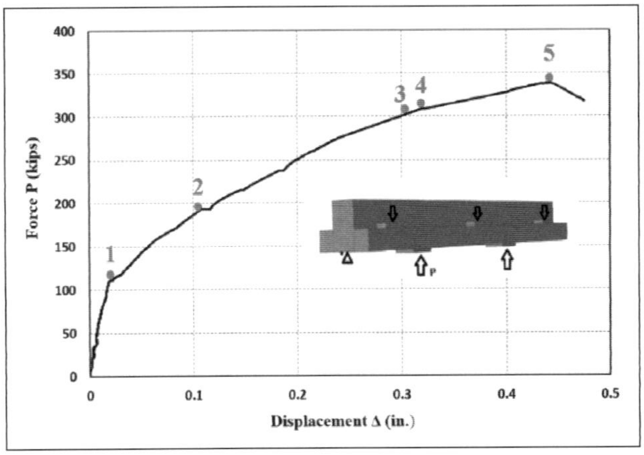


**Figure 5.** Force–displacement curve of specimen ITBC-0-T-2M. (1 = First flexure crack at 109 kips, 2 = First flexural shear crack at 190 kips, 3 = Yielding of transverse rebars at 308 kips, 4 = Diagonal crack appears in the ledge at the end face at 309 kips, 5 = Peak load at 338 kips)

the shear reinforcements, followed by the failure of the hanger stirrups, which resulted in wide crack at both end faces of the specimen. The north end and south end of the test specimen at failure are shown in Figure 6, *c* and *d*, respectively. The total number of shear cracks observed in the specimen at peak reaction force was 74, and the maximum flexural shear crack width at peak was measured as 0.020 in.

## *Specimen ITBC-30-T-2M*

The reaction force–displacement curve for the specimen ITBC-30-T-2M is shown in Figure 7. As shown in Figure 8*a*, flexural crack 1 was first observed at a load of 128 kips. Subsequently, flexural shear cracks (Crack 2, Crack 3, and Crack 4) were observed at a load of 187 kips. When the loads applied by the actuators were progressively increased, the transverse stirrups in the overhang part started yielding at a support reaction of 345 kips. At this loading stage, the strains in the shear stirrups (NS9, SS9, and NS1) reached greater than 0.0023, which is the yielding strain of the rebar. The flexural shear crack pattern observed at yielding of the transverse stirrups at north and south ends are shown in Figure 8, *b* and *c*, respectively. The maximum flexural shear crack width was measured as 0.013 in.

As shown in Figure 8*d*, diagonal cracks 5 and 6 were observed on the end face of the ledge at a support reaction of 346 kips and 358 kips, respectively. Furthermore, crack 7 occurred between the ledge and the web on the end face of the specimen at a support reaction of 381 kips, as shown in Figure 8*e*. The width of crack 7 increased significantly at a peak support reaction of 384 kips, and the load-carrying capacity of the specimen started to drop. The peak support reaction of 384 kips corresponds to an individual actuator load of 256 kips. The failure mode of the specimen was primarily attributed to the shear failure caused by the yielding

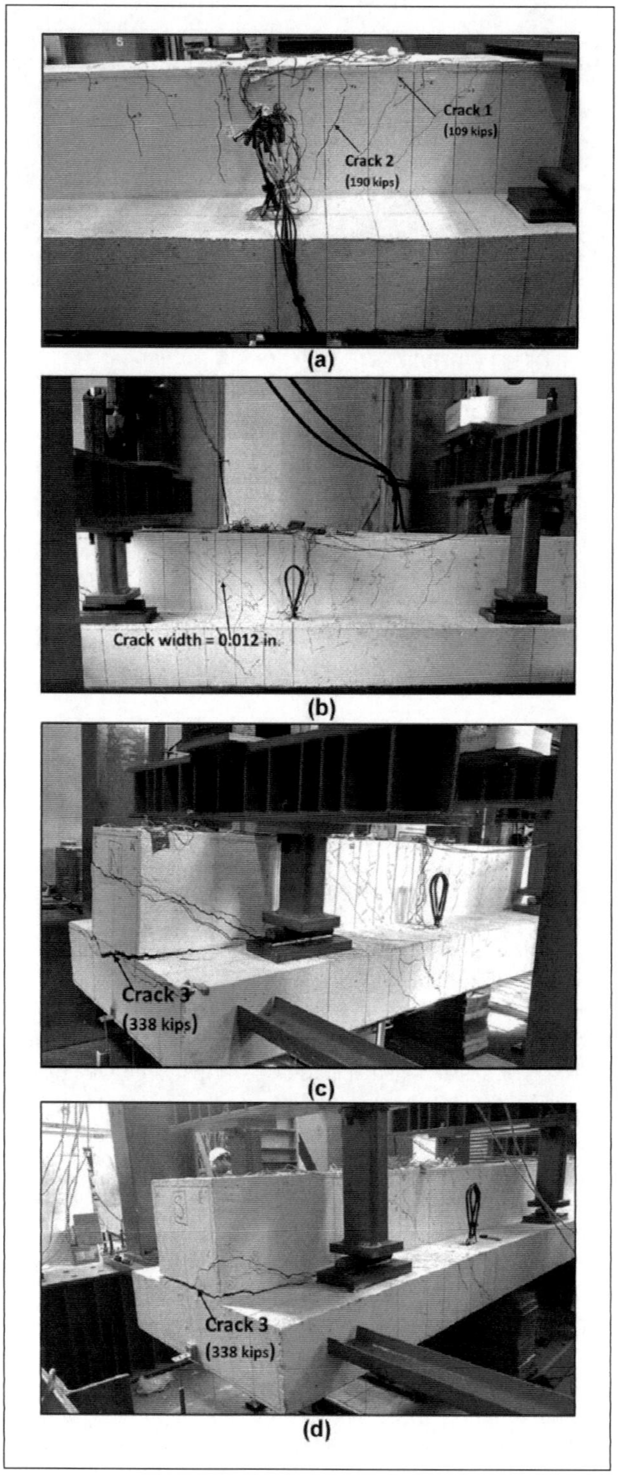


**Figure 6.** Cracking of specimen ITBC-0-T-2M.
(*a*) Flexure and flexure shear cracks, (*b*) Flexural shear crack at yielding of transverse stirrups on north face, (*c*) North face of the specimen at failure, (*d*) South face of the specimen at failure.

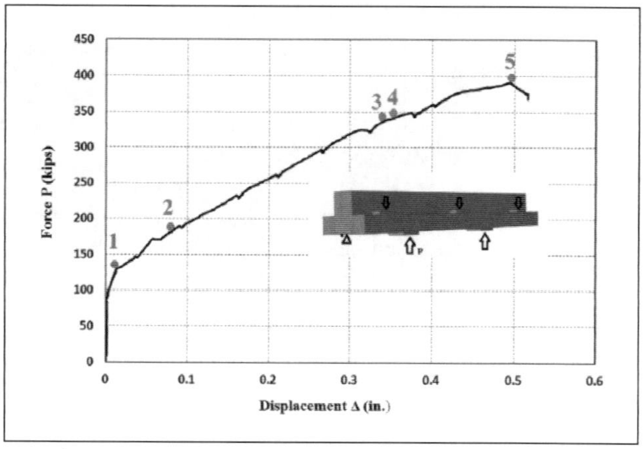


**Figure 7.** Force–displacement curve of specimen ITBC-30-T-2M. (1 = First flexure crack at 128 kips, 2 = First flexural shear crack at 187 kips, 3 = Yielding of transverse rebars at 345 kips, 4 = Diagonal crack appears in the ledge at the end face at 346 kips, 5 = Peak load at 384 kips)

of the shear reinforcements, followed by the failure of the hanger stirrups at both ends of the specimen. The north end of the test specimen at failure is shown in Figure 8*f*. The total number of shear cracks observed in the specimen at peak reaction was 102, and the maximum width of flexural shear crack at peak was measured as 0.023 in.

## *Specimen ITBC-30-S-2M*

As shown in Figure 9, the reaction force obtained from the load cell was plotted against the displacement obtained from the LVDT to obtain the complete load–displacement curve for specimen ITBC-30-S-2M. As shown in Figure 10*a*, flexural crack 1 was first observed at a load of 127 kips. Subsequently, flexural shear cracks (Crack 2, Crack 3, Crack 4, and Crack 5) were observed at a load of 190 kips. The loads applied by the actuators were progressively increased and the strains in the shear stirrups (S bars) were closely monitored. The transverse stirrups in the overhang part started yielding at a support reaction of 350 kips. At this loading stage, the strains in the shear stirrups (NS1 and NS9) reached greater than 0.0023, which is the yielding strain of the rebar. The flexural shear crack pattern observed at yielding of the transverse stirrups at north and south ends are shown in Figure 10, *b* and *c*, respectively. The maximum flexural shear crack width was measured as 0.010 in.

As shown in Figure 10*d*, diagonal cracks 6 and 7 were observed on the end face of the ledge at a support reaction of 344 kips and 356 kips, respectively. The shear reinforcements in overhang portion (NS8, NS7, SS7, SS8, and SS1) were yielded at a support reaction of 350 kips to 379 kips. Furthermore, crack 8 occurred between the ledge and the web on the end face of the specimen at a support reaction of 379 kips, as shown in Figure 10*e*. The width of crack 8 increased significantly at a peak support reaction of 381 kips, and the load-carrying capacity of the specimen started to drop. The peak support reaction of 381 kips corresponds

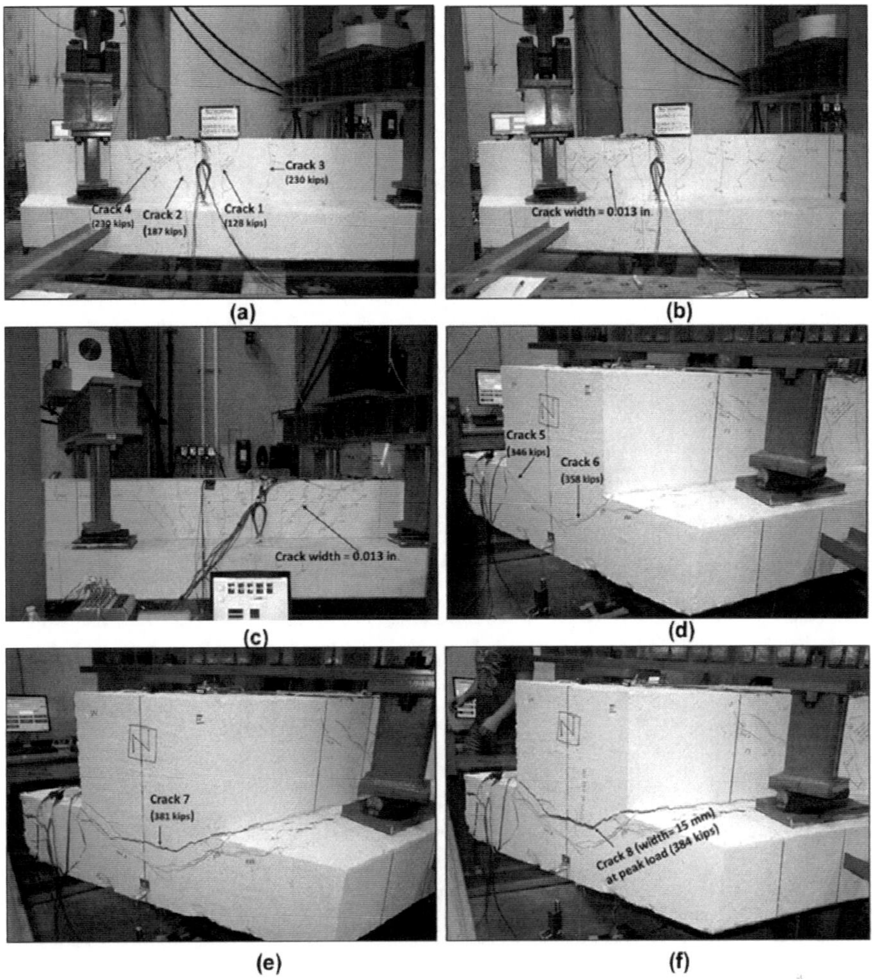


**Figure 8.** Cracking of specimen ITBC-30-T-2M.
(*a*) Flexure and flexure shear cracks, (*b*) Flexural shear crack at yielding on north face, (*c*) Flexural shear crack at yielding on south face, (*d*) Diagonal crack at the end face of the ledge, (*e*) Cracks between the ledge and the web in the end face of the specimen, (*f*) North face of the specimen at failure.

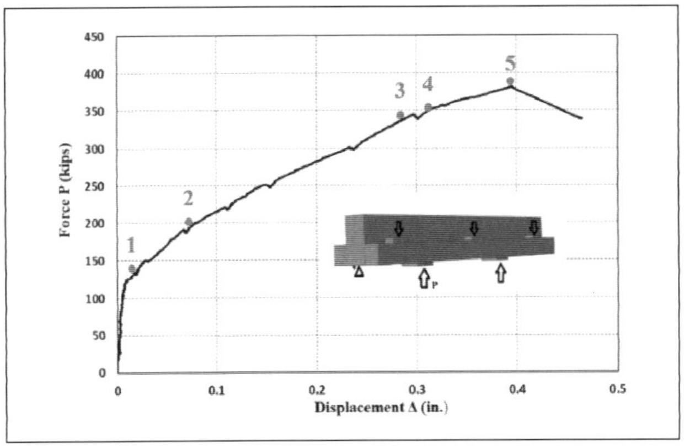


**Figure 9.** Force–displacement curve of specimen ITBC-30-S-2M.
(1 = First flexure crack at 127 kips, 2 = First flexural shear crack at 190 kips, 3 = Diagonal crack appears in the ledge at the end face at 344 kips, 4 = Yielding of transverse rebars at 350 kips, 5 = Peak load at 381 kips).

to an individual actuator load of 254 kips. The failure mode of the specimen was primarily attributed to the shear failure caused by the yielding of the shear reinforcements, followed by the failure of the hanger stirrups at both ends of the specimen. The north end of the test specimen at failure is shown in Figure 10*f*. The total number of shear cracks observed in the specimen at peak reaction was 72. The maximum flexural shear crack width at peak was measured as 0.019 in.

Table 2 summarizes the primary test results of three ITBCs with traditional and skew reinforcing. From the observed results, it is evident that there is no significant difference between the specimens with two types of reinforcing detailing in terms of strength. This is because the yielding of transverse stirrups happened at a reaction force of 346 kips and 350 kips, whereas the peak reaction force was observed at 384 kips and 381 kips for specimens ITBC-30-T-2M and

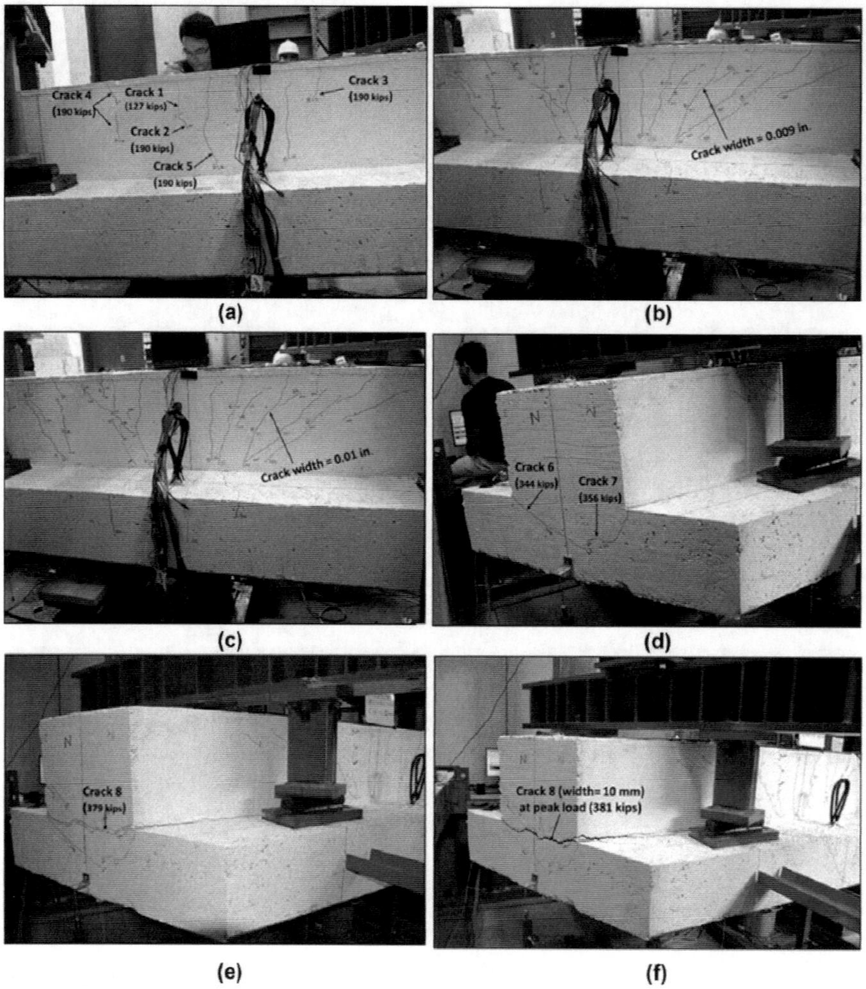


**Figure 10.** Cracking of specimen ITBC-30-S-2M.
(*a*) Flexure and flexure shear cracks, (*b*) Flexural shear crack at yielding on north face, (*c*) Flexural shear crack at yielding on south face, (*d*) Diagonal crack at the end face of the ledge, (e) Cracks between the ledge and the web in the end face of the specimen, (*f*) North face of the specimen at failure.

ITBC-30-S-2M, respectively. The maximum diagonal shear crack width observed at yielding was 0.013 in. and 0.010 in. for 30-degree skew ITBC with traditional and skew reinforcing, respectively. Moreover, as shown in Table 2, the number of shear cracks observed at peak reaction force in the ITBC with traditional reinforcing was more than that of the ITBC with skew reinforcing. This better performance may be attributable to the even spacing of hanger and ledge rebars in the skew ITBC. It can be inferred from the above discussion that from a serviceability and cracking point of view, the ITBC with skew transverse reinforcing performed comparatively better than that with traditional transverse reinforcing.

To compare the test results of these three specimens, the normalized reaction force–deflection curves are plotted in Figure 11. The reaction force obtained from the loadcell

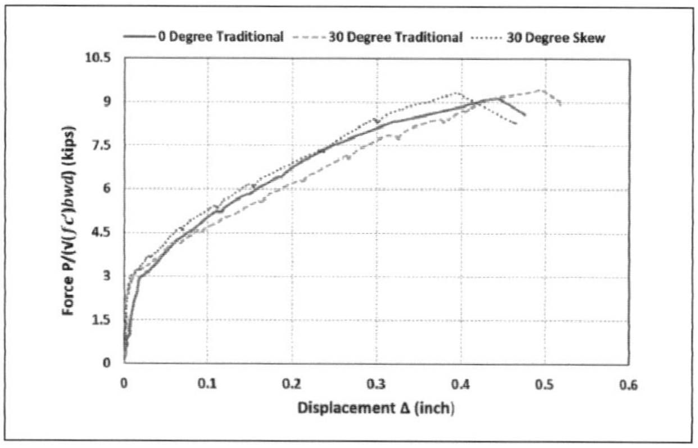


**Figure 11.** Normalized force–displacement curve of specimens ITBC-0-T-2M, ITBC-30-T-2M, and ITBC-30-S-2M.

**Table 2.** Summary of Primary Test Results

| Test result | ITBC test specimens | | |
|---|---|---|---|
| | ITBC-0-T-2M | ITBC-30-T-2M | ITBC-30-S-2M |
| Flexural cracking (kips) | 109 | 128 | 127 |
| Flexural shear cracking (kips) | 190 | 187 | 190 |
| Yielding of transverse reinforcement (kips) | 308 | 345 | 350 |
| Peak reaction force (kips) | 338 | 384 | 381 |
| Maximum flexural shear crack width at yielding reaction force (in.) | 0.012 | 0.013 | 0.010 |
| Total number of shear cracks observed at peak reaction force (#) | 74 | 102 | 72 |
| Maximum flexural shear crack width at the peak reaction force (in.) | 0.020 | 0.023 | 0.019 |

was normalized with $\sqrt{f_c'}b_w d$ and plotted with the displacement obtained from the LVDT, as the shear strength of an inverted-T section is associated with the compressive strength of the concrete. Figure 11 shows that the capacities of 30-degree skew inverted-T specimens with traditional transverse reinforcement and skew transverse reinforcement are almost the same. But the displacement corresponding to the peak force of specimen ITBC-30-T-2M is somewhat greater than that of specimen ITBC-30-S-2M. This behavior can be attributed to the former having a greater quantity of hanger and ledge reinforcements than the latter. The initial stiffness of specimen ITBC-0-T-2M is relatively lower than the other two. This is because support settlement was not recorded during the test, and the displacement which is plotted against the reaction force is not net displacement.

## Conclusion and Future Study

Three skew ITBC specimens with two types of transverse reinforcement configuration were tested with a constant shear span to depth ratio of 1.4. Based on the test results, it is recommended that skew ITBC with skew reinforcing may be a better alternative to traditional reinforcing, because the traditional detailing of transverse reinforcement in the skew ITBC introduces complexity to the design and construction process. The serviceability performance of skew reinforcing is comparatively better than traditional reinforcing because of the even distribution of hanger and ledge rebars. Nevertheless, this paper is limited to three ITBCs only. The researchers at University of Houston will be performing more physical tests of ITBC specimens with 45-degree and 60-degree skew angles. Additional parameters that were not considered in the proposed test matrix will be investigated in detail with the help of a 3D FE analysis model to establish sufficient databases. After thorough research, a set of design guidelines can be established for the design of the skew ITBC with skew reinforcements.

### Acknowledgments

The authors would like to express their gratitude to the Texas Department of Transportation (TxDOT) for their financial support and collaborative efforts for this project. This research is supported by TxDOT for Grant 0-6905. The authors would like to specifically thank the contributions of the project supervisory committee which consists of: Chris Glancy, RTI (Research and Technology Implementation Office); Michael Carlson, HOU (Houston District); Bobby Bari, HOU (Houston District); Walter "Ray" Fisher, DAL (Dallas District); Courtney Holle, BRG (Bridge Division); Aaron Garza, BRG (Bridge Division); Hector Garcia, DOT (Department of Transportation) and Andrew Smyth, DOT (Department of Transportation). Finally, the authors appreciate the rest of the support staff at the Thomas T.C. Hsu Structural Lab at UH (University of Houston) and the many other researchers who helped with instrumentation, testing, and analysis of results.

### References

1. TxDOT. *Bridge Design Manual-LRFD*. Texas Department of Transportation (TxDOT), Austin, TX., 2015.
2. AASHTO. *LRFD Bridge Design Specifications. Customary U.S. Units, 7th Edition, with 2015 Interim Revisions*. American Association of Highway and Transportation Officials (AASHTO), Washington, D.C., 2014.
3. ACI 318. Building Code Requirements for Structural Concrete (ACI 318-14) and Commentary. Farmington Hills, 2014.
4. Zhu, R. H., H. Dhonde, and T. T. C. Hsu. Crack Control for Ledges in Inverted 'T' Bent Caps. *No. Research Report 0-1854-5*. University of Houston, Department of Civil & Environmental Engineering, 2003.
5. Ley, M. T., K. A. Riding, S. Bae, and J. E. Breen. Experimental verification of strut-and-tie model design method. *ACI Structural Journal*, Vol. 104, No. 6, 2007, p. 749.
6. Larson, N., E. F. Gomez, D. Garber, O. Bayrak, and W. Ghannoum. Strength and Serviceability Design of Reinforced Concrete Inverted-T Beams. *Report No. FHWA/TX-13/0-6416-1*, Center for Transportation Research, The University of Texas at Austin, Austin, Tx, 2013, p. 234.
7. Mirza, S. A., and R. W. Furlong Serviceability Behavior and Failure Mechanisms of Concrete Inverted T Beams Bridge

Bentcaps, *Journal of American Concrete Institute*, Vol. July–August 1983 Proceedings, No. 4, 1983, pp. 294–304.
8. Mirza, S. A., and R. W. Furlong Strength Criteria for Concrete Inverted T Girders. *Journal of Structural Engineering, ASCE*, Vol. 109, No. 8, Aug 1983, pp. 1836–1853.
9. Mirza, S. A., and R. W. Furlong The Design of Reinforced and Prestressed Concrete Inverted T Beams for Bridge Structures, *PCI Journal*, Vol. 31, No. 3, May–June 1986, pp. 157–163.
10. Galal, K., and M Sekar, Rehabilitation of RC inverted-T girders using Anchored CFRP sheets. *Composites Part B: Engineering*, Vol. 39, No. 4, 2008, pp. 604–617.
11. Kuo, W. W., T. T. Hsu, and S. J. Hwang. Shear Strength of Reinforced Concrete Beams. *ACI Structural Journal*, Vol. 111, No. 4, 2014, p. 809.

*The Standing Committee on Concrete Bridges peer-reviewed this paper (18-04680).*

*The contents of this paper reflect the views of the authors, who are responsible for the facts and the accuracy of the data presented here. The contents do not necessarily reflect the official views or policies of TxDOT.*

TRR
JOURNAL OF THE TRANSPORTATION RESEARCH BOARD

Article

Transportation Research Record
2018, Vol. 2672(41) 75–86

DOI: 10.1177/0361198118767420
journals.sagepub.com/home/trr
SAGE

# Modified Elastic Dynamic Analysis (EDA) for Seismic Demand on In-Span Hinge Shear Keys in Multi-Frame Bridges

**Masoud Mehr[1,2] and Arash E. Zaghi[1]**

**Abstract**
Long concrete box-girder bridges are typically constructed in multiple frames that are separated by in-span hinges. Shear keys, located at the in-span hinges, help preserve the transverse integrity of the bridge frames. To date, no reliable method other than nonlinear time history analysis exists to estimate the seismic force demands on in-span shear keys. Methods such as pushover and elastic dynamic analysis (EDA) do not provide accurate estimations. In this study, a rational and reliable analysis method was developed for obtaining the seismic demand on in-span shear keys of multi-frame bridges. A large number of time history analyses were performed on two- to five-frame bridge models with single- and two-column bents that were designed according to California Department of Transportation (Caltrans) Seismic Design Criteria. The results show that high-frequency modes of vibration of the superstructure significantly contribute to the shear key force demands. These modes may also cause transverse yielding in columns. It was established that a modified EDA method may be used to approximate the shear key force demands. In the proposed modified method, the modal forces are reduced separately by the corresponding modal displacement ductility before performing modal combination. This method accounts for the nonlinear response under high-frequency modes of vibration.

Multi-frame bridges are composed of individual segments, called frames. Figure 1 shows a typical multi-frame bridge. The multi-frame construction, when used for long bridges, eases post-tensioning of the superstructure and reduces the unfavorable effects of creep deformations. In addition, this method of construction accommodates thermal expansion and contraction in the superstructure without inducing large forces in supporting columns (*1*). To ensure the seismic integrity of multi-frame bridges, the frames are connected to each other by in-span hinges, where shear key elements, generally made of concrete blocks or steel pipes called pipe-seat extenders (*2*), are placed (*3*). In-span hinges permit relative movement of adjacent frames in the longitudinal direction.

The advantages of the multi-frame bridge system make it an appealing and practical option for long bridges. However, the lack of a reliable method of analysis for the estimation of the seismic force demands on shear key elements may lead to costly or unsafe design of shear keys.

Design codes such as AASHTO LRFD Bridge Design Specifications (*4*) and AASHTO Guide Specifications for LRFD Seismic Bridge Design (*5*) do not currently provide guidance for establishing the seismic design forces for in-span shear key elements. The California Department of Transportation (Caltrans) Seismic Design Criteria (SDC) 1.7 (*6*), states that

Determining the earthquake force demand on shear keys is difficult since the magnitude is dependent on how much relative displacement occurs between the frames. Forces generated with EDA (Elastic Dynamic Analysis), i.e. response spectral analysis, should not be used to size shear keys. EDA overestimates the resistance provided by the bents and may predict force demands on the shear keys that differ significantly from the actual forces.

SDC 1.7 allows shear transfer between adjacent frames if the ratio of the natural period of transverse vibration of the stiffer frame to that of the more flexible frame is larger than 0.7. While no specific design formulation is recommended, the upper bound design force is suggested to be the lateral capacity of the weaker frame. In cases where

[1]University of Connecticut, Storrs, CT
[2]WSP USA, Glastonbury, CT

**Corresponding Author:**
Address correspondence to Masoud Mehr: m.mehrraoufi@gmail.com

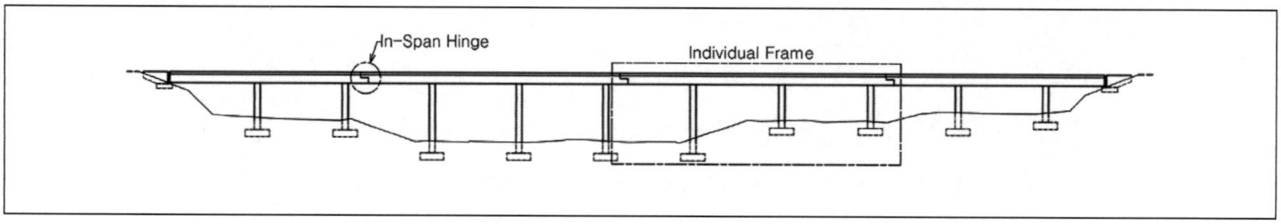


**Figure 1.** Schematics of a multi-frame bridge.

the ratio of the periods of adjacent frames is smaller than 0.7, the SDC suggests limiting the capacity of shear keys to prevent transfer of large lateral forces to the stiffer frame by ensuring the yielding of shear keys.

Several studies have examined the longitudinal seismic response of multi-frame bridges. The primary goal of these studies was to determine the minimum gap size and seat width necessary to avoid significant pounding and unseating of adjacent frames (*7–11*). Other studies investigated the design of hinge restrainers as a retrofit measure for bridges with inadequate seat widths (*12–15*). Recent studies have focused on analyzing the fragility of multi-frame bridges with horizontal curve and/or integral abutments (*16–17*). Due to the coupling of transverse and rotational displacements of individual frames, as well as the interactions of these frames, the transverse response of multi-frame bridges is more complex than their longitudinal response (*18*). The complexity of the transverse response of multi-frame bridges is compounded by in-plane flexibility of the superstructure (*19*).

Mehr and Zaghi (*20*) investigated the transverse seismic response of multi-frame bridges and concluded that the in-plane flexibility of superstructures and the high-frequency modes associated with the in-plane deformations of superstructures play a significant role in the seismic response of multi-frame bridges. Mehr and Zaghi (*21*) also studied the effects of geometric and ground motion characteristics on the seismic demand on in-span hinges shear key elements. They found that, in general, the time of occurrence of the maximum shear key forces in a time history analysis does not coincide with the time of the maximum transverse displacement or acceleration on the superstructure. The effects of non-uniform stiffness of adjacent frames on shear key force diminish as the number of frames increases. They observed that the shear key forces may be reduced by placing the in-span hinges on the side of the bent with lower stiffness. Soft soils result in large shear key force demands in two-column bent bridges, whereas shear key forces in single-column bent bridges are less sensitive to the soil type. The impact on shear key elements due to the transverse gap closure may increase the forces by an average factor of 2.5. The effects of non-uniform base excitation may increase shear key forces by approximately 1.25 times in soil types C, D, and E (*21*).

This study aims to develop a reliable, easy-to-implement formulation that enables seismic demand analysis of shear key forces. A set of prototype multi-frame bridges was designed in accordance with Caltrans SDC 1.7, and extensive nonlinear time history (NTH) analyses were conducted. This was followed by a series of pushover and elastic dynamic analyses (EDA). Next, rigid modal analysis was conducted, and the shear key force was compared with that of NTH analysis. These three methods of analysis failed in accuracy. The concept of modal ductility was introduced as a reliable method for modification of individual modal responses before combining the modal effects. The proposed modified EDA estimates the in-span shear key forces more accurately than other methods of analysis.

## Analytical Simulations

A refined NTH analysis is essential to estimate consistently and accurately the force demands on shear keys. In this study, 1,848 sets of NTH, EDA, and pushover analyses were conducted on 56 model bridges with varied geometries under a suite of 33 ground motions of different magnitudes and soil types. To ensure realistic results, the prototype models were carefully designed following the procedure recommended by Caltrans SDC (*6*). The following section provides a brief overview of the methodology used for the analytical studies. Comprehensive information about the modeling assumptions, ground motion selection, and design of the prototype bridges are presented in Mehr and Zaghi (*20, 22*).

### *Geometry of the Prototype Bridges*

Fifty-six multi-frame bridges with a variety of realistic geometries were defined. Figure 2 illustrates four sets of two-, three-, four-, and five-frame straight bridges with no skew. Bridge span lengths of 110 ft (33.6 m) were chosen to represent short-span constructions (not shown in the figure) while lengths of 200 ft (61 m) were selected to represent long-span bridges. The in-span hinges were placed at one-fifth of the span length. Figure 2 also shows the in-span hinges, consisting of two L-shaped solid diaphragms at the ends of the adjacent frames. The

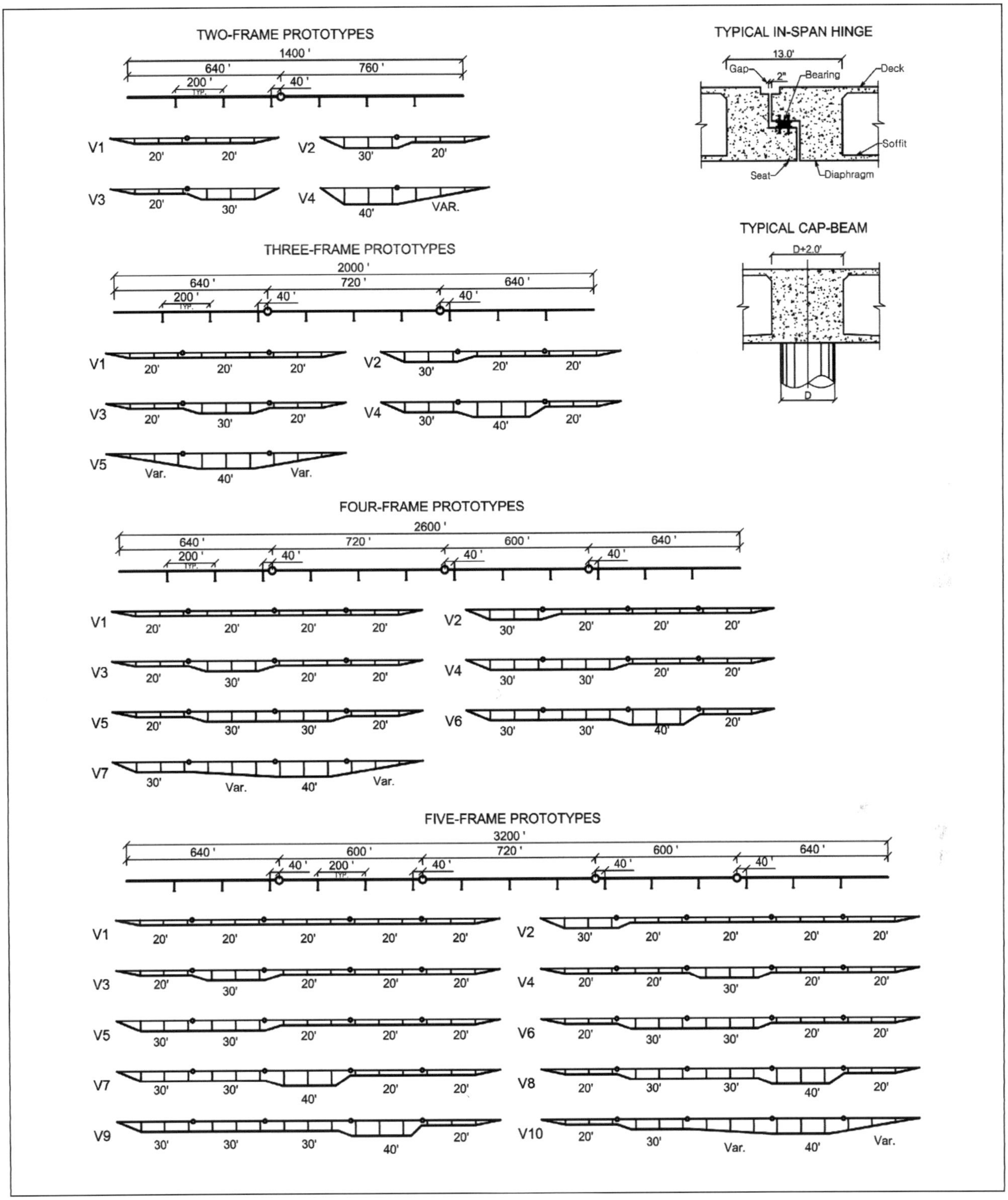


**Figure 2.** Prototype sets, showing geometries and valley shapes.

in-span shear keys were assumed to be made of 15 × 15-in, and 15 × 15-in concrete blocks, for single-column and two-column prototype bridges, respectively. As specified by AASHTO, the longitudinal gap size was calculated as 2.0 in (*4*). The transverse gap size around the shear key blocks was assumed to be zero to isolate the effect of transverse pounding. The effect of transverse pounding was investigated in a previous study (*21*).

The superstructure system was a concrete multi-cell box girder. The long-span prototypes had superstructure widths of 40 ft (12.2 m) for single-column bents and 64 ft (19.5 m) for double-column bents, respectively. The short-span bridge models exclusively used a 40-ft wide superstructure. The geometric proportions of the superstructures and the thicknesses of the deck slab, soffits, and webs were determined based on the specifications of Caltrans MTD-10-20 (*23*). Two substructure systems were studied: single-column extended pile-shaft bents and two-column bents with flexural pins at the base of the columns. To investigate the effect of varied column heights, the long-span prototypes included 26 realistic valley shapes, that is, ground profiles. The column heights are shown in Figure 2. The column heights are measured from the soffit to the top of the ground level for the single-column bent and to the top of the footings for the two-column bents. Except for one of the prototypes in each subset, the column height within each frame was kept constant. For the short-span bridge prototypes, the column height was constant along the bridge. In both single-column and two-column systems, the columns are integrated into the concrete cap beams. These cap beams are 2 ft (0.6 m) wider than the diameter of the columns. The abutments were seat-type, with a weak backwall and two exterior sacrificial shear keys that were sized to yield under a design earthquake (*6*).

## Design of the Prototype Bridges

The design of the prototype bridges followed Caltrans SDC 1.7 (*6*). Each prototype was designed for 11 different acceleration response spectrums (ARS) selected from Appendix B of SDC 1.7. This set included three hazard levels, namely "moderate" (PGA = 0.2 g, Mw = 6.25–6.75), "large" (PGA = 0.4 g, Mw = 7.0–7.5), and "severe" (PGA = 0.6 g, Mw = 7.75–8.25), for soil types B, C, and D. Only moderate and large hazard levels were considered for soil type E. The inclusion of different ARSs ensures that the findings of this study are applicable to varying site specifications and hazard levels. The design procedure is presented in detail by Mehr and Zaghi (*20, 22*).

## Analytical Modeling of the Prototype Bridges

Three-dimensional nonlinear spine models were created with OpenSees v2.4.4 (*24*). Since the superstructure, diaphragm, and cap beam elements are designed to remain elastic in an earthquake, they were modeled using "elasticBeamColumn" elements that have the corresponding flexural, shear, and torsional stiffness values. Two multi-component assemblies were created to capture the behavior of abutments and in-span hinges. These assemblies included the performance of the bearings, shear keys, longitudinal and transverse gaps, and backwalls by using several nonlinear ZzeroLength" elements. The effect of longitudinal pounding was captured by the uniaxial material model "ImpactMaterial" (*25*). The elastomeric bearings were modeled using the uniaxial material "Elastic" in the horizontal directions. The uniaxial material "Elastic-no-Tension" was used in the vertical direction to allow uplift.

Zero-length elements with uniaxial material "Elastic" were used to model the in-span shear keys. The elastic in-span shear keys enabled capturing the upper bound values of shear key demands. The transverse stiffness of the in-span shear keys was calculated per Megalley (*26*). As opposed to the in-span shear keys, the two exterior shear keys at the abutments were modeled with elastoplastic behavior according to Bozorgzadeh, Megalley, Ashford, and Restrepo (*27*). The uniaxial material "Steel02" in series with the material "ImpactMaterial" was used for the exterior shear keys. The yield capacity of the abutment shear keys was assumed to be equal to the dead load reaction at the abutment (*6*). "HyperbolicGapMaterial" was used to model the soil backfill at the abutment backwall.

Nonlinear "forceBeamColumn" elements with "Mid-Distance" integration points were used to model the columns to control the length and location of plastic hinges. A nonlinear fiber section with aggregated torsional stiffness was implemented into the element. SDC 1.7 provided the specifications used to determine the plastic hinge length. The unconfined concrete strength, $f'_c$, was assumed to be 5.0 ksi. The parameters of uniaxial material "Concrete04" were calculated using the Mander model for confined concrete (*28*). Longitudinal bars were modeled using the uniaxial material "Reinforcing Steel" with 60-ksi yield strength. The global and local modes of vibration were captured by assigning translational and rotational masses to the nodes of the superstructure. To account for P-delta effects, the "Pdelta" coordinate transformation object of OpenSees was used. Rayleigh damping at 5% was used (*29, 30*). The behavior of the individual column elements was confirmed using the results of shake table tests performed by Zaghi and Saiidi (*31*).

## Ground Motions

The time history ground motions were selected from the PEER NGA database (*32*). Three biaxial ground motions with comparable magnitude, intensity level, and soil type were selected for each of the design spectrums, that is, ARSs, mentioned in the Design of Prototype Bridges section. SeismoMatch spectral matching software (*33*) generated a library of 33 spectrum-matched biaxial acceleration histories. For the design of ordinary bridges, Caltrans SDC 1.7 does not require the consideration of the combined effects of vertical and horizontal components of ground motions (*6*). Thus, the vertical components of the ground motions were not included in the analyses. It should be noted that the vertical components may affect the response of bridges in near-field earthquakes by changing the axial forces of columns. Further details about the process of selection and spectrum matching of the ground motions is presented in Mehr and Zaghi (*20*).

## Methods for Estimating Shear Force Demands

A correlational analysis was performed to determine the relationship between the reference shear key force obtained from NTH analysis and shear key force calculated from: (a) nonlinear static analysis (pushover), (b) elastic dynamic analysis (EDA), and (c) rigid modal analysis. Other refined methods based on these general methods or their combinations were also investigated. Reliability and accuracy were the two major criteria considered to evaluate the suitability of each method. Reliability refers to the consistency of the results across the set of prototypes and hazard levels. Accuracy refers to the degree of correlation between the predicted force and the reference shear key force. The mean absolute percentage error (MAPE) as shown in Equation 1 is a standard statistical measure of accuracy. This measure is used to quantify reliability and accuracy of the results for each seismic hazard level (*34*).

$$MAPE = \frac{1}{n}\sum_{i=1}^{n}\left|\frac{Y_i - X_i}{Y_i}\right| \times 100 \quad (1)$$

where $Y$ is the reference shear key force and $X$ is the shear key force from the method of interest, and $n$ is the total number of data points. A lower MAPE reflects a more accurate method. Scatter plots in the following figures show the relationship of the determined shear key force to the reference shear key force and MAPE for each analysis method.

### Static Nonlinear "Pushover" Analysis

To perform the pushover analyses, the design displacements for the in-span hinges were found using EDA. The in-span hinge nodes were pushed to their corresponding design displacements. The shear key forces were obtained using two methods: (a) the difference of the absolute shear forces in adjacent superstructure elements, $|V1|-|V2|$, where $V1$ is the shear force in one frame and $V2$ is the shear force in an adjacent frame; and (b) one-quarter of the absolute values of $V1$ and $V2$, which is equal to the push force, $P$, divided by four, $P/4 = (|V1| + |V2|)/4$. The later method assumes that the shear key force is the average of the shear forces in the adjacent superstructure elements. Figures 3*a* and *b* show the reference shear key forces versus $|V1|-|V2|$ and $P/4$, respectively. The method based on $|V1|-|V2|$ performed poorly for the two-column bridges. This may be because of the small values of $|V1|-|V2|$ when the shear forces in the left and right superstructures are comparable. The method based on $P/4$ is more reliable than the method based on $|V1|-|V2|$; however, the method based on P/4 is not accurate enough. It was concluded that pushover analysis does not determine reliable and accurate shear key force demands because it does not capture the effects of high-frequency modes of vibration of the superstructure; therefore, pushover analysis should not be used for this purpose.

### Elastic Dynamic Analysis (EDA)

As opposed to the pushover method, the EDA method captures the effects of high-frequency modes of vibration. The ground motion response spectrums were used and the total responses, including shear key forces, were obtained using complete quadratic combination (CQC) of the first 50 modes, achieving a 97% mass participation ratio. The correlations between shear key force from EDA and the reference shear key force from NTH are presented in Figure 4*a*. Under moderate ground motions, where the structures remain practically elastic, there is a relatively good correlation. However, under large and severe motions, the reference shear key forces are significantly smaller than the EDA shear key forces. To address the over-prediction of shear key forces under large ground motions by the EDA method, the elastic spectrums were replaced with inelastic spectrums (*35*). The shear key forces that resulted from this method are compared with the reference shear key force in Figure 4*b*. The use of inelastic spectrums resulted in underprediction of shear key forces under moderate and severe motions, as most of the high-frequency modes of vibration that contribute to shear key forces are elastic. However, the use of a constant-ductility inelastic spectrum suppresses the effects of these modes.

The number of dynamic modes correlates with the mass participation ratio. Reaching a large ratio, typically more then 0.9, in the EDA method means effects of high-frequency modes of vibration are included in the structural response. For practical purposes, the shear key force corresponding to the mass participation ratio of 0.97 was assumed as the maximum shear key force response from EDA. The statistical distribution of the required mass participation ratio to achieve 90% of the

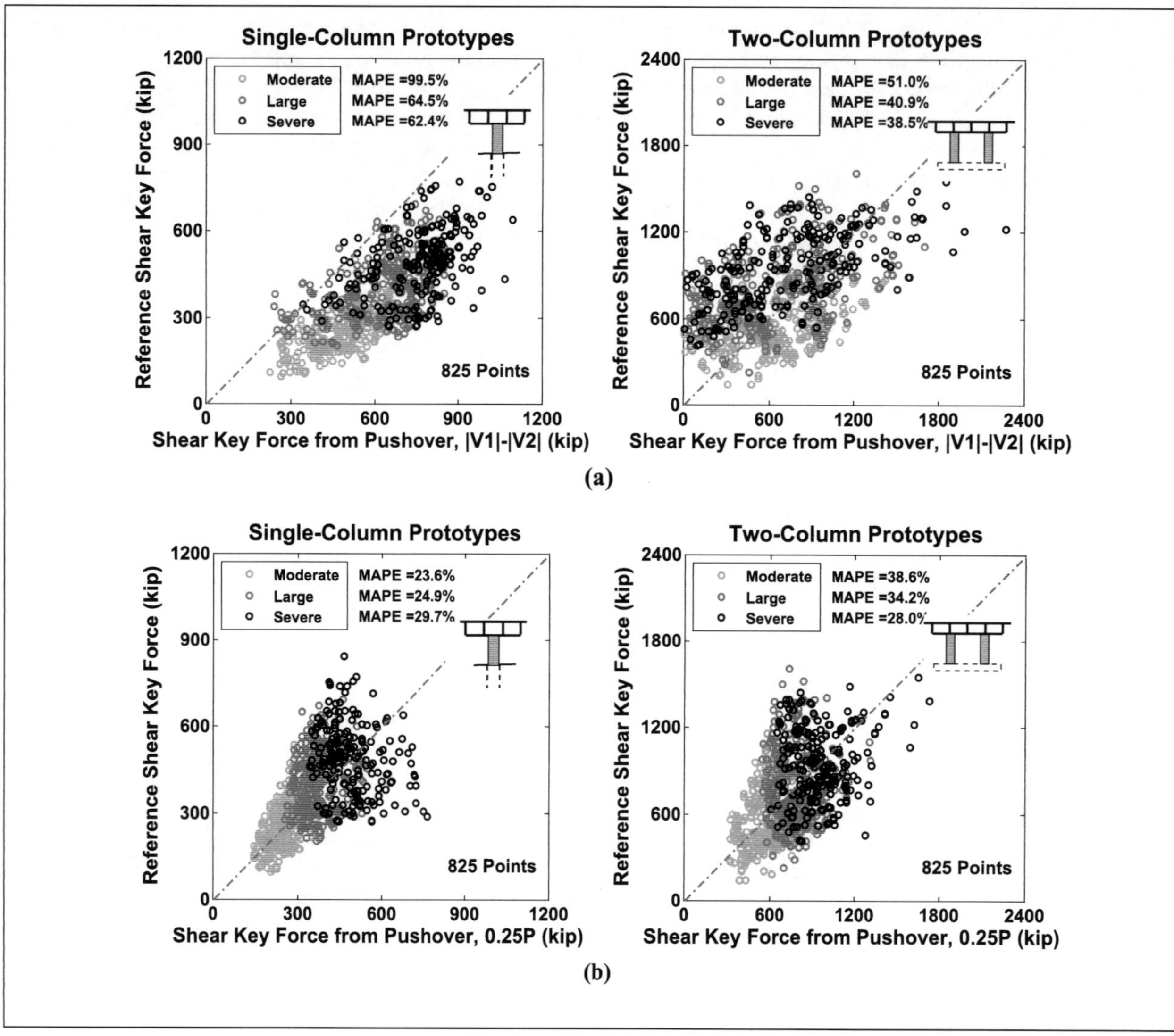


**Figure 3.** Scatter plots (*a*) of $|V1|–|V2|$ from pushover analyses with the reference shear key forces, (*b*) ($P/4$) from pushover analyses with the reference shear key forces.

maximum EDA shear key force for single- and two-column bents is shown in Figure 5*a* and *b*. Likewise, the minimum mass participation ratios needed to achieve 90% of the maximum displacement at in-span hinges are shown in Figure 5*c* and *d*. The figures show that: (a) a dynamic mode that does not participate in the displacement response may significantly participate in the shear key force, and (b) mobilizing a mass participation ratio of 0.9 may not be sufficient for estimation of the maximum elastic shear key forces. This data suggests that for the calculation of shear key forces, the target mass participation ratio needs to be larger than 0.95. Particularly in two-column bridges with a stiffer superstructure, modes with a small mass participation ratio can mobilize large shear key forces.

### *Other Methods*

In addition to the methods discussed above, three others were investigated: (a) EDA modified with displacement ductility, (b) the rigid dynamic mode method, a simple method used by the Nuclear Regulatory Commission to capture dynamic effects of high-frequency modes of vibration (*36*), and (c) a combination of pushover and rigid dynamic mode. All three methods proved unsuccessful. The EDA method, modified with displacement ductility, led to under-prediction of shear key forces. The rigid dynamic mode method lacked accuracy; in this method, only inertial effects are considered. The final method produced relatively better results, but ultimately lacked accuracy. The details of these analyses are presented in Mehr and Zaghi (*22*).

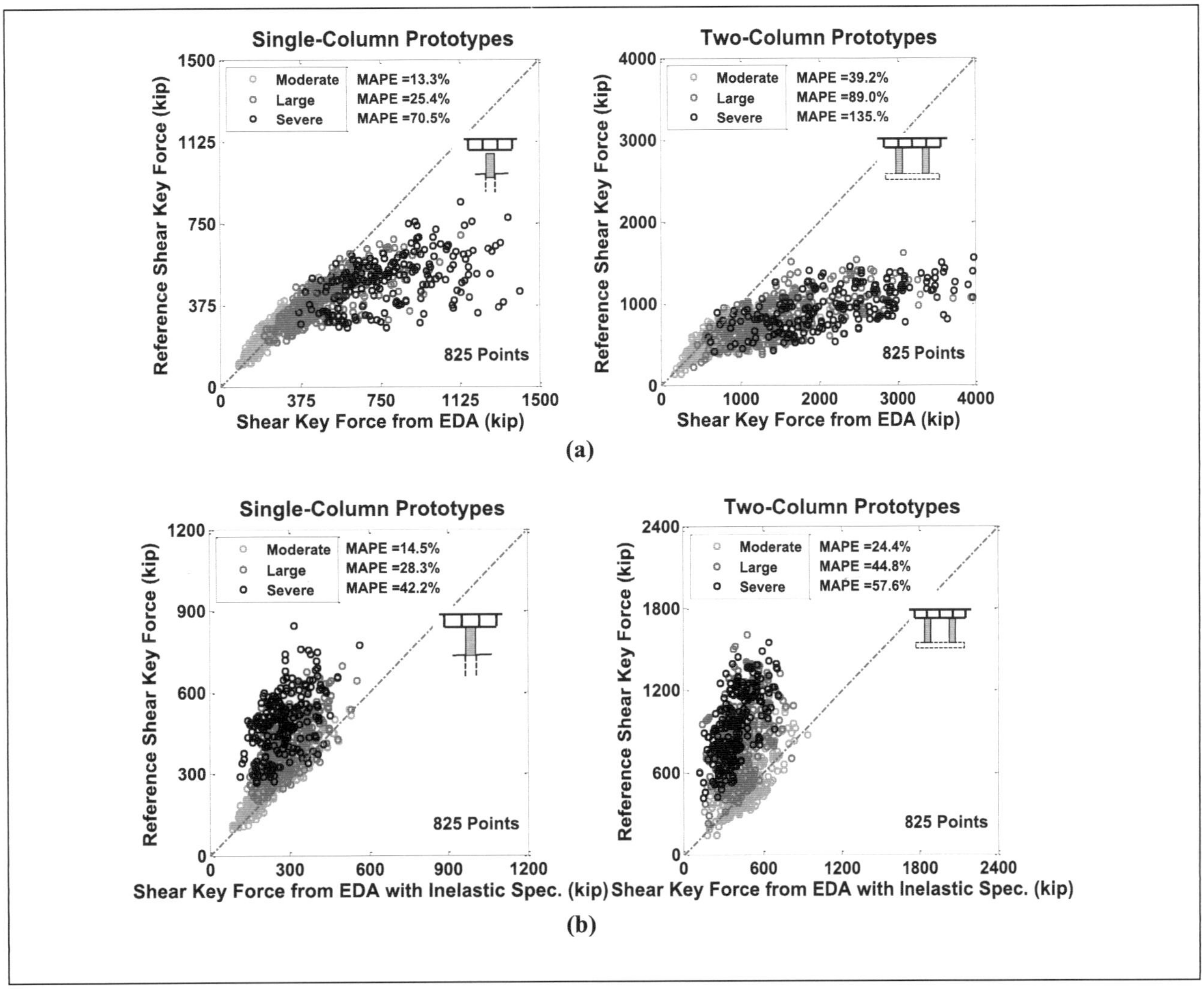


**Figure 4.** (*a*) Correlation of forces from EDA with reference shear key forces; and (*b*) Correlation of forces from EDA with inelastic spectrum with reference shear key forces.

## Proposed Method for Estimating Shear Key Demands

Results presented earlier show that the uniform modification of modal shear key forces leads to significant errors. In this study, a refined method was developed to capture the effects of the nonlinearities at certain modes. The proposed method, referred to as MEDA here, uses the concept of EDA and modifies individual modal forces using "modal displacement ductility." Modal displacement ductility will be introduced in the following section. The modified forces are then combined using a standard method, such as square root of sum of squares (SRSS) or complete quadratic combination (CQC). The proposed method preserves the effects of high-frequency modes of vibration, as well as nonlinearity of low-frequency modes.

### *Modal Displacement Ductility*

Recent studies on building systems have shown that structures typically become inelastic at the first mode of vibration, but remain elastic at higher modes in response to ground motions (*37*). In this study, it was discovered that in multi-frame bridges high-frequency modes can generate large inelastic responses. This observation led to the development of the concept of modal displacement ductility for multi-frame bridges. Modal displacement ductility, $\mu_{D,n}$, is defined as the maximum displacement ductility of columns under spectral displacement of one mode (Figure 6, Equation 2).

$$\mu_{D,n} = max\left|\Delta_{(i)D,n}/\Delta^{col}_{(i)Y}\right| \geq 1.0 \qquad (2)$$

where $\Delta_{(i)D,n}$ is the spectral displacement of the *i*th column under *n*th mode of vibration, and $\Delta^{col}_{(i)Y}$ is the

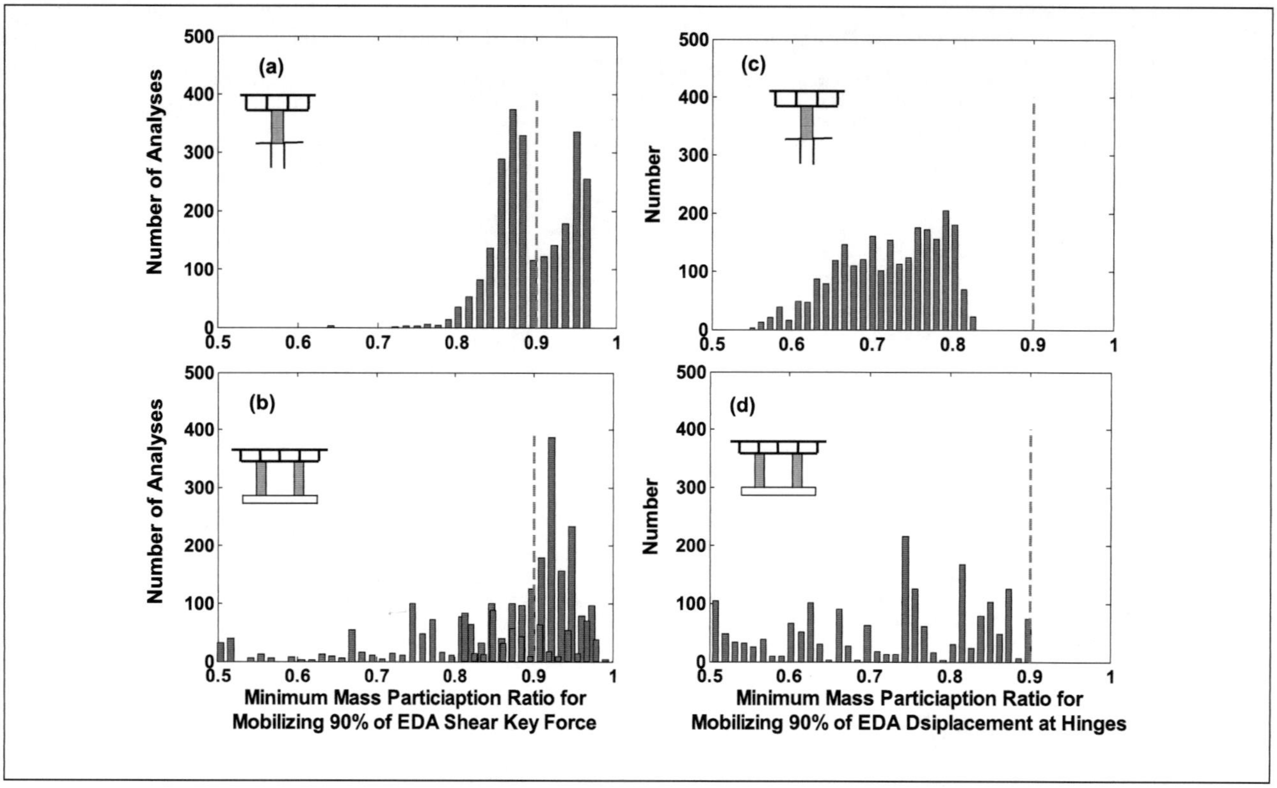


**Figure 5.** Distribution of the mass participation ratios for mobilizing 90% of the maximum shear key forces: (a) single-column prototypes, (b) two-column prototypes; Distribution of the mass participation ratios for mobilizing 90% of the maximum transverse displacement at hinges: (c) single-column prototypes, (d) two-column prototypes.

idealized yield displacement of the $i$th column according to Caltrans SDC 1.7, Sec. 2.2.4. Ductility values less than 1.0 indicate an elastic response, in which case $\mu_{D,n}$ is reset to 1.0. The $\Delta_{(i)D,n}$ can be calculated by Equation 3, or by using analytical packages such as SAP2000 (*37*).

$$\Delta_{(i)D,n} = \Gamma_n \phi_{(i)D,n} S_{a,n} / \omega_n^2 \quad (3)$$

where $\Gamma_n$ is the modal participation factor, $\phi_{(i)D,n}$ is the mode shape at the $i$th column, $S_{a,n}$ is the spectral acceleration and $\omega_n$ is the frequency of the $n$th mode.

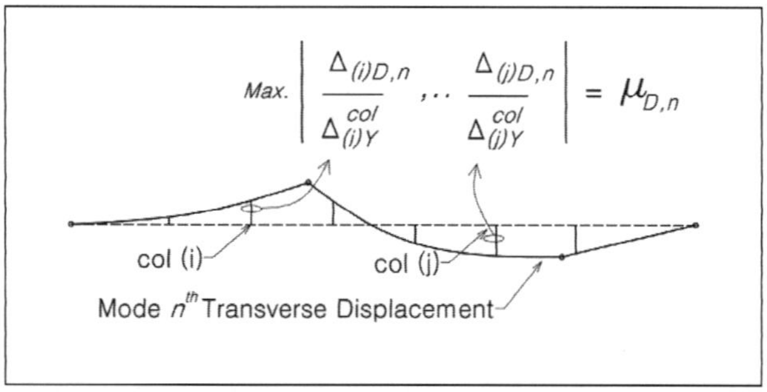


**Figure 6.** Illustration of modal displacement ductility ($\mu_{D,n}$).

For the set of prototype bridges, the values of modal displacement ductility for different modal periods are shown in Figure 7. For the single-column system, the inelastic modes, that is, $\mu_{D,n} > 1.0$, were the 1st to 4th modes with periods between 1.0–2.8 sec. For the two-column system, modes as high as the 22nd, with periods between 0.5–2.2 sec, may contribute to an inelastic response. This data demonstrates the importance of accounting for the nonlinearity of different modes. In the proposed method, the values of modal displacement ductilities were used for modification of modal shear key forces.

## *Modal Shear Forces*

Similar to the modal displacements, the modal shear key forces can be calculated using Equation 4 (*38*).

$$V_n^{sk} = \Gamma_n \phi_{(sk)V,n} S_{a,n} / \omega_n^2 \quad (4)$$

where $V_n^{sk}$ is the spectral shear key force obtained from the $n$th mode of vibration. $\phi_{(sk)V,n}$ is the modal shear force quantity at shear key $sk$, which can be found through Eigenvector analysis.

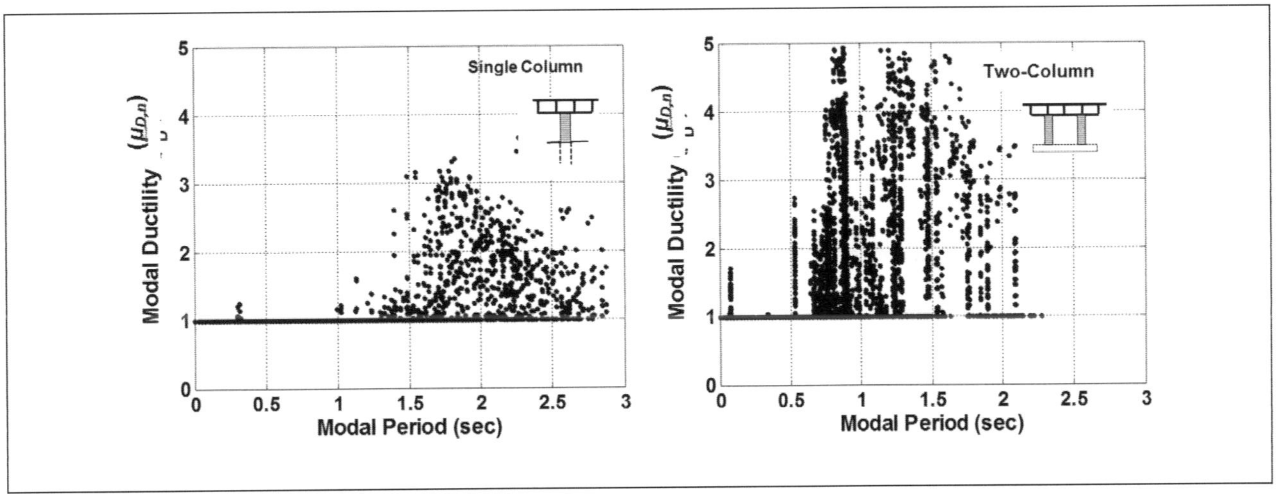


**Figure 7.** Transverse modal ductility ratios for different modal periods.

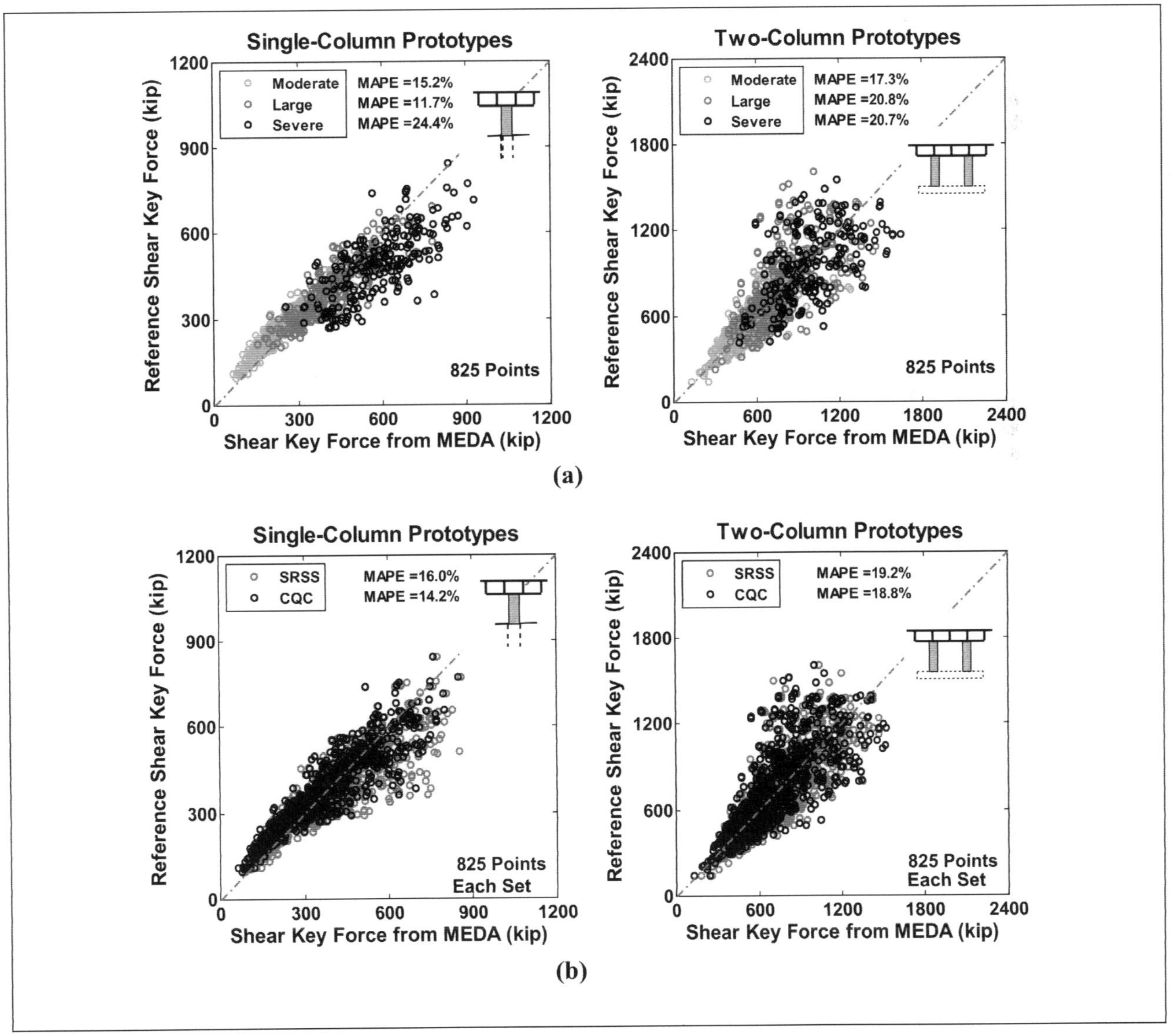


**Figure 8.** (*a*) Correlation of MEDA forces with reference shear key forces for all prototypes, (*b*) Comparing the SRSS and CQC combination methods in MEDA for all prototypes.

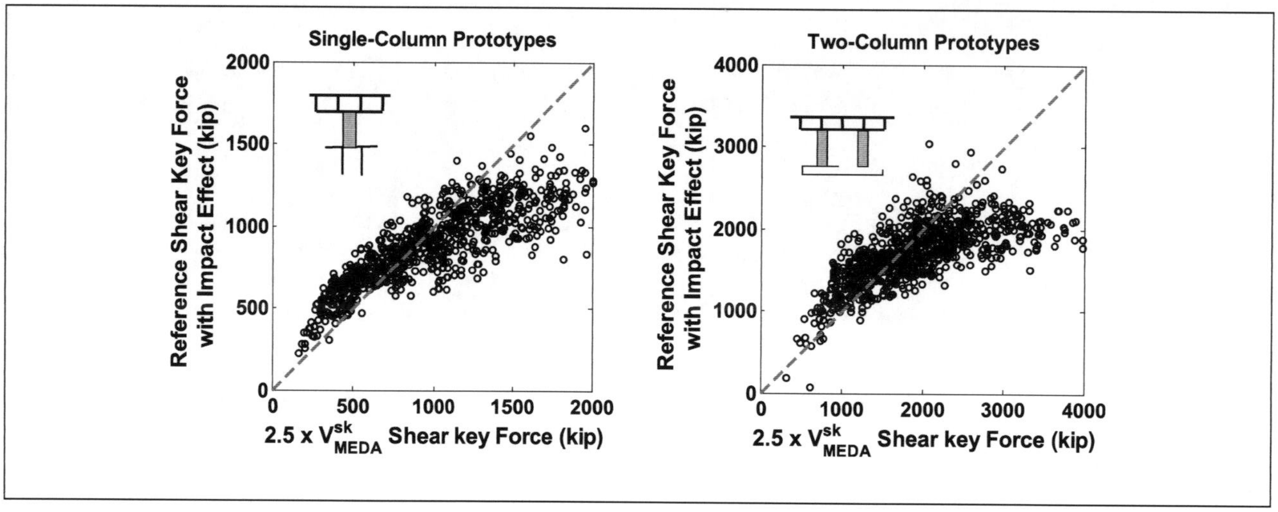


**Figure 9.** Correlation of design forces with NTH shear key forces including impact effects.

Based on the concept of equal displacements, inelastic forces can be estimated by dividing the elastic force by displacement ductility. In the method proposed in this study, this concept is implemented to modify the participation of each mode in shear key force by the ductility of that mode as presented in Equation 5.

$$\bar{V}_n^{sk} = V_n^{sk}/\mu_{D,n} \quad (5)$$

Finally, the modified modal forces are combined using a standard method such as square root of sum of squares (SRSS) or complete quadratic combination (CQC) to find the total shear key force.

Figure 8*a* shows the shear key forces obtained from the MEDA compared with the reference shear key force. The MEDA method is a reliable method with both modal combination methods (SRSS and CQC) yielding similar results (Figure 8*b*).

The MEDA method does not include the effect of the impact on shear key elements. The complexity of the impact phenomenon makes it impractical for development of a formula. An impact factor of 2.5, which is equal to the median of impact factors, was suggested by Mehr and Zaghi (*21*). Figure 9 shows the correlation of the factored MEDA shear key forces versus the reference shear key forces, including transverse impact effects.

Additionally, a modification factor of 1.25 was proposed to account for the effect of non-uniform base excitation for soil types C, D, and E (*21*). This modification factor, along with the impact factor, can then be implemented into Equation 6 to determine the design seismic shear demand on in-span hinge shear keys.

$$V_u^{sk} = \text{Factor1} \times \text{Factor2} \times V_{MEDA}^{sk} \leq V_{up}^{sk} \quad (6)$$

$$\text{Factor1} = 2.5\,(\text{Impact Factor}) \quad (7)$$

$$\text{Factor 2} = \begin{cases} 1.0 \text{ Soil Type A, and B} \\ 1.25 \text{ Soil Type C, D, and E} \end{cases} \quad (8)$$

$$V_{up}^{sk} = 0.8 \times min\left(V_o^{Left\,Frame}, V_o^{Right\,Frame}\right)\,(\text{Upper Bound}) \quad (9)$$

where $V_o^{frame}$ is the summation of the transverse overstrength shear capacities of the columns in each frame as defined in SDC 1.7 Sec. 2.3.2.1. This study revealed that the maximum shear key forces, including impact effects, do not exceed 80% of the transverse capacity of the weaker frame adjacent to the hinge. Currently, SDC 1.7 assumes 100% of capacity of the weaker frame as the upper bound force. The mean value was found to be about 35% of the transverse capacity of the weaker frame.

## Conclusions

A modified elastic dynamic analysis method (MEDA) was developed to determine in-span hinge shear key force demands. The effect of the nonlinear behavior is considered through the modification of individual modes by the modal ductility proposed in this study. To develop, and verify the proposed method, approximately 1,848 NTH analyses were conducted on a set of 56 box-girder prototype bridges designed in accordance with the seismic design criteria of Caltrans. This method presents reliable and accurate results. Effects of the transverse impact and non-uniform base excitation from previous studies were included in the proposed method.

It was found that the in-plane vibration of the superstructure significantly contributes to the shear key forces. Thus, assuming that the superstructure is rigid in-plane may yield inaccurate estimation of shear key force demands. The standard pushover method fails to estimate shear key force demands. The standard EDA method significantly over-predicts the shear key forces, as it ignores nonlinearities in structural response. Conversely, using EDA with an inelastic response spectrum leads to significant under-prediction of shear key forces. Similarly, modifying EDA forces with one response factor, such as ductility or response factor under-predicts the forces.

The reduction of elastic forces should only be applied to inelastic modes. The analyses showed that in multi-frame bridges higher modes may present more nonlinearity than the first mode. Individual elastic modal forces should be divided by the ductility ratio of the corresponding mode before modal combination. For accurate estimation of elastic shear key forces in EDA and MEDA, it is necessary to have a larger mass participation ratio than is necessary for estimation of displacements. The upper bound shear key force was found to be approximately 80% of the transverse overstrength capacity of the weaker frame connected by the in-span hinge.

## Acknowledgments

This research was funded by the California Department of Transportation (Caltrans). Special thanks are due to Charles Sikorsky, the Caltrans Research Program Manager, for his significant support and advice throughout the project. Daryoush Tavatli, Marc Friedheim, Toorak Zokaie, Mark Mahan, Tom Ostrom, and Mike Keever of Caltrans are thanked for their critical feedback, comments, and suggestions. Saad El-Azazi is thanked for his early support on the projects. Former students Alicia Echevarria, Amanda McBride, and Kevin Zmetra are thanked for their assistance.

## References

1. Hube, M. A., and K. M. Mosalam. *PEER 2008/103: Experimental and Computational Evaluation of Current and Innovative In-Span Hinge Details in Reinforced Concrete Box-Girder Bridges*. Pacific Earthquake Engineering Center (PEER), Berkeley, CA, 2008.
2. Caltrans. *xs7-070: Restrainer Details for Pipe Extender/Shear Key (New). Bridge Standard Detail Sheets, Division of Engineering Services (DES)*. California Department of Transportation, Sacramento, CA, 2014. http://www.dot.ca.gov/hq/esc/techpubs/manual/bridgemanuals/bridge-standard-detail-sheets/
3. Yashinsky, M. *Recent Change to Seismic Design Practice in California*. Caltrans Office of Earthquake Engineering, Sacramento, CA, 2013.
4. AASHTO. *AASHTO LRFD Bridge Design Specifications*, 7th, Interim 2015 ed. American Association of State Highway and Transportation Officials, Washington, DC, 2014.
5. AASHTO. *LRFD Seismic Bridge Design*, 2nd ed. American Association of State and Transportation Officials, Washington, DC, 2011.
6. Caltrans. *Seismic Design Criteria Version 1.7*. California Department of Transportation, Sacramento, CA, 2013. http://www.dot.ca.gov/hq/esc/earthquake_engineering/sdc/documents/Seismic-Design-Criteria-(SDC-1.7-Full-Version,-OEE-Release).pdf
7. Fenves, G. L., and M. Ellery. *PEER 1998/08: Behavior and Failure Analysis of a Multiple-Frame Highway Bridge in the 1994 Northridge Earthquake*. Pacific Earthquake Engineering Research Center (PEER), Berkeley, California, 1998.
8. Hao, H., and N. Chouw. Seismic Design of Bridges for Prevention of Girder Pounding. *Electronic Journal of Structural Engineering (EJSE)*, Vol. 8, 2008, pp. 133–141.
9. Singh, S. P. *Earthquake Analysis and Response of Multi-Frame Bridges*. PhD dissertation. University of California, Berkeley, CA, 1994.
10. DesRoches, R., and S. Muthukumar. Implications of Seismic Pounding on the Longitudinal Response of Multi-Span Bridges - an Analytical Perspective. *Earthquake Engineering and Engineering Vibration*, Vol. 3, No. 1, 2004, pp. 57–65.
11. Shrestha, B., H. Hao and K. Bi. *Pounding and Unseating Damage Mitigation on Bridge Structures subjected to Spatially Varying Ground Motions Using Restrainers and Rubber Bumpers*. Australian Earthquake Engineering Society Conference, Tasmania, 2013.
12. DesRoches, R., and G. L. Fenves. Simplified Restrainer Design Procedure for Multi-Frame Bridges. *Earthquake Spectra*, Vol. 17, No. 4, 2001, pp. 551–567.
13. DesRoches, R., and G. L. Fenves. *MCEER-0013: Design Procedures for Hinge Restrainers and Hinge Seat Width for Multiple-frame Bridges*. Multidisciplinary Center for Earthquake Engineering Research (MCEER), University of California, Berkeley, CA, 1998. http://mceer.buffalo.edu/publications/catalog/reports/Design-Procedures-for-Hinge-Restrainers-and-Hinge-Seat-Width-for-Multiple-frame-Bridges-MCEER-98-0013.html
14. Trochalakis, P., M. O. Eberhard and J. F. Stanton. Design of Seismic Restrainers for In-Span Hinges. *Journal of Structural Engineering*, Vol. 123, No. 4, 1997, pp. 469–478.
15. Tegos, I. A., and O. G. Markogiannaki. *Analytical Investigation on the use of FRP Materials as Seismic Restrainers in Concrete Bridges*. Bridge Conference, Istanbul, 2014.
16. Jeon, J.-S., R. DesRoches, T. Kim, and E. Choi. Geometric Parameters Affecting Seismic Fragilities of Curved Multi-Frame Concrete Box-Girder Bridges with Integral Abutments. *Engineering Structures*, Vol. 122, No. 1, 2016, pp. 121–143.
17. Pahlavan, H., B. Zakeri, G. Ghodrati Amiri, and M. Shianfar. Probabilistic Vulnerability Assessment of Horizontally Curved Multi-Frame RC Box-Girder Highway Bridges. *Journal of Performance of Constructed Facilities*, Vol. 30, No. 3, 2016.
18. Priestly, M. J., F. Seible, and G. M. Calvi. *Seismic Design and Retrofit of Bridges*, 1st ed. John Wiley & Sons, Inc., New York, 1996.
19. Akbari, R., and S. Maalek. Adequacy of the Seismic Analysis Methods for Single-Column-Bent Viaducts Considering Regularity and Higher Modes Effects. *Journal of Vibration and Control*, Vol. 16, 2010, pp. 827–852.

20. Mehr, M., and A. E. Zaghi. Seismic Response of Multi-Frame Bridge. *Journal of Bulletin of Earthquake Engineering*, Vol. 14, No. 4, 2016, pp. 1219–1243.
21. Mehr, M., and A. Zaghi. Contributing Factors to Seismic Force Demand on In-Span Shear Keys in Multi-Frame Bridges. *Structures and Infrastructure Engineering*, 2018. (Under review).
22. Mehr, M., and A. E. Zaghi. *Development of a Rational Design Method for Shear Keys at In-Span Hinges in Multi-Frame Highway Bridges, PART-1: A Rational Seismic Analysis Method for In-Span Hinge Shear Keys*. California Department of Transportation, Sacramento, CA, 2014.
23. Caltrans. *Memo to Designers 10-20: Deck and Soffit Slabs*. California Department of Transportation (Caltrans), Sacramento, CA, 2008.
24. McKenna, F., and G. L. Fenves. *Open System for Earthquake Simulation Framework (OpenSees) Ver. 2.4.4*. Pacific Earthquake Engineering Research Center (PEER), University of California, Berkeley, CA, 2014. http://opensees.berkeley.edu/
25. Muthukumar, S., and R. DesRoches. A Hertz Contact Model with Non-Linear Damping For Pounding Simulation. *Earthquake Engineering & Structural Dynamics*, Vol. 7, No. 35, 2006, pp. 811–828.
26. Megalley, S. H., P. F. Silva and F. Seible. *Seismic Response of Sacrificial Exterior Keys in Bridge Abutments*. Report no. SSRP–2001/23. Department of Structural Engineering, University of California, San Diego, La Jolla, CA, 2002.
27. Bozorgzadeh, A., S. H. Megalley, S. Ashford, and J. I. Restrepo. *Seismic Response of Sacrificial Exterior Keys in Bridge Abutments*. Department of Structural Engineering, University of California, San Diego, La Jolla, CA, 2007.
28. Mander, J. B., M. J. Priestley, and R. Park. Theoretical Stress-Strain Model for Confined Concrete. *Journal of Structural Engineering*, Vol. 114, No. 8, 1988.
29. Chopra, A. K. *Dynamics of Structures*, 2nd ed. Prentice Hall, New Jersey, 2001.
30. Aviram, A., K. R. Mackie and B. Stojadinovic. *Guideline for Nonlinear Analysis of Bridge Structures in California*. Pacific Earthquake Engineering Research Center, PEER, Berkeley, CA, 2008.
31. Zaghi, A. E., and M. S. Saiidi. *CCEER-10-01: Seismic Design of Pipe-Pin Connections in Concrete Bridges*. Center for Civil Engineering Earthquake Research (CCEER), Reno, Nevada, 2010.
32. PEER. *Ground Motion Database*. Pacific Earthquake Engineering Research Center, 2011.
33. SeismoMatch- A computer program to match a specific target response spectrum. Seismosoft, Pavia, Italy, 2011, www.seismosoft.com.
34. Montgomery, D. C., G. C. Runger and N. F. Hubele. *Engineering Statistics*. 5th ed. John Wiley & Sons, 2010.
35. Krawinkler, H., and A. A. Nassar. *Seismic Design Based on Ductility and Cumulative Damage Demands and Capacities*. Elsevier Applied Science, New York, 1992.
36. Gupta, A. K. *Modal Combination in Response Spectrum Method*. Eighth World Conference on Earthquake Engineering, San Francisco, 1984.
37. NIST. GCR12-917-20: *Tentative Framework for Development of Advanced Seismic Design Criteria for New Buildings*. National Institute of Standards and Technology (NIST), U.S. Department of Commerce, Gaithersburg, MD, 2012.
38. CSI. *Analysis Reference Manual for SAP2000, ETABS, SAFE and CSiBridge*. Computers and Structures Inc. (CSI), Berkeley, California, 2013.


*The Standing Committee on Seismic Design and Performance of Bridges peer-reviewed this paper (18-00719).*

Article

TRR
JOURNAL OF THE TRANSPORTATION RESEARCH BOARD

Transportation Research Record
2018, Vol. 2672(41) 87–100
© National Academy of Sciences:
Transportation Research Board 2018
Article reuse guidelines:
sagepub.com/journals-permissions
DOI: 10.1177/0361198118781148
journals.sagepub.com/home/trr
SAGE

# Impact of Specialized Hauling Vehicles on Load Rating Older, Bridge-Class, Reinforced Concrete Box Culverts

**William D. Lawson[1], Hoyoung Seo[1], James G. Surles[2], and Stephen M. Morse[3]**

**Abstract**

This paper describes the comparison of load ratings associated with application of three live load models recognized by AASHTO—AASHTO legal loads, the notional rating load including single-unit specialized hauling vehicles (SHVs), and the HL-93 design tandem live load—versus load ratings associated with application of the typical HS-20 standard truck. The test bed for this study was a statistically representative sample of Texas' older bridge-class reinforced concrete box culvert structures. Rating factors were determined using the load factor rating method with demands calculated from a production-simplified, calibrated, two-dimensional soil–structure interaction model using linear elastic constitutive models for both concrete and soil. The study was motivated in part by research which showed that SHVs create force effects significantly greater than those from the HS-20 truck (for bridges proper), and recent federal policy mandating that states load rate their bridges for SHVs. Findings from this study indicate the standard HS-20 truck, and *not* SHVs or other legal or design loads, is the critical model for most culvert load rating applications. In particular, operating rating factors calculated from both the AASHTO legal loads and SHV models tend to be higher than corresponding rating factors calculated using the HS-20 standard truck, most of the time. The response is explained primarily by considering the relatively short span length of culvert structures and the load-attenuating benefit of cover soil above the culvert top slab. More detailed exploration of rating variables suggests interactions between culvert geometry, cover soil thickness, and the various types of applied vehicle loads.

This paper describes impacts resulting from application of specialized hauling vehicles (SHVs) and other non-HS20 load patterns on load ratings for older, bridge-class, cast-in-place, reinforced concrete box culvert structures. More specifically, this paper compares culvert load ratings calculated using the standard HS-20 truck with load ratings that were calculated based on application of three types of other recognized vehicle load models: (1) AASHTO legal loads, (2) the notional rating load including single-unit SHVs that meet Federal Bridge Formula B, and (3) the HL-93 design live load per LRFD design specifications.

This paper is motivated by 2013 FHWA policy guidance that requires states to include SHVs in their load rating and posting analyses of all bridges by the year 2017 to 2022 (*1*). SHVs are closely spaced multi-axle single-unit trucks (4–7 axles) introduced by the trucking industry in the last decade (*2*). These short wheel base trucks—dump trucks, ready-mix concrete trucks, tank trucks, solid-waste trucks, and so forth—can carry up to 80,000 lbs and still meet the requirements of the Federal Bridge Formula B. However, SHVs "…cause force effects that exceed the stresses induced by HS20 in bridges by up to 22% and by the Type 3, 3S2, or 3-3 posting vehicles by over 50% in certain cases" (*3*), especially "bridges with a shorter span and shorter loading length such as transverse floor beams" (*1*). Evaluation of this issue lies at the confluence of three domains of transportation policy and operations, and these are bridge inspection, truck size and weight, and culvert structures.

## *Load Rating*

Load rating is a component of the bridge inspection process and consists of determining the live load-carrying capacity of a bridge structure. As used in this paper, multi-span box culverts are bridge-class provided the centerline length of all spans combined is greater than 20 feet (*4*). The AASHTO *Manual for Bridge Evaluation, 2nd Edition* (*MBE*), Section 6, provides technical guidance on load rating and posting of existing bridges (*5*) and identifies three methods for load rating: load and resistance factor rating (LRFR), load factor rating (LFR), and allowable stress rating (ASR).

[1]Department of Civil, Environmental and Construction Engineering, Texas Tech University, Lubbock, TX
[2]Texas Tech University, Lubbock, TX
[3]Civil & Environmental Engineering, Michigan Technological University, Houghton, MI

**Corresponding Author:**
Address correspondence to William D. Lawson: william.d.lawson@ttu.edu

The *MBE* Section 6, Part A presents LRFR procedures for load rating bridges consistent with the load and resistance factor design philosophy. The *MBE* Section 6, Part B provides a choice of procedures for load rating bridges using either ASR or LFR. Both approaches accommodate two rating levels, the inventory rating level associated with the design capacity and operating rating level associated with the ultimate capacity.

Inasmuch as this research focuses on the rating of *older* bridge-class culverts, the LFR method is used, and the rating equation summarizes the principles of load rating under this method, Equation 1.

$$RF = \frac{C - A_1 D}{A_2 L(1+I)} \quad (1)$$

The rating factor, *RF*, is essentially a live load factor of safety for a particular section in a structure calculated from the capacity, *C*, minus the factored dead load, $A_1 D$, and divided by the factored live load including impact, $A_2 L\ (1 + I)$. Per the LFR method as specified in the AASHTO *Standard Specifications for Highway Bridges* (*6*), the analysis considers three potential failure modes: moment (flexure), shear, and axial thrust. If the rating factor is greater than 1.0, the section can carry the applied live load at that section; if less than 1.0, the section does not have adequate capacity. The rating for the section is calculated by multiplying the rating factor by the nominal tonnage of the load rating truck. The lowest rating from all sections governs the load rating for the structure (*5*).

Technical guidance for load rating is written primarily with traditional bridges in mind, although significant portions of this policy can be extrapolated to bridge-class culverts. The *MBE* Section 6 does not specifically mention culverts, neither Part A nor Part B. However, the 2013 Interim Revisions to the *MBE* add a completely new article, Section 6A.4.12, on rating of reinforced concrete box culverts using the LRFR method. Further, the 2014 Interim Revisions to the *MBE* provide additional guidance on live loads for LRFR rating of culverts. In contrast, the *MBE* Section 6B (ASR, LFR) does not currently provide any specific load rating guidance for culverts apart from brief statements in the 2016 Interim Revisions on assigned load ratings and on potential overconservatism associated with simplified modeling approaches (*5*).

## *Vehicle Live Load Models*

The evolution of live load design models for bridges dates to the 1920s (*7*). Relative to live load configuration, the *MBE* features four dominant truck load models and these include the HS-20 standard truck load, AASHTO legal loads, the notional rating load including single-unit SHVs, and the HL-93 design live load (Figure 1). Of course, other truck load combinations exist, sometimes called exclusion vehicles, but these four models are primary for load rating.

The standard HS-20 truck was introduced in the 1940s to represent an idealized tractor-semi-trailer combination (*7*). First identified in 1944, the HS-20 truck features three axles and a total weight of 72,000 lbs, as shown in Figure 1*a*. The *MBE*, Section 6 Part B, specifies that the HS-20 truck be used to determine the extreme live load force effect in the basic rating equation under ASR and LFR (*5*). State DOTs, such as Texas, also specify the HS-20 truck be used to rate existing structures per the LFR method (*8*).

The AASHTO legal loads, also referred to as "AASHTO Routine Commercial Traffic legal loads" are a set of three vehicle configurations (Figure 1*b*) identified in the 1980s that were found to be typical of the legal load trucks operating in many states (*9*). AASHTO Type 3 is representative of a three-axle, single-unit truck. Type 3S2 is representative of a five-axle tractor-semitrailer, that is, the common "18-wheeler." Type 3-3 is representative of a six-axle truck–trailer combination. The *MBE* Part 6A specifies that the AASHTO legal loads be used for LRFR rating of bridges for routine legal commercial traffic, with the ratings suitable for use as a basis for decision-making related to load posting or bridge strengthening. Similarly, the *MBE* Part 6B specifies the AASHTO legal loads be used for LFR or ASR rating for bridge posting considerations (*5*).

The third truck load group, AASHTO Specialized Hauling Vehicle legal loads (SHVs), represents a class of legal vehicles not initially incorporated within the set of typical AASHTO legal loads, and thus requires special consideration for load rating/posting decisions (*3*). Figure 1*c* identifies the notional rating load (NRL) that is considered as a screening load model for single-unit trucks that meet Formula B. Also included are four single-unit Formula B legal loads that represent the worst four-axle (SU4), worst five-axle (SU5), worst six-axle (SU6), and worst seven-axle (SU7) trucks identified for this group. The *MBE* Part 6A specifies that the NRL/SHV loads be used in addition to the AASHTO legal loads for LRFR ratings suitable for bridge posting considerations. Likewise, the *MBE* Part 6B also specifies the NRL/ SHV loads be used for LFR or ASR ratings associated with bridge posting.

The final live load model, the HL-93 design live load (Figure 1*d*), is a family of three live load models adopted for use in the LRFD design of bridges. The first—identified as "Design Truck"—is none other than the standard HS-20 truck, discussed previously. The second HL-93 load—identified as "design tandem"—generally corresponds to what had previously been identified as "alternate military loading" (*7*). The third HL-93 load depicts a train of two HS-20 trucks spaced at least 50ft apart with loads reduced to 90%. This HL-93 family of loads was found to best fit the exclusion vehicles of the Interstate design era and thus served as the basis for a new live load model that would achieve more uniform and consistent safety of bridges (*7*). The *MBE* Part 6A specifies that the HL-93 family of loads be used for the design load rating of bridges. The *MBE* Part 6B does not mention HL-93 loads other than a statement in the 2016 Interim Revisions that

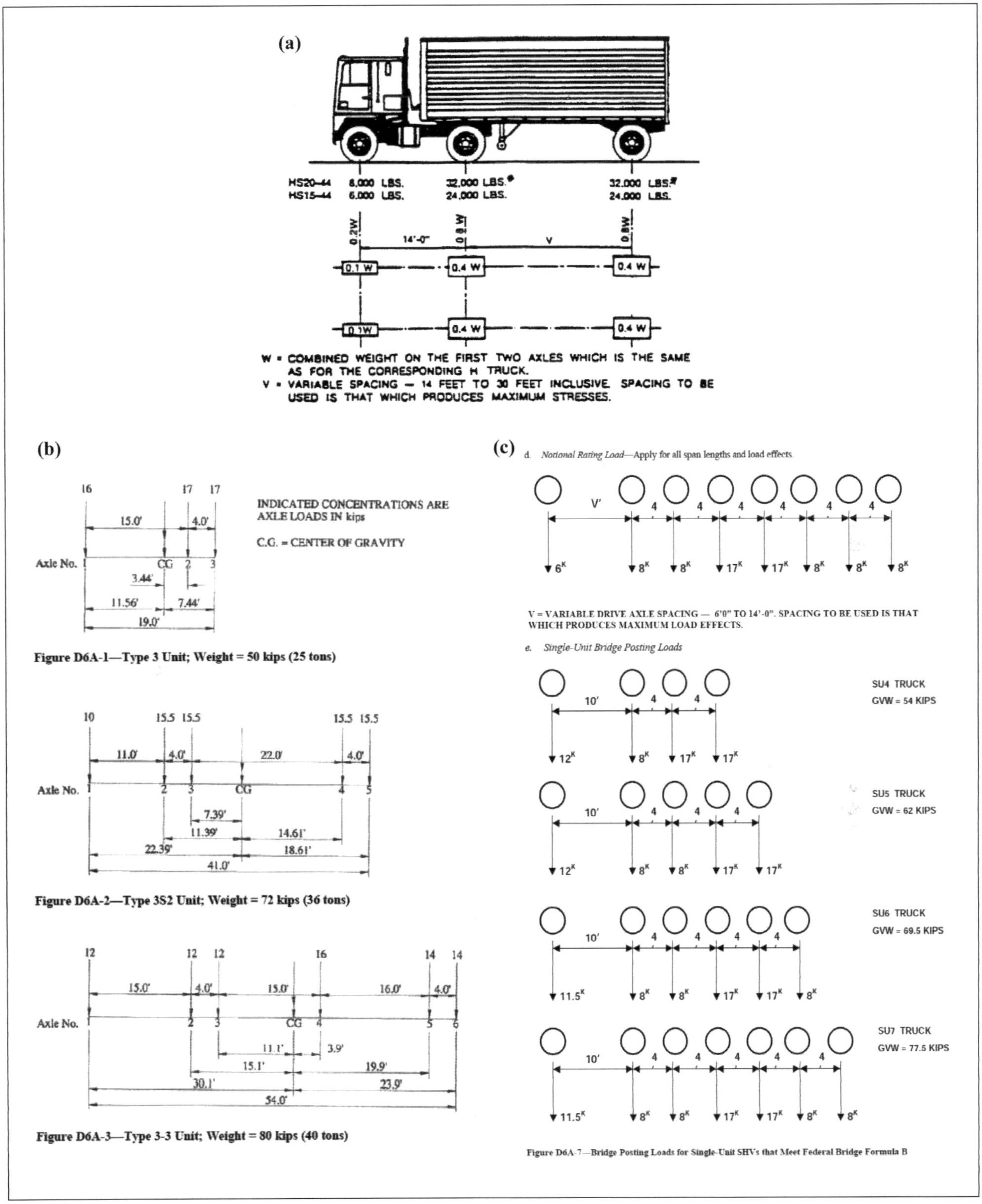


**Figure 1.** (continued)

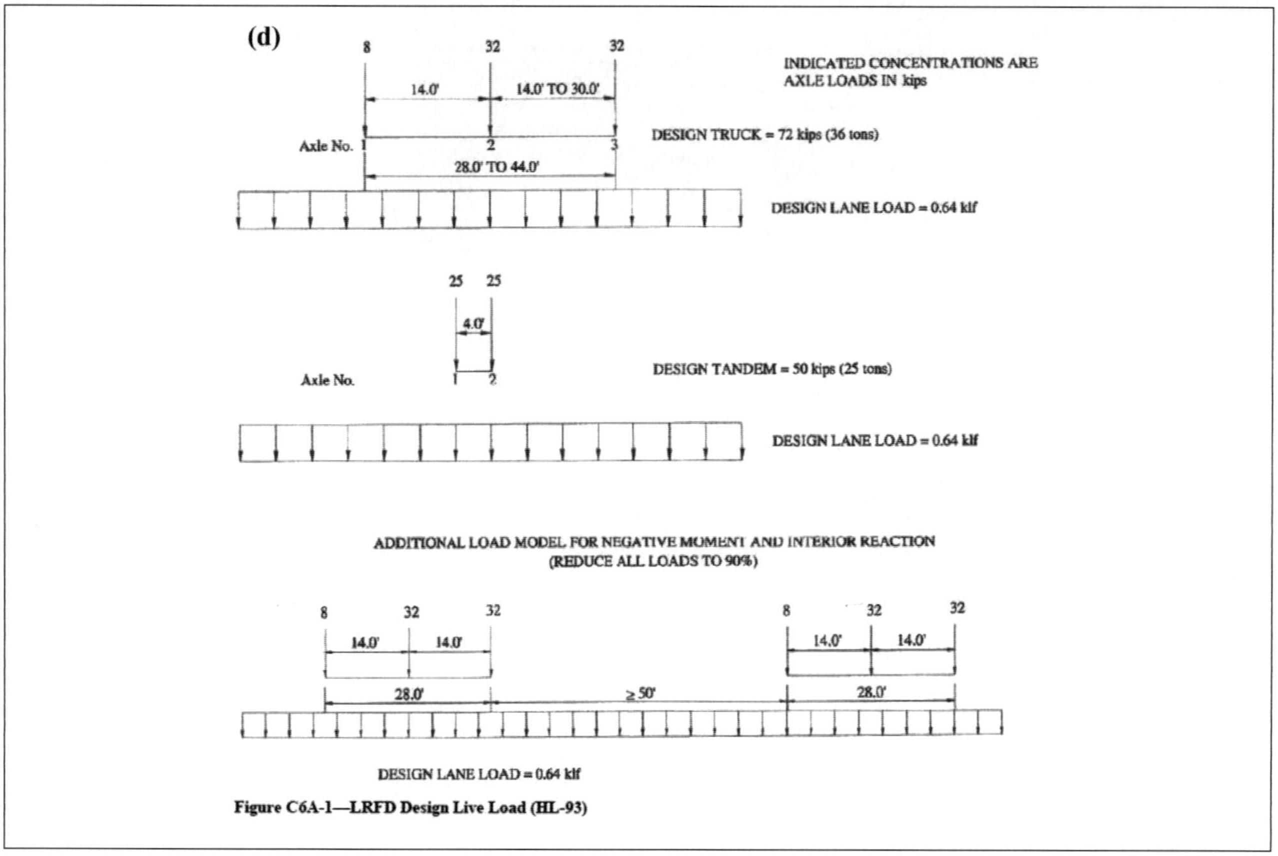


**Figure 1.** Live load design models identified in the AASHTO *MBE* (*5*). From *AASHTO Manual for Bridge Evaluation, 2nd Edition*, Copyright 2011, by the American Association of State Highway and Transportation Officials, Washington, DC. Used by permission: (*a*) HS-20 Standard Truck load, (*b*) AASHTO Legal Loads, (*c*) Notional Rating Load/ Single-unit SHVs and (*d*) HL-93 LRFD Design Truck Load.

bridges having an HL-93 operating rating factor greater than 1.0 need not be rated for SHVs.

In addition to live load configuration, the *MBE* Section 6A addresses other aspects of bridge live load including dynamic load allowance, lane load models, multiple presence probabilities, and calibrated live load factors for a variety of conditions associated with the LRFR rating method. Likewise, the *MBE* Section 6B addresses these aspects for ASR and LFR. This technical guidance may be said to directly apply to bridges proper, and the commentary makes a point to say that "specifications are calibrated documents in which the loads, load factors and design methods are part of the whole and should not be separated" (*5*). The *MBE*'s treatment of these other aspects of load rating specific to culverts primarily exists in the 2013 Interim Revisions for LRFR rating of culverts (Section 6A). The *MBE* Section 6B does not provide this same information for the LFR and ASR rating methods. Here the culvert load rating engineer is directed to the AASHTO *Standard Specifications* (*6*) for guidance on the "other" rating issues.

## *Culvert Structures*

The number of in-service bridge-class culverts in the U.S. is large. National Bridge Inventory (NBI) 2015 data identify 136,971 bridge-class culverts out of a total of 611,845 highway bridges in the U.S., or 22 percent of the total inventory (*10*). Further, 32 states record more than 1,000 culverts in their bridge inventory, with the proportion of culverts ranging from 3% to 55%. The Texas Department of Transportation (TxDOT) is solidly invested in this issue, having an inventory of 19,594 bridge-class culverts—by far the largest of any state—accounting for 14% of the bridge-class culverts in the U.S. (*10*).

The makeup of a state's culvert inventory can be quite diverse, encompassing various structure geometries and both flexible and rigid material types. NBI data (*10*) identify ten types of materials for bridge-class culverts, with the primary types being concrete (67%) or continuous concrete (21%). Texas would actually claim to have a very consistent inventory, with 97% of Texas' older culverts being multi-span, cast-in-place reinforced concrete boxes.

Culverts have a long service life as evidenced by NBI data which show in-service culverts in the U.S. dating to 1900 (*10*). In Texas, of 13,409 on-system bridge-class concrete culverts, 10,846 (81%) were built *prior* to 1980. Roughly one-third of Texas' culverts were constructed in years leading up to and following World War II, and about half of Texas' culverts were constructed during the Interstate Highway era.

### Overview of this Paper

This paper argues that bridge-class culverts are a significant and important subset of highway bridges, and these culverts must be load-rated. Owing to the size of its bridge-class culvert inventory Texas is at the forefront of this issue, and the state's population of reinforced concrete box culverts can serve as a suitable test bed for evaluating how SHV and other alternative vehicle loads influence load rating factors for older bridge-class culverts, especially compared with common practice where rating factors are calculated based on application of the standard HS-20 truck. Accurate load rating of older, in-service culverts serves multiple purposes including avoiding the unnecessary replacement of such structures, which is directly related to infrastructure sustainability and resiliency (*11*).

## Method

The work described here represents an empirical study that compared load rating results for a statistically representative sample of Texas' older bridge-class culverts across a range of vehicle loads. This section provides details of the research method including the culvert sample, the load rating process, and details of the applied loads.

### Population and Sample

The population of interest for this study consisted of older, bridge-class reinforced concrete box culvert structures. TxDOT's inventory includes 13,409 on-system culverts, and those built prior to 1980 number 10,846 (81%), almost all of which (97%) are reinforced concrete boxes. This compares with 88% of the national inventory of bridge-class culverts being concrete or continuous concrete. While not a statistically random sample *per se*, the claim of this study is that Texas' population of on-system culvert structures is both numerically significant and reasonably descriptive of the national population of older, bridge-class concrete box culvert structures.

A simple random sample of Texas' culvert population was drawn at two levels. First, a sample of 1,000 culverts was identified and rated for an approved set of rating variables including the HS-20 standard truck load. Results from the sample (1,000 culverts) are representative of the population (10,846 culverts) with a nominal margin of error of ± 3% for the pass/fail rate. Second, within the sample but still statistically representative of the population, a simple random subsample of 100 culverts was identified for comparative analyses of alternative vehicle loads. The nominal margin of error associated with results from the subsample relative to the population of Texas culverts is ± 10%.

Table 1 provides descriptive information about the sample and subsample of Texas culverts. As introduced in Table 1, "design segment" refers to that portion of a culvert structure having a unique design. Because the culverts of interest for this study are older, during their long service lives many of these culverts were upgraded to improve hydraulic capacity (lengthened), traffic capacity (widened), or both—in some cases years or decades after the original culvert was built. These major modifications were usually done under designs and specifications which differed from the original structure, and in fact, about half of Texas' older culverts comprise more than one design segment. Thus the segment is the unit of analysis for load rating, with multi-segment culverts being controlled by the segment having the lowest rating. The data in Table 1 and the charts which follow reflect the individual design segments that comprise both the sample and the subsample.

Other descriptive data in Table 1 such as year built, number of spans, span length, and box height are self-explanatory. Cover soil thickness, aggregated into three categories, nominally describes the amount of material (soil and pavement) above the top slab of the culvert structure.

Typical design practice for Texas culverts was to use standard designs, with the MBC family of standards used prior to WWII, and the MC family of standards used during the Interstate Highway era. Two generations of MBC standards exist. Though not consistently shown on all archive drawings, the earliest MBC designs date to the mid-1920s, feature a non-haunched cross-section, and identify the design load as a "typical 15-ton truck." Later MBC standards date to the mid-1930s, feature a haunched cross-section, and a typical note identifies the designs were based on a live load of "two 15-ton trucks with 25% impact distributed in accordance with AASHO [1931] specifications." The 15-ton trucks are interpreted to mean the H15 truck load (*7*). The MC family of standards feature unhaunched cross-sections. Original design years range from 1949 to 1958, and general notes on the MC standards identify the design loading as "H-20 or H20 S16 in accordance with 1957 AASHO specifications."

The soils surrounding the culverts were classified as low stiffness, medium stiffness, and high stiffness. The surface pavement structure type was obtained from Texas' pavement management information system, the predominant categories being as identified. Collectively, these data lay out the characteristics of Texas' population of older concrete box culverts and also show how closely the subsample matches the sample.

**Table 1.** Descriptive Information for Sample and Subsample of Texas Culvert Structures

| Parameter | Sample (1,000 culverts) | | Subsample (100 culverts) | |
|---|---|---|---|---|
| | Mean/Median | Range | Mean/Median | Range |
| (1) | (2) | (3) | (4) | (5) |
| Design segments | 1,385 total | – | 139 total | – |
| Year built | 1955.5/1958 | 1923–1979 | 1954/1956 | 1925–1979 |
| No. of spans | 4.1/4.0 | 1–30 | 4.2/4.0 | 1–15 |
| Span length (ft) | 7.4/7.0 | 3–10 | 7.0/7.0 | 4–10 |
| Box height (ft) | 5.8/6.0 | 2–12 | 5.6/5.0 | 2–10 |
| Cover soil thickness | Direct (0–2ft): 77.3% | – | Direct (0–2ft): 83.4% | – |
| | Low (2–6ft): 18.1% | – | Low (2–6ft): 14.4% | – |
| | Deep (>6ft): 4.6% | – | Deep (>6ft): 2.2% | – |
| Design family | MC std.: 63.1% | – | MC std.: 61.9% | – |
| | MBC std.: 17.5% | – | MBC std.: 24.5% | – |
| | Other stds: 19.4% | – | Other stds: 13.6% | – |
| Soil type/ stiffness | Low: 36.3% | – | Low: 30.2% | – |
| | Medium: 58.0% | – | Medium: 63.3% | – |
| | High: 5.7% | – | High: 6.5% | – |
| Pavement type | Seal Coat: 38.4% | – | Seal Coat: 41.0% | – |
| | Thin ACP: 20.1% | – | Thin ACP: 20.1% | – |
| | Int. ACP: 26.5% | – | Int. ACP: 30.2% | – |
| | Other: 15.0% | – | Other: 8.7% | – |

## *Load Rating Procedure*

The load rating procedure used for this study was developed through a series of research projects performed over the past 10 years that were sponsored by TxDOT. More specifically, the rating analyses relied on a calibrated two-dimensional, soil–structure interaction, finite element model to calculate the load demands. This production-simplified approach used linear elastic constitutive models for both soil and concrete where the effective stiffness of external culvert members (i.e., culvert top slab, exterior walls, bottom slab) was modeled using $0.5I_{gross}$ and the effective stiffness of internal culvert members (i.e., interior walls) was modeled using $1.0I_{gross}$. Strength reduction factors for capacity followed AASHTO policy (*6*). Load spreading in the out-of-plane direction was simulated using a rectangular prism with dimensions 1.15 × cover soil depth as per guidance from the AASHTO *LRFD Bridge Design Specifications* (*12*). The overall model was calibrated based on data from full-scale instrumented culvert load tests (*13*), and fine-tuned to optimize accuracy and precision of the predicted load demands through a series of parametric and other analytical studies (*14–18*).

The first step to achieving reliability in culvert load rating was to make sure that personnel could consistently produce repeatable results. The actual load rating process required extensive and direct interaction with TxDOT's bridge (culvert) files and utilized a seven-step approach. TxDOT's *Culvert Rating Guide, 2nd Edition* (*19*) and CULVLR 2.0 load rating program (*20*) —both recently updated/revised—were created to accomplish this work.

## *Applied Vehicle Loads*

As has been noted, this study compared culvert load ratings calculated using the standard HS-20 truck with load ratings that were calculated based on application of three other groups of vehicle loads. The study modeled axle weights and axle spacings as per guidance in the *MBE* and shown in Figure 1, with the following adjustments:

- This study relied on a two-dimensional finite element model of unit width (i.e., 1.0ft), thus the axle loads were converted to wheel loads (divided by 2).
- The finite element model directly modeled wheel loads (in-plane) for all cover soil depths. The prismatic function (out-of-plane) accounted for two-wheel interaction when the soil cover depth exceeded the AASHTO-specified interaction depth.
- Analyses for NRL/SHVs did not include SU6, as this particular axle configuration is reasonably covered by the other models.
- The *MBE* associates the HL-93 design load family with LRFD/LRFR and does not identify this model for LFD/LFR. However, TxDOT's historic usage of exclusion vehicles very similar to the HL-93 design tandem (e.g., the 24-ton alternate military loading) are such that the HL-93 design tandem (25-ton)—hereafter

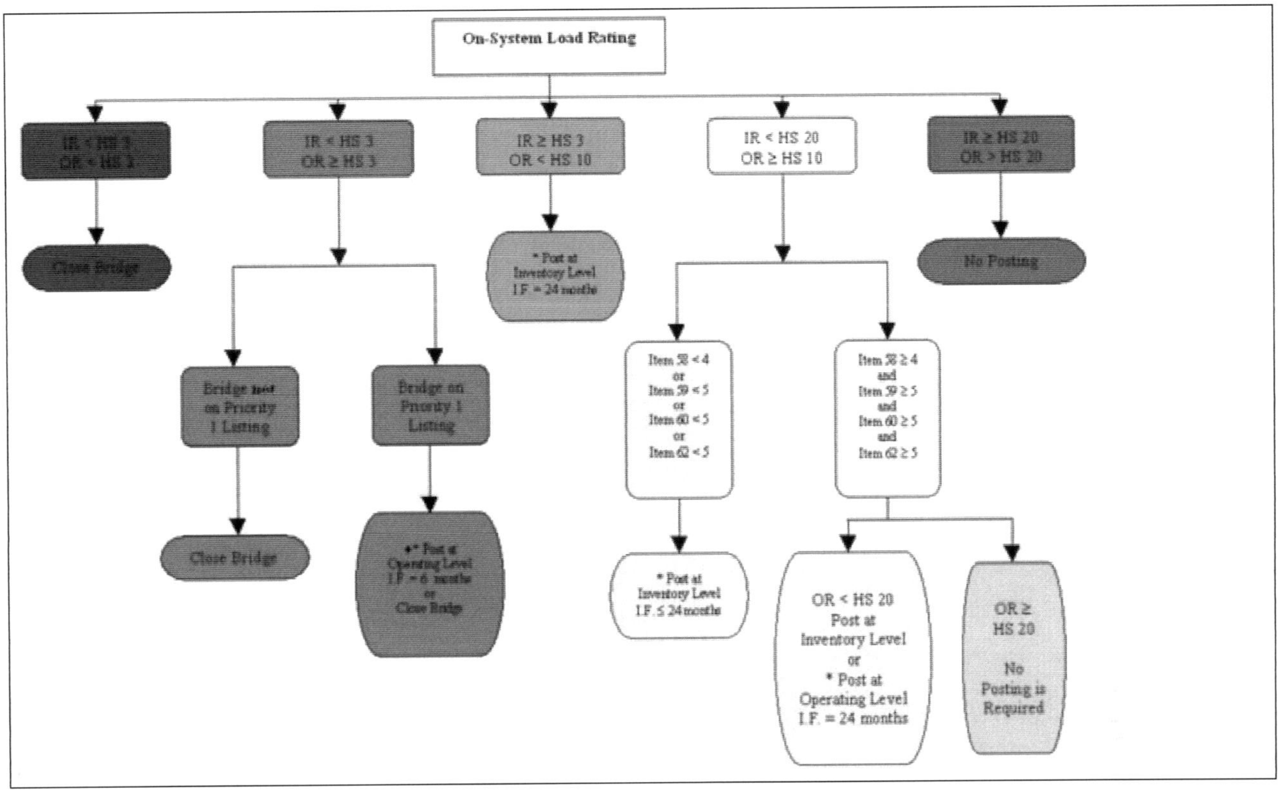


**Figure 2.** TxDOT on-system load posting guidelines (*21*).

"HL-93 design tandem"—is included as one of the live load models in this analysis.

- The HS-20 truck and the NRL each show one variable axle spacing, V. Because of the relatively short spans of culverts, the variable axle spacing in each case was fixed to the minimum value.
- All ratings modeled only the live load configuration (i.e., the design truck) and the impact factor. Owing to the absence of specific guidance in the *MBE* Section 6B, and for the sake of uniformity and clarity, lane loads and multiple presence factors were not considered in the rating calculations.
- Comparative load ratings did not include the load-attenuation benefit of pavement; that is, the pavement was neglected.

In all other respects, the culvert load rating variables were kept constant at the approved values as per the primary research study (*14*).

## Results and Discussion

### *Subsample vs. Population*

The statistical underpinnings of this study rely on the premise that a random subsample of 100 TxDOT culverts is statistically representative of Texas' population of pre-1980, on-system culverts, and by implication is also reasonably descriptive of the *nation's* population of older, bridge-class culvert structures.

In support of this claim, Figure 2 presents TxDOT's load posting guidelines as per NBI guidance and published in TxDOT's *Bridge Inspection Manual* (*21*). This chart reflects posting thresholds for load ratings—that is, the inventory rating (IR) and operating rating (OR) in units of tons—where the load rating is the rating factor (from Equation 1) multiplied by the nominal tonnage of the rating truck. Because typical load rating analyses are based on the standard HS-20 truck, an IR = 10 means the inventory rating factor was approximately 10/20 = 0.50, with similar calculations for the other identified rating thresholds. As an aid to interpretation, load posting categories are color-coded such that "green" and "light green" indicate results where posting would not be required, and colors other than green represent varying degrees of posting severity.

Figure 3 presents load rating/posting results for the subsample (100 culverts) and sample (1,000 culverts) of TxDOT's culvert population, color-coded in a manner consistent with Figure 2 to align with TxDOT's load posting guidance. Rating factors were calculated per AASHTO *MBE* policy for LFR using a live load factor of 2.17 for inventory level and 1.3 for operating level (*5*). All calculations were performed for ratable culvert segments using the approved set of rating variables including pavement.

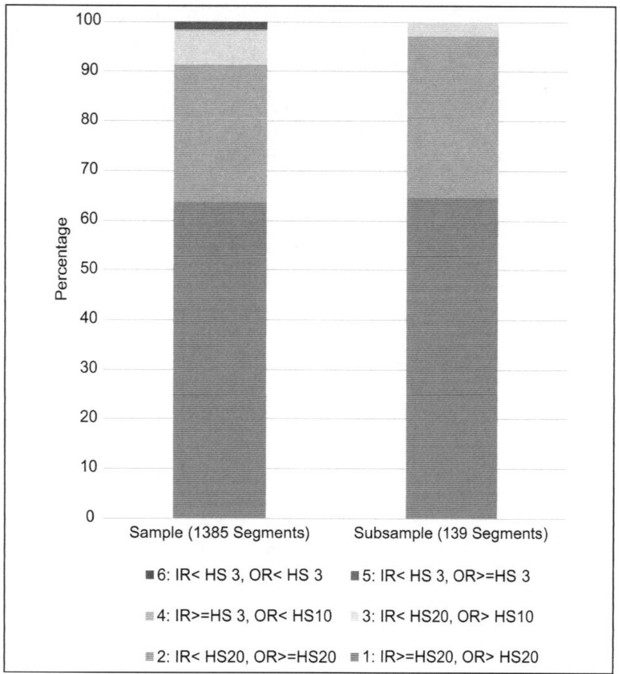


**Figure 3.** Load rating/posting results for sample and subsample of Texas culverts.

These summary-level data indicate Texas' population of older culverts is reasonably "healthy," with 97% of segments in the culvert subsample and 91% in the culvert sample rating "in the green," that is, not requiring posting. Very few culvert segments (less than 2%) required the strictest levels of posting and/or replacement. This level of performance is consistent with a serviceable culvert inventory evidenced by NBI inspection/condition ratings for the sample which range from 4/9 to 8/9, with most culverts (63% ±3%) scoring 7/9 or better (*14*).

Focusing on the 1,000-culvert sample (*14*), the mode of failure was flexure (moment) for 99% of rated segments (1372/1385), whereas shear capacity controlled for only for 1% of rated segments (13/1385). As to the location of maximum distress, in about 90% of cases the top region of the culvert is where the critical sections occurred. These included the top slab corners and midspan (51% ± 3.0%), the top exterior wall (37% ± 2.9%), and the top interior walls (1% ± 0.7%). This modeled behavior is consistent with the observation that as live load enters the culvert from the ground surface, the top region of the structure will experience the greatest live load intensities; thus, demands will be higher in this region, so it is reasonable that this area should most often control the rating. Likewise, the lowest percentage of critical sections occurred in the bottom culvert slab (2% ± 0.8%), which was expected. The exterior walls showed a significant portion of controlling sections in the top as already noted, and at the bottom (8% ± 1.6%). Similarly for the 100-culvert subsample (139 segments), the critical section occurred in the top region of the culvert for about 92% to 94% of cases, for all applied live load models.

## Summary of Load Ratings by Vehicle Load Type

Load rating results for the nine vehicle loads evaluated in this study appear in Figure 4, in two ways. Figure 4*a* presents culvert load posting data (similar to Figure 3) associated with all vehicle loads. The *y*-axis identifies the percentage of culvert segments rated (out of 139) associated with each load posting category. Using the HS-20 live load as an example, about 41% of the culvert segments have an OR > HS-20 and IR ≥ HS-20, about 46% of segments have an OR ≥ HS-20 and IR < HS-20, and about 13% of segments have an OR > HS-10 and IR < HS-20. By comparison these values are 80%, 17% and 3% when the live load model is SU7.

Figure 4*b* presents box plots of the culvert operating rating factors (ORFs) associated with these same vehicle loads. It is to be recalled that these comparative load ratings do not include the benefit of pavement, and thus the results differ slightly from Figure 3.

The most striking observation from these data (Figure 4*a*) is the strong percentage of culvert segments that rate sufficiently high so as to not require posting. All the AASHTO legal load and SHV/NRL live load models produce culvert ratings above 95% not requiring posting, and this is without the load-attenuating benefit of pavement. Such a high level of performance was not expected given published findings which suggested SHVs would produce force effects for bridges proper (not culverts) up to 22% larger than the HS-20 load (*3*).

The box plot (Figure 4*b*) illustrates the ORFs associated with each load type. The horizontal line in each box is the median ORF (values are superimposed for clarity), the bottom and top of each box represent the first and third quartiles, whiskers indicate the data range, and asterisks indicate statistical outliers. These data were transformed into the log10 scale to achieve constant variance and normal error distribution necessary to satisfy analysis of variance (ANOVA) assumptions, and Tukey pairwise comparisons were made. The Tukey analysis revealed some clear separations among the vehicle load types. The combination vehicles, AASHTO Type 3-3 and Type 3S2, yield the highest average ORF values. This is in contrast to HS-20 standard truck (also a combination vehicle) which yields the lowest average ORF values. SHV types SU4, SU5, SU7, NRL, and AASHTO Type 3 (all representative of single-unit straight trucks) follow and, as a group, are not significantly different from one another. Lastly, the HL93 design tandem (a tandem axle) shows a relatively high average ORF not significantly different from the HS-20 standard truck.

In summary, both the AASHTO legal loads and the SHVs on average produce higher ratings compared with the HS-20 load, and per the rating equation this is indicative of lower force effects. More will be said about this later.

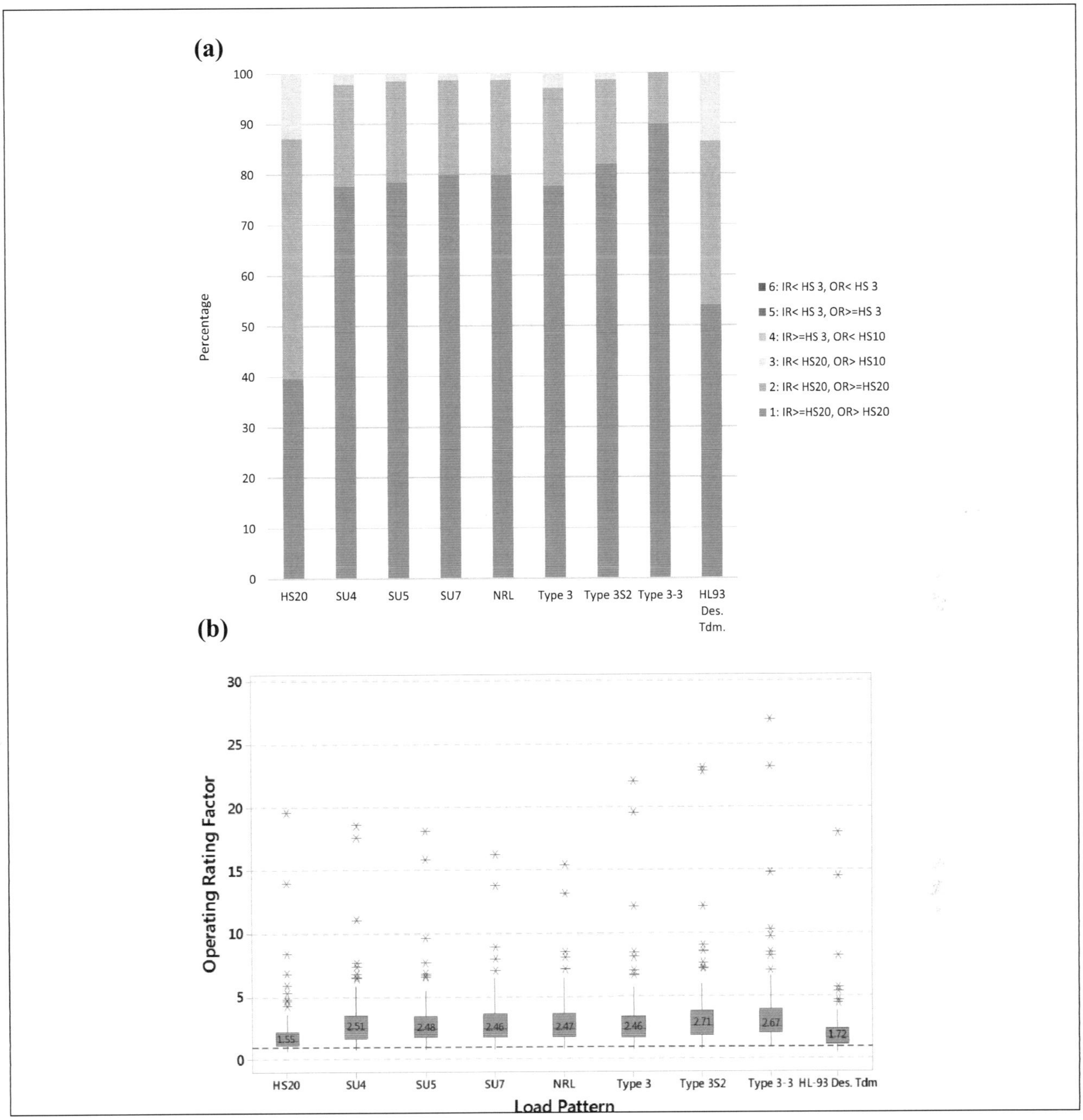


**Figure 4.** Load rating factors associated with nine vehicle loads for 100 culverts: (*a*) Load rating results by load posting performance and (*b*) Load rating results by operating rating factor (ORF).

## *Alternative Vehicle Loads vs. the HS-20 Standard Load*

To better understand the influence of non-HS-20 live load models on culvert load ratings, it is helpful to directly compare ORFs associated with each alternative load with the ORFs obtained using the standard HS-20 load. Figure 5 provides a scatterplot of such a comparison for the SU7 model (worst seven-axle SHV), and this plot is reasonably illustrative of all the alternative vehicle loads. Figure 5*a* includes all ORF data for the comparison, and Figure 5*b* "zooms in" to show that portion of the scatterplot where most data exist.

Focusing on Figure 5*b*, the horizontal axis is the culvert ORF associated with HS-20 load and the vertical axis is the culvert ORF associated with the SU7 load. The 45-degree line is the solid light gray line such that data points above this line represent culvert segments where the SU7 ORF is greater than the HS-20 ORF and data points below this line are those cases where the SU7 ORF is less than the HS-20 ORF. The solid red line is the statistical regression line for the SU7/HS-20 ORF

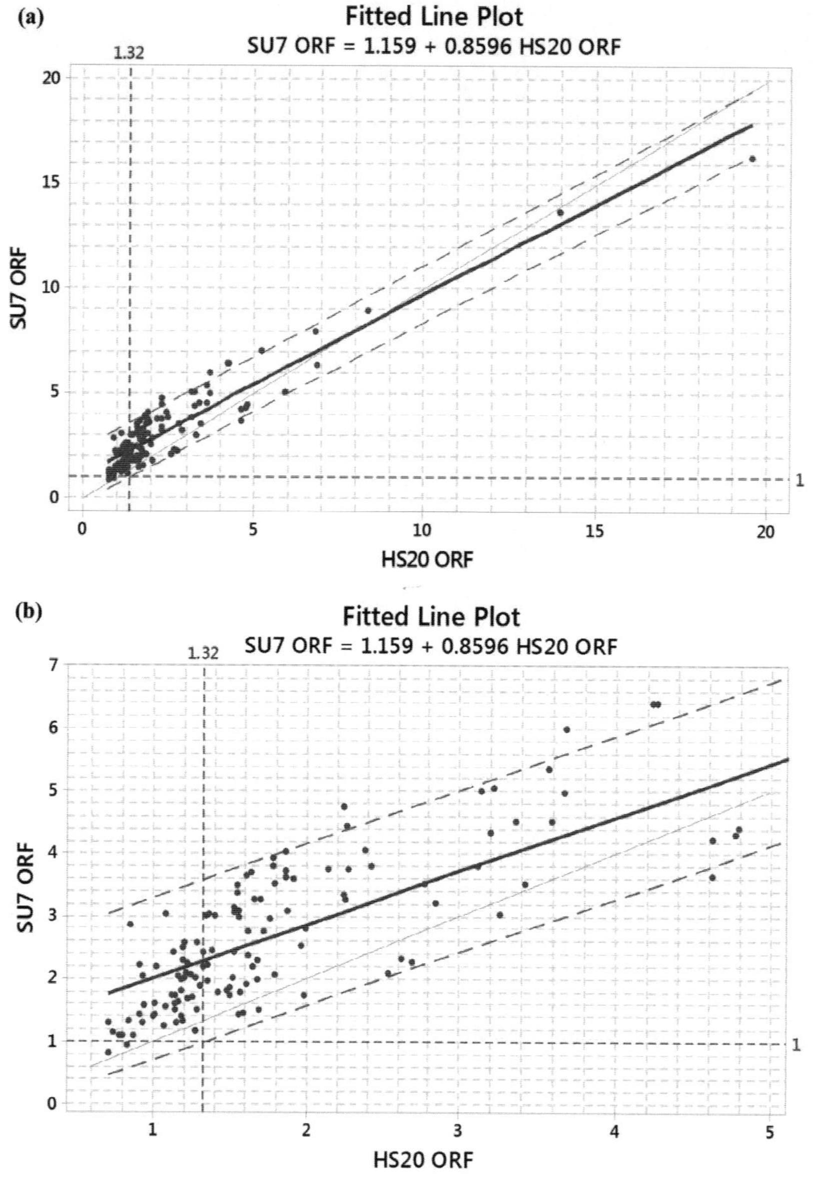


**Figure 5.** Comparison of SU7 ORFs vs HS-20 ORFs: (*a*) Scatterplot showing all ORF data for the subsample and (*b*) Scatterplot zoomed in to show higher-density ORF data.

relationship, with the equation for the line shown on the chart. Further, the dashed green lines represent the upper- and lower-bound 90% prediction interval for the regression line.

Notice the dashed red lines on Figure 5*b*. The horizontal dashed red line corresponds to an SU7 ORF of 1.0, and this ORF threshold is typically considered the cutoff for load posting. That is, culverts (bridges) with ORF values greater than or equal to 1.0 do not require posting, but culverts with ORF values less than 1.0 do require posting.

The vertical dashed red line exactly intersects both the horizontal dashed red line (SU7 ORF = 1.0) and the lower-bound prediction limit for the SU7/HS-20 ORF regression line. Further, this particular vertical line occurs at the HS-20 ORF of 1.32 (identified on the figure). The interpretation of this point of intersection is that, for the data provided, when the HS-20 ORF is known to be equal to or greater than 1.32, there is 95% confidence that the SU7 ORF will be equal to or greater than 1.0. Space limitations preclude showing all charts, but Table 2 summarizes the analyses for all alternative vehicle loads considered in this study.

Table 2 highlights various details about the relationships between ORFs associated with alternative vehicle loads and ORFs for the standard HS-20 truck. The threshold values for each alternative vehicle load are presented in Column 2. This column shows that HS-20 ORF thresholds for all AASHTO legal loads and the SHVs are relatively low, not unlike that for

**Table 2.** Comparison of Alternative Load ORF Values vs. HS20 ORF Values

| Alternate vehicle load | HS-20 ORF threshold (Alt. Load ORF ≥ 1) | Rated segments with alt. load ORF < HS-20 ORF | | | Rated segments with HS-20 ORF ≥ 1 but alt. load ORF < 1 |
|---|---|---|---|---|---|
| | | No. | % | 95% CI | No. |
| (1) | (2) | (3) | (4) | (5) | (6) |
| SU4 | 1.32 | 8 | 5.8 | 2.5–11.0 | – |
| SU5 | 1.28 | 11 | 7.9 | 4.0–13.7 | – |
| SU7 | 1.32 | 17 | 12.2 | 7.3–18.9 | – |
| NRL | 1.31 | 18 | 12.9 | 7.9–19.7 | – |
| Type 3 | 1.23 | – | – | 0.0–2.6 | – |
| Type 3S2 | 1.25 | – | – | 0.0–2.6 | – |
| Type 3-3 | 1.05 | – | – | 0.0–2.6 | – |
| HL93 design tandem | 2.05 | 70 | 50.4 | 41.8–58.9 | 5 |

the SU7 model in Figure 5. However, the HS-20 ORF threshold associated with the HL-93 design tandem load is significantly higher.

Columns 3 through 5 illustrate the relationships further. Column 3 identifies the number of rated segments, out of 139 total segments in the 100-culvert subsample, where the ORF associated with the alternative vehicle load is less than the ORF for the HS-20 standard truck. Column 4 expresses this number as a percentage, and Column 5 identifies the 95% confidence interval range associated with this percentage. These data reveal that none of the AASHTO legal loads produce ORFs less than those for the HS-20 standard truck. However, about 6–13% of ORFs resulting from SHVs are lower than the corresponding HS-20 ORFs. Further, ORFs resulting from the HL-93 design tandem are evenly split relative to the HS-20 ORFs, with 70 less and 69 greater than the HS-20 value. In other words, the regression model for the HL-93 design tandem very closely matches the 45-degree line.

Another question of interest is, for those rated segments where the ORF associated with the HS-20 standard truck was greater than or equal to 1.0, how often were the corresponding ORFs for alternative vehicle loads less than 1.0? Column 6 reveals that none of the AASHTO legal loads and none of the SHVs produced any ORFs less than 1.0 when the corresponding HS-20 ORF was greater than or equal to 1.0. However, the ORF associated with the HL-93 design tandem was less than 1.0 in five instances where the corresponding ORF for the HS-20 truck was 1.0 or greater.

This type of analysis establishes the statistical basis to identify thresholds where comparative load rating analyses might not be required. That is, the analysis identifies known ORFs (e.g., from HS-20 loads) where the regulatory authority can be reasonably confident that proposed alternative vehicles will not produce a load rating with an ORF less than 1.0. For this study, these threshold ORFs appear in Column 2 of Table 2. Such an approach is not uncommon in load rating policy statements. For example, an FHWA policy memorandum on load rating for emergency vehicles identifies just such a rating threshold for an AASHTO Type 3 model (*22*).

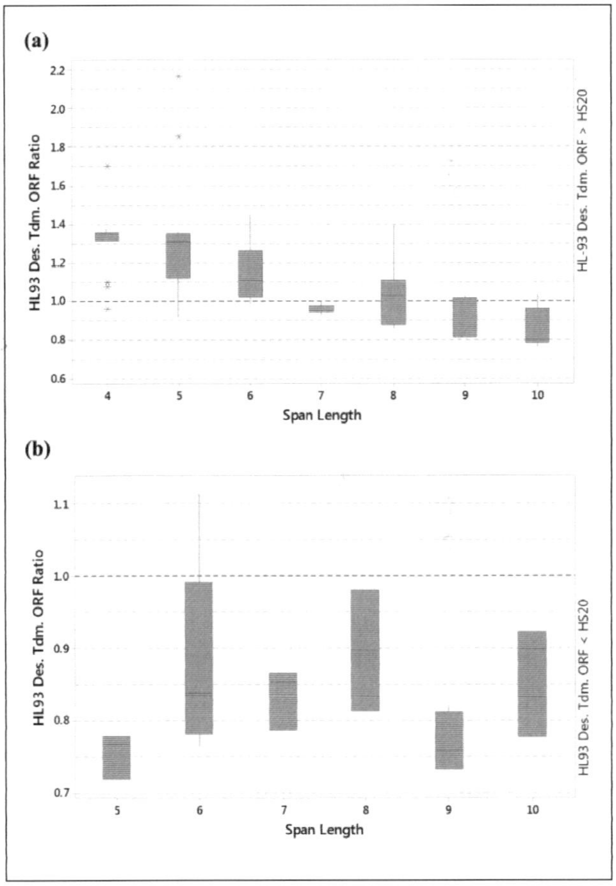


**Figure 6.** Interaction between span length and cover soil thickness, HL-93 design tandem load: (*a*) Span length vs. ORF ratio, direct traffic condition and (*b*) Span length vs. ORF ratio, low-fill condition.

## *Influence of Culvert Variables*

What factors cause the load rating behavior associated with alternative vehicle loads as depicted in Figure 4 and Figure 5? While it is beyond the scope of this paper to delve into

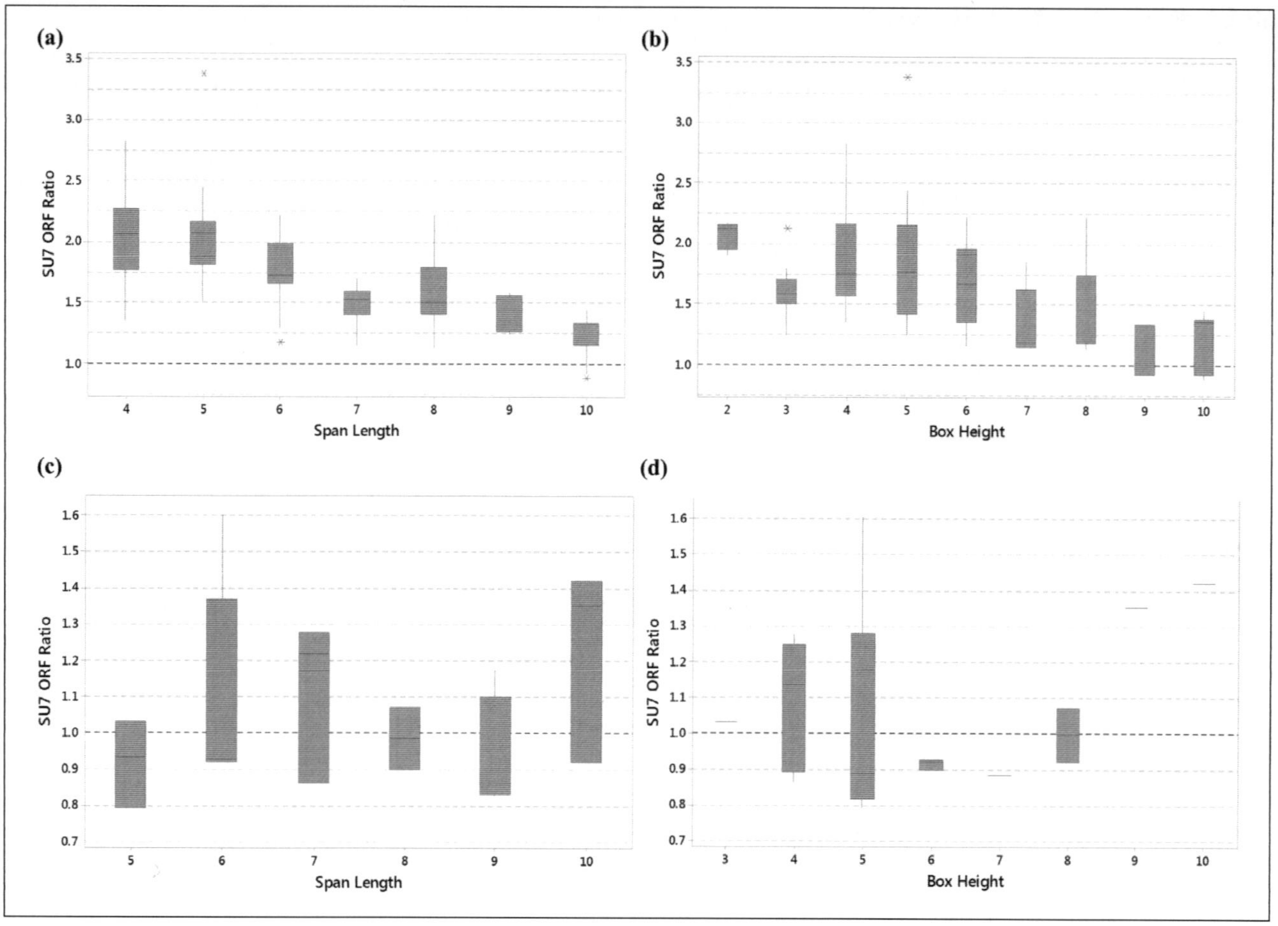


**Figure 7.** Influence of span length and box height, direct traffic and low-fill culverts, SU7 load: (*a*) Span length vs. ORF ratio, direct traffic condition, (*b*) Box height vs. ORF ratio, direct traffic condition, (*c*) Span length vs. ORF ratio, low-fill condition, and (*d*) Box height vs. ORF ratio, low-fill condition.

structural analysis details, the empirical load rating results do provide insight. ANOVA for all models over the full set of culvert parameters indicated that cover soil thickness, span length, and box height tended to influence the load rating results, in some cases with interactions.

First, consider the influence of cover soil depth, which interacts with both the span length and box height variables. Recall that Texas' population of older, bridge-class culvert structures strongly comprises "direct traffic" culverts, that is, culvert segments where the thickness of material (both soil and pavement) above the top of the top slab is in the range of 0.0 to 2.0ft. Per Table 1, about 83% of Texas culverts are in the "direct traffic" category, and about 14% are "low fill" (i.e., 2 to 6ft of cover soil). The HL-93 design tandem load—expressed in terms of a ratio relative to HS-20 (i.e., HL-93 design tandem ORF/HS-20 ORF)—illustrates the interaction (Figure 6).

The primary observation from Figure 6 is that a clear, well-defined, statistically significant relationship exists for span length under the direct traffic condition, whereas no such relationship is evident for the low-fill condition. Relative to span length (direct traffic), as span length increases the ORF ratio decreases. For spans of 6ft or less, the HL-93 design tandem ORF is greater than the HS-20 ORF, that is, the HS-20 ORFs would be considered "conservative." However, for span lengths 7ft and greater, the opposite occurs—the HL-93 design tandem ORF becomes less than the HS-20 ORF, indicating that the HL-93 design tandem load would control the rating. This behavior is consistent with intuition that suggests the short HL-93 design tandem axle spacing (4ft total) would create more severe force effects for longer spans. However, this behavior does not materialize for low-fill culverts, the inference being that thicker cover soil attenuates the effect.

Consider the influence of span length and box height, this time illustrated for the SU7 model under the direct traffic condition (Figure 7, *a* and *b*) and for the low-fill condition (Figure 7, *c* and *d*). Again, the box plots are expressed in terms of a ratio relative to HS-20 (i.e., SU7 ORF/HS-20 ORF). The primary observation from Figure 7, *a* and *b*—which are specific to direct traffic culverts—is that a well-defined relationship exists for span length under the SU7 model similar to that for the HL-93 design tandem: as span length increases the ORF ratio decreases. However, in this case, for all spans represented, the SU7 ORF is greater than the HS-20 ORF, that is, the HS-20 ORFs would be considered "conservative." A similar but less-clear trend exists relative to box height: as box height increases the ORF ratio decreases and this trend holds for all box heights in the sample.

Again, this behavior is consistent with intuition that suggests the short wheel base SU7 model would create more severe force effects for longer spans. However, given the typical span of the SU7's heaviest axles (4ft) combined with relatively short culvert spans (compared with bridges), the full intensity of the SU7 force effects is not manifested. Rather, the much heavier axle loads associated with the HS-20 standard truck tend to control the load rating for most culvert applications.

Likewise, Figure 7, *c* and *d* illustrate the attenuating effect of cover soil. Notwithstanding the fact that the dataset for the low-fill condition (cover soil depth varies from 2ft to 6ft) is limited, these charts suggest no defined pattern for the identified ORF ratio.

## Conclusion

This study has presented empirical evidence from a statistically representative sample of Texas' older, bridge-class reinforced concrete box culvert structures that indicates the standard HS-20 truck is the critical model for most culvert load rating applications. In particular, ORFs calculated from both the AASHTO legal loads and SHV models tend to be higher than corresponding rating factors calculated using the HS-20 standard truck, most of the time. This finding is initially surprising given that documented load rating studies for bridges identify that SHVs tend to produce higher force effects compared with the HS-20 model. The response is explained primarily by considering the relatively short span length of culvert structures and the load-attenuating benefit of cover soil above the top slab. More detailed exploration of rating variables suggests interactions between culvert geometry, cover soil thickness, and the various types of alternative vehicle loads. But collectively, the findings of this study suggest that concrete bridge-class box culverts which rate acceptably for the HS-20 standard load will also tend to rate well for the alternative vehicle loads currently specified in policy.

### Author Contributions

The authors confirm contribution to the paper as follows: study conception and design– Lawson, Seo, Surles, Morse; data collection– Morse, Seo, Surles; analysis and interpretation of results– Surles, Lawson, Seo; draft manuscript preparation– Lawson, Seo, Surles. All authors reviewed the results and approved the final version of the manuscript.

### References

1. Federal Highway Administration. *Memorandum: Load Rating of Specialized Hauling Vehicles*. 2013. https://www.fhwa.dot.gov/bridge/loadrating/131115.cfm/. Accessed July 27, 2017.
2. Federal Highway Administration. *Questions and Answers: Load Rating of Specialized Hauling Vehicles*. 2014. https://www.fhwa.dot.gov/bridge/loadrating/shv_qa.pdf/. Accessed July 27, 2017.
3. Sivakumar, B., F. Moses, G. Fu, and M. Ghosn. *Legal Truck Loads and AASHTO Legal Loads for Posting*. NCHRP Report 575. Transportation Research Board, Washington, D.C., 2007.
4. Code of Federal Regulations. *23 CFR 650 Subpart C National Bridge Inspection Standards*. 2016. https://www.gpo.gov/fdsys/pkg/CFR-2016-title23-vol1/pdf/CFR-2016-title23-vol1.pdf. Accessed September 16, 2016.
5. Association of State Highway and Transportation Officials. *Manual for Bridge Evaluation, 2nd Edition, with 2011, 2013, 2014, 2015, and 2016 Interim Revisions*. American Association of State Highway and Transportation Officials, Washington, D.C., 2011.
6. American Association of State Highway and Transportation Officials. *Standard Specifications for Highway Bridges*, 17th ed. American Association of State Highway and Transportation Officials, Washington, D.C., 2002.
7. Kulicki, J. M., and D. R. Mertz. Evolution of Vehicular Live Load Models during the Interstate Highway Design Era and Beyond. In *50 Years of Interstate Structures: Past, Present and Future. Transportation Research Circular E-C104*. Transportation Research Board, Washington, D.C., 2006.
8. Texas Department of Transportation. *Bridge Design Manual – LRFD*. Texas Department of Transportation, Austin, 2015.
9. Committee for the Truck Weight Study. *Truck Weight Limits: Issues and Options*. Special Report 225. Transportation Research Board, Washington, D.C., 1990.
10. Federal Highway Administration. *Bridge Tables/Download NBI ASCII Files*. 2016. http://www.fhwa.dot.gov/bridge/nbi/ascii.cfm. Accessed July 30, 2016.
11. National Academy of Engineering. *Restore and Improve Urban Infrastructure NAE Grand Challenges for Engineering*. National Academy of Engineering, Washington, D.C., 2017. http://www.engineeringchallenges.org/challenges/infrastructure.aspx. Accessed July 27, 2017.
12. American Asssociation of State Highway and Transportation Officials. *LRFD Bridge Design Specifications, U.S. Customary Units*, 7th ed. American Asssociation of State Highway and Transportation Officials, Washington D.C., 2014.
13. Lawson, W. D., T. A. Wood, C. D. Newhouse, and P. W. Jayawickrama. *Evaluating Existing Culverts for Load Capacity Allowing for Soil Structure Interaction*. Report No. FHWA/TX-10/0-5849-1. Texas Department of Transportation, Austin, 2010.
14. Lawson, W., H. Seo, J. Surles, S. Morse, P. Jayawickrama, S. Bae, T. Wood, S. Gunerathne, B. Afolabi, M. Mousavi, and A. Javid. *Load Rating TxDOT's pre-1980 In-Service Culverts: Model Enhancements, Procedure Improvements, and Results for 1,000 Structures*. FHWA/TX-16-88-4XXIA001-R1. DRAFT Report. Texas Tech University, Lubbock, 2016.
15. Seo, H., T. A. Wood, A. H. Javid, and W. D. Lawson. Simplified, System-Level Pavement- Stiffness Model for

Box Culvert Load Rating Applications. *Journal of Bridge Engineering*, Vol. 22, No. 10, 2017, p.04017066. https://doi.org/10.1061/(ASCE)BE.1943-5592.0001098.
16. Wood, T. A., W. D. Lawson, J. G. Surles, P. W. Jayawickrama, and H. Y. Seo. Improved Load Rating of Reinforced Concrete Box Culverts using Depth-Calibrated Live Load Attenuation. *Journal of Bridge Engineering*, Vol. 21, No. 12, 2016, p. 04016095. https://doi.org/10.1061/(ASCE)BE.1943-5592.0000967.
17. Wood, T. A., W. D. Lawson, and P. W. Jayawickrama. Influence of Cover Soil Depth on RC Box Culvert Load Rating. *Transportation Research Record: Journal of the Transportation Research Board*, 2015. 2511: 63–71.
18. Wood, T. A., W. D. Lawson, P. W. Jayawickrama, and C. D. Newhouse. Evaluation of Production Models for Load Rating Reinforced Concrete Box Culverts. *Journal of Bridge Engineering*, Vol. 20, No. 1, 2014, p. 04014057. https://doi.org/10.1061/(ASCE)BE.1943-5592.0000638.
19. Lawson, W. D., H. Seo, J. G. Surles, and S. M. Morse. *Culvert Rating Guide*, 2nd ed. Texas Department of Transportation, Austin, 2017.
20. Morse, S. M., H. Seo, J. G. Surles, and W. D. Lawson. *CULVLR 2.0 Culvert Load Rating Software*. FHWA/TX-17-88-7XXIA001-P4. Texas Tech University, Lubbock, 2017.
21 Texas Department of Transportation. *Bridge Inspection Manual*. Texas Department of Transportation, Austin, 2013.
22. Federal Highway Administration. *Memorandum: Load Rating for the FAST Act's Emergency Vehicles*. 2016. https://www.fhwa.dot.gov/bridge/loadrating/161103.cfm/. Accessed July 27, 2017.

*The Standing Committee on Culverts and Hydraulic Structures (AFF70) peer-reviewed this paper (18-01521).*

TRR
JOURNAL OF THE TRANSPORTATION RESEARCH BOARD

Article

Transportation Research Record
2018, Vol. 2672(41) 101–112

DOI: 10.1177/0361198118790640
journals.sagepub.com/home/trr
SAGE

# Nondestructive Evaluation of Non-Strand Defects in Stay Cable Specimens

**Madhu M. Karthik[1], Tevfik Terzioglu[2], Casey Jones[2], Stefan Hurlebaus[3], and Mary Beth D. Hueste[3]**

**Abstract**

Stay cable systems are used extensively for the construction of long-span bridges owing to their inherent advantages. In most cable-stayed bridges the steel strands within the main tension elements (MTEs) are grouted to provide a corrosion protection system. However, there could be segregated grout, voids, and water accumulation/infiltration within these voids in the ducts due to poor quality of the grouting material and grouting techniques. These locations may act as initiators for the corrosion of steel strands within the ducts, which can have serious implications on the load carrying capacity and safety of cable-stayed bridges. In this investigation, several nondestructive evaluation (NDE) techniques are evaluated to determine their ability to detect grout defects within the anchorage regions, end caps, and MTEs of the stay cable system. Four stay cable specimen mock-ups with known grout defects and voids were fabricated. NDE methods that were part of this investigation included ground penetrating radar, infrared thermography, electrical capacitance tomography, impact echo, sounding, ultrasonic tomography, ultrasonic pulse velocity, low frequency ultrasound, visual testing, and borescope. While a few NDE methods could identify voids and water infiltration defects within the high-density polyethylene (HDPE) MTEs, it is evident that there is still a need for a method that is effective in detecting grout defects within the metal ducts embedded in the anchorage region and metal MTEs. Although visual testing and borescope methods are effective in identifying the defects, prior information about the location of the defects is generally required.

Stay cable systems have been widely used for long-span bridges. A cable-stayed bridge consists of a bridge deck supported by a system of inclined cables, or stays, passing over or attached to towers located at the main abutments or piers. The cables typically consist of a group of strands, which in turn are made from a set of wires. Due to economy in weight, material, and cost, as well as its versatile concept that lends itself to a wide variety of geometric configurations, modern cable-stayed bridges have been successfully used in the United States since the late 1970s. Although cable-stayed bridges are not a new development, their application has not been widespread until recently.

Corrosion of strands in cable-stayed bridges is a major problem. Many different protection systems have been used to provide long-term durability and corrosion resistance to the stay cables. However, almost all these methods have experienced problems. In a comprehensive survey of cable-stayed bridges around the world, Watson and Stafford (*1*) concluded that nearly 200 cable-stayed bridges are in danger due to potential corrosion problems in stay cables. Several other researchers claimed that Watson and Stafford overstated the problem (*2*). Although Watson and Stafford's conclusions were controversial, it started a necessary dialog on the corrosion of stay cables and the principal cause of cable deterioration.

Various factors such as misalignment of anchorages, excessive vibrations under live load, and excessive use of de-icing salts on the bridge deck contributed to the failure of wires in the Kohlbrand Bridge in Germany (*3*). More than 500 broken wires were detected after an inspection of the Lake Maracaibo Bridge in Venezuela due to the lack of maintenance and failure to replace the neoprene boots that protected the anchorage area (*3*). Telang et al. (*4*) and Mehrabi (*5*) present an investigation report of the condition assessment of the Luling Bridge, in Luling, Louisiana. Hands-on inspection and nondestructive testing of the free length of the main tension

[1]Civil Engineering, Indian Institute of Technology, Palakkad, India
[2]Texas A&M Transportation Institute, College Station, TX
[3]Zachry Department of Civil Engineering, Texas A&M University, College Station, TX

**Corresponding Author:**
Address correspondence to Madhu M. Karthik: madhu@iitpkd.ac.in

elements (MTEs) revealed a variety of damage conditions and anomalies including longitudinal and transverse splits, bulges and holes in the polyethylene pipe, exposure and degradation of grout filler and steel wires, tape damage, and grout voids and delamination.

Strand defects in cable-stayed bridges arise due to poor corrosion protection systems, which are typically provided by grouting the ducts. Although grouting materials and construction practices have considerably improved over the years, existing cable-stayed bridges have exhibited a considerable degree of grout defects within the MTEs. Poor grouting practices and poor grout quality can create detrimental grout conditions such as voids, water infiltration and segregation of grout, which can then lead to corrosion of steel strands within the MTEs. Therefore, it is important to identify the presence and severity of grout defects so that preventative actions can be taken to avoid any potential damage to the strands.

Four mock-up specimens representing stay cable specimens were constructed to evaluate possible nondestructive evaluation (NDE) technologies that could be used for the condition assessment of stay cable systems. Both strand and grout defects were introduced into the mock-up specimens. The location of each of these defects was documented and used to evaluate the effectiveness of the NDE technology in identifying the location and severity of grout defects.

The various techniques used for the evaluation of grout defects in the stay cable specimens, include ground penetrating radar (GPR), infrared thermography (IRT), electrical capacitance tomography (ECT), impact echo (IE), sounding, ultrasonic tomography (UST), ultrasonic pulse velocity (UPV), low frequency ultrasound (LFUT), visual testing (VT), and borescope. The NDE techniques are compared with the actual defects, and their effectiveness in identifying the grout defects is discussed. Further details of the stay cable specimens, location of defects, construction of the specimens, and the various NDE techniques used for evaluating the grout defects are discussed in detail in Hurlebaus et al. (*6*).

## Description of Stay Cable Specimens

Four stay cable specimen mock-ups were constructed at the Texas A&M University RELLIS Campus in Bryan, Texas, by taking into account various typical details of the MTE of cable-stayed bridges. Each of the four stay cable specimens were constructed with its own unique combination of sheathing material and corrosion protection system. Additionally, detrimental conditions were implemented along the length of the specimens, both in the free spans of the MTEs and the anchorage regions. To simplify construction and inspection, the mock-up specimens were constructed horizontally.

Figure 1 shows the plan, elevation, and end-sections of a typical stay cable specimen. The two anchorages represent the concrete deck and pylon anchorage regions of cable-stayed bridges. The free span of all four stay cable specimens are approximately 17 ft long, which are restrained at both ends by the concrete anchorages. The concrete anchorages are 2 ft thick and 6 ft long. The anchorage blocks for each specimen are dimensionally identical and contain steel reinforcement detail that is typical in an actual cable-stayed bridge. For the MTE, metal and high-density polyethylene (HDPE) ducts having a nominal inner diameter of 4.0 in. and outer diameter of 4.5 in. are used. Each system contains 19 seven-wire 0.6 in. diameter strands.

After construction of the anchorage regions and placement of the ducts, the strand defects were placed and documented prior to grouting. Stay cable specimen 1 (SC1) and stay cable specimen 2 (SC2) have low-relaxation parallel strands, and grout was used as the corrosion protection system. Steel and HDPE ducts were used for SC1 and SC2, respectively. Stay cable specimen 3 (SC3) has ungrouted, greased and sheathed strands in HDPE duct, and stay cable specimen 4 (SC4) has epoxy-coated strands encased within HDPE ducts. Although SC3 and SC4 have no grout defects, they are part of a larger study to identify NDE technologies capable of detecting both grout and strand defects (which is not a subject of discussion in this manuscript). For the ungrouted specimens, the top three strands in each duct were stressed to a load of approximately 12 kips to ensure that the tendons were taut, resulting in minimum deflection, and sagging of the tendons.

Both strand and grout defects were implemented into each of the four stay cable specimens. To facilitate the placement of the various defects, each mock-up specimen was divided into segments that were approximately 3 ft long. The specimens contained two defect locations within the anchorage regions, and six defect locations along the free spans. As strand defects are not part of this investigation, only the grout defects are described further in detail. Table 1 provides the location of the grout defects and a description of all the grout defects in the stay cable specimens, including the level of severity for each defect. The grout defects were implemented in SC1 and SC2 at their respective locations, and the remaining sections were grouted completely with normal grout. Further details about the stay cable specimen construction, defect fabrication, and defect placement may be found in Hurlebaus et al. (*6*).

## Evaluation of Grout Defects in Stay Cable Specimens

The grout defects in the stay cable specimens were inspected using electromagnetic methods, mechanical wave and vibration methods, and visual methods.

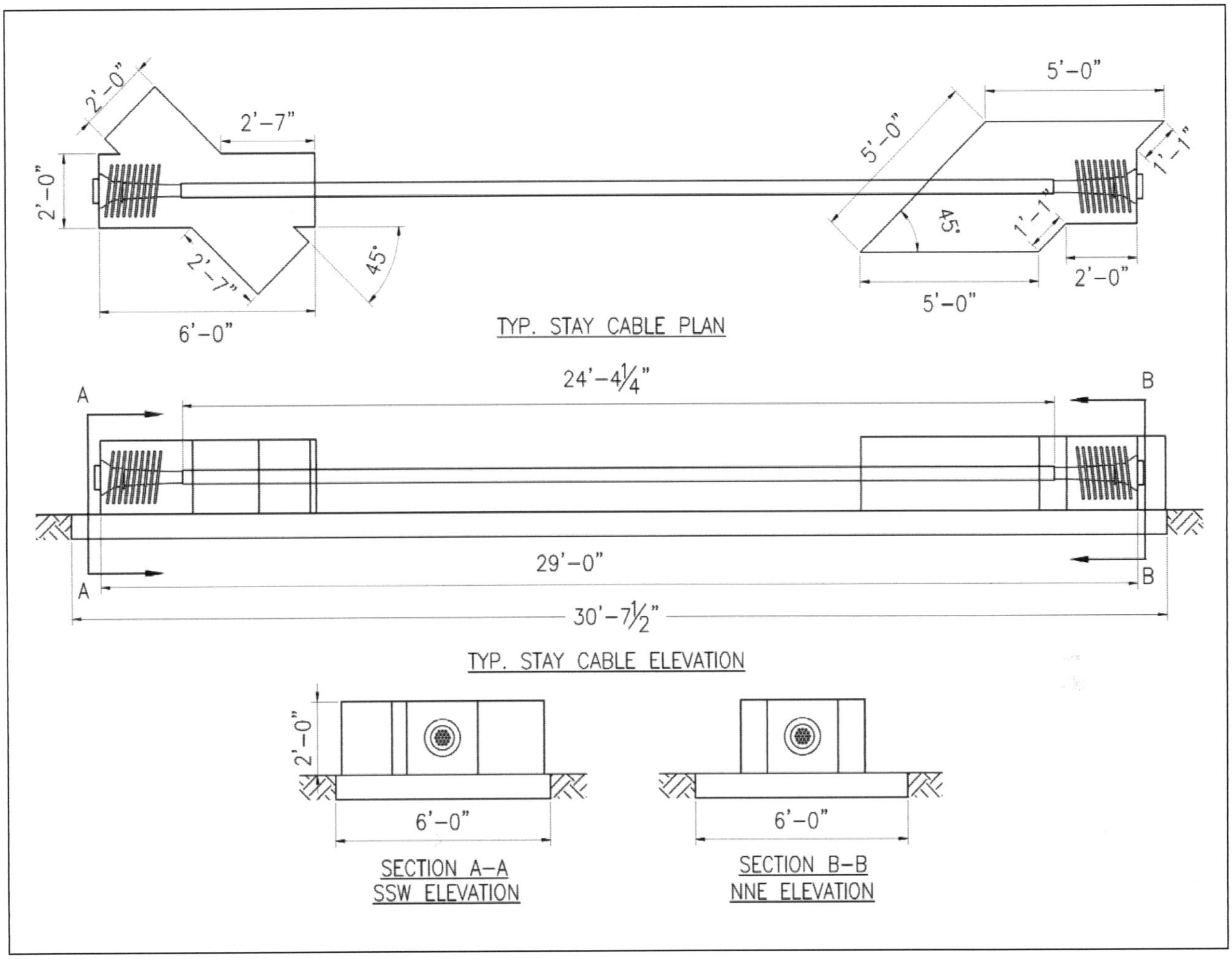


**Figure 1.** Geometric details of stay cable mock-up specimens.

**Table 1.** Location and Description of Defects in the Stay Cable Specimens

| Defect location | SC1 | SC2 | SC3 | SC4 |
|---|---|---|---|---|
| A–B | – | – | V4 | V4 |
| B–C | – | – | V4 | V4 |
| C–D | – | – | V4 | V4 |
| D–E | W2 | W2 | V4 | V4 |
| E–F | – | – | V4 | V4 |
| F–G | V4 | V4 | V4 | V4 |
| G–H | – | – | V4 | V4 |
| H–I | GS1 | GS1 | V4 | V4 |
| I–J | W1 | V1 | V4 | V4 |
| J–K | W1 | V1 | V4 | V4 |

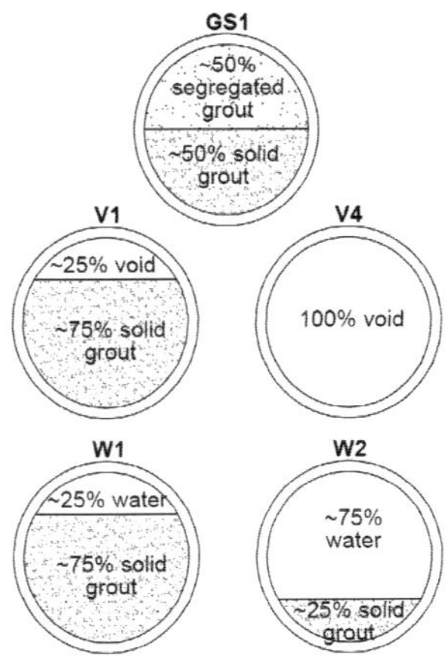


*Note:* – = no grout defect. Refer to Figure 2a for defect location.

## Ground Penetrating Radar (GPR)

GPR is an electromagnetic technique, in which electromagnetic pulses are emitted from a transmitter and the reflected pulses are received using receivers. A variation in the material's electrical conductivity and dielectric permittivity results in a difference in the reflected pulses. By analyzing the reflected pulses, it may be possible to identify anomalies in the structure.

A StructureScan Mini HR GPR unit with an operating frequency of 2.6 GHz was used in this investigation. This frequency allows for good resolution and a scan depth of up to 16 in. As the survey wheels on the device move along the length of the structure being investigated, the data is recorded. A 2.0 in. grid system was created for inspecting grout defects in the anchorage regions of the stay cable specimens. The scan was performed along vertical and horizontal lines, and later combined in the RADAN software to generate 3-D models of the anchorage regions. To enable the inspection of the MTEs, temporary wooden supports were installed on either side of the MTEs to allow the survey wheels to turn as the device moved.

Figure 2 presents a comparison of the GPR testing results of the MTEs of the stay cable specimens with the actual defects. Figure 2*b* shows the inspection results of the metal duct in specimen SC1. It is evident that this method was unable to detect the air voids and the water infiltration defects within the metal duct. Figure 2*c* shows the GPR inspection results of the HDPE duct in specimen SC2. As seen in the figure, GPR signal over regions with air void and water infiltration were distinct with scattered bands when compared with the other regions with solid grout. However, the air voids and water infiltration could not be differentiated from each other. Figure 2*d,e* present inspection results from specimens SC3 and SC4, respectively. The GPR signals do not show solid bands along the length of the duct, leading to the conclusion that specimens SC3 and SC4 may be either completely void or filled with water. Owing to the rebound of the radar signal from the internal prestressing strands, it was not possible to determine the size of the void or water infiltration for any specimen.

The GPR technique was also used for the inspection of the anchorage regions. However, due to the large amounts of reinforcing steel in the anchorage regions, it

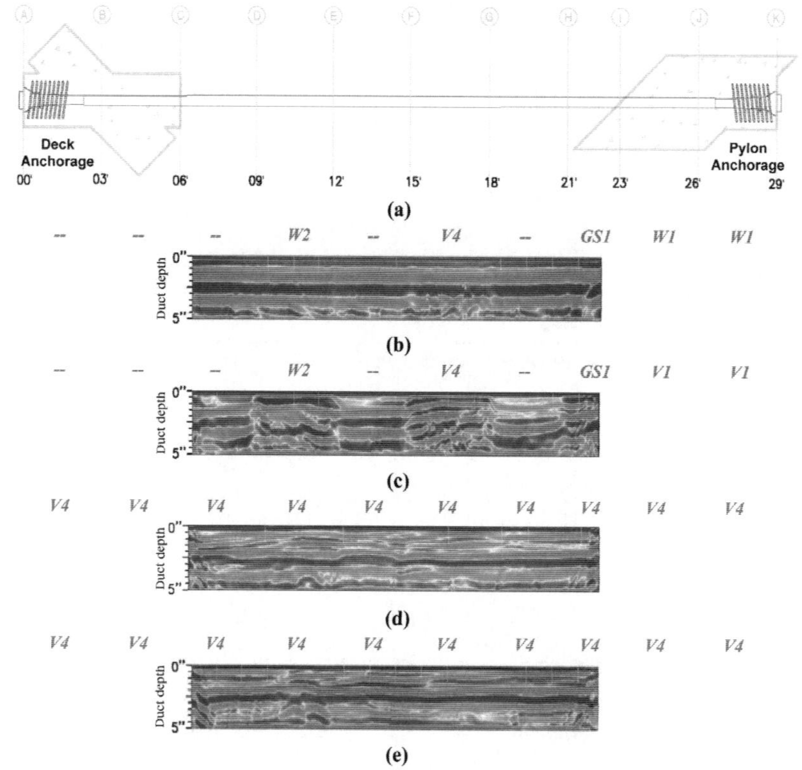


**Figure 2.** GPR inspection of stay cable specimens: (*a*) segments for defect placement within stay cable specimens; comparison of defects with GPR results from MTEs of (*b*) SC1, (*c*) SC2, (*d*) SC3, and (e) SC4.

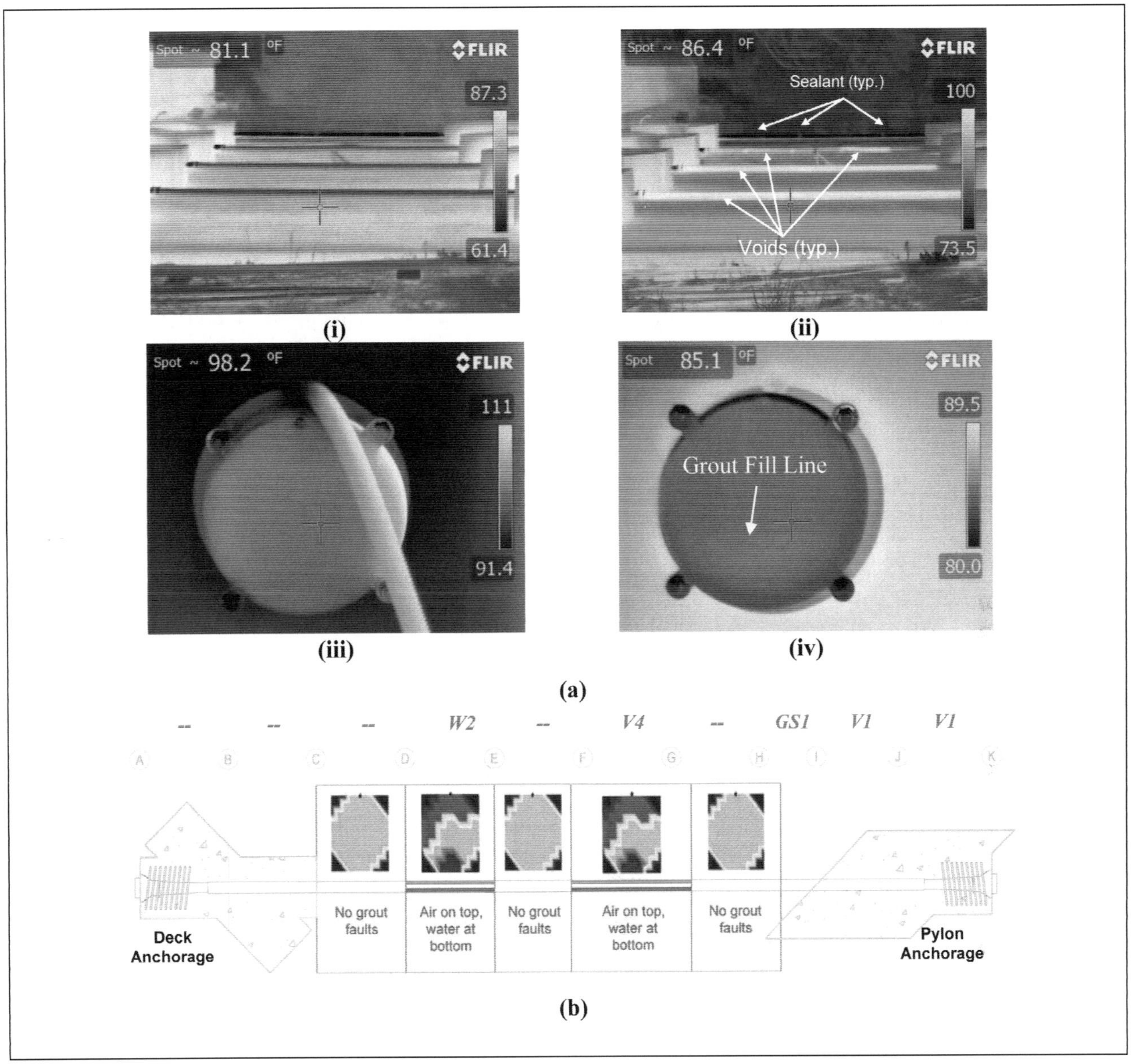


**Figure 3.** IRT and ECT inspection of stay cable specimens: (*a*) IRT images of (*i*) stay cable specimens approx. 7 min after sunrise, (*ii*) stay cable specimens approx. 100 min after sunrise, (*iii*) end cap region at deck anchorage of SC2, and (*iv*) end cap region at pylon anchorage of SC2; (*b*) comparison of results from ECT with the defect key of SC2.

was unable to even locate the metal ducts in the deck and pylon anchorage regions. Owing to its poor performance in detecting grout defects in the anchorage regions, the results are not presented herein.

## *Infrared Thermography (IRT)*

IRT inspection is an electromagnetic technique and uses the principle of examining the emissivity of individual materials within the object being inspected. Depending on the emissivity, each material releases or absorbs heat at different rates, and the differential temperatures during this transition period can provide valuable information about the inner composition of the object. However, it is important to perform IRT during times of the day when atmospheric temperature gradients are high, thus forcing the object being inspected to heat or cool in order to reach equilibrium with the surrounding environment.

Passive IRT was used to inspect defects in the free spans of the MTEs, the anchorage regions, and end caps of the stay cable specimens. Note that active IRT was not pursued in this investigation, as it is impractical to use an external source of heat to inspect large regions of in-service structures. A high-performance, forward-looking infrared (FLIR) T640 thermal imaging camera was used for the IRT inspection. The accuracy of the camera was calibrated within ± 3.6°F or ± 2% of the measurement at 77°F nominal temperature. To enable the inference of IRT inspection results, infrared images were captured at regular intervals starting before sunrise during the heating-up period. An appropriate scale was chosen to enable the detection of defects.

Figure 3*a*(*i,ii*) presents the infrared images of the MTEs of the stay cable specimens. In both the images, specimen SC1 is at the back of the image and SC4 is in the front. The voided regions of the stay cable specimens are not evident in Figure 3*a*(*i*) but may be clearly distinguished from the completely grouted sections in Figure 3*a*(*ii*), which were taken roughly 100 minutes after sunrise. The water defect in section D–E of specimen SC2 appears to be similar to the air void in section F–E, indicating that water infiltration and air voids may not be differentiated in the infrared images.

It is important to note that the voided regions in specimen SC1 were not identifiable in the infrared images. This may be attributed to the low emissivity of steel, meaning that steel reflects significantly more heat than the HDPE ducts. This characteristic results in steel ducts absorbing significantly less heat and as a result being continually closer to thermal equilibrium with the environment, thus not allowing thermal differential between the voids and the fully grouted regions. However, the sealant between grouting defects in the metal duct may be identified in all the images, as it is a polyurethane compound which absorbs practically no heat and remains at a constant temperature despite environmental temperature changes.

The infrared images of the anchorage regions did not provide any discernable information (and hence are not shown) and were unable to even locate the metal ducts within the anchorage regions of the stay cable specimens. From this it is evident that the IRT inspection technique is not applicable in inspecting defects within the metal ducts embedded in the deck and pylon anchorage regions.

Figure 3*a*(*iii,iv*) shows infrared images of the end caps of the specimen SC2 in the deck and pylon anchorage regions. The infrared image of the end cap in the deck anchorage does not show the presence of any differential temperature, indicating that the end cap is either completely filled with grout or completely empty. However, the infrared image of the end cap in the pylon anchorage shows a discernable grout fill line, indicating that the end cap is partially filled.

### *Electrical Capacitance Tomography (ECT)*

ECT is an electromagnetic technique and has the advantage of being a safe, fast, and relatively inexpensive technique. ECT obtains capacitance data from multi-electrode sensors, and by several iterations makes permittivity images of sections. ECT has been primarily applied to detect the air pockets in oil flow. However, there are inherent problems with the image resolution which is a major roadblock in the application of this method. In this investigation, ECT was used to detect grout defects in the free spans of the MTEs.

The inspection process involved identifying the beginning and ending points of voids, water, and bleeding grout in real time. ECT is not applicable to metal ducts, therefore only SC2 with grouted HDPE ducts was part of this investigation. Figure 3*b* shows a comparison of the results from ECT for SC2 with the defect key. Although ECT was able to identify the location of the water and air defects with good accuracy, the cross-sectional slices indicated the presence of both water and air at locations having only water infiltration (section D–E) and air void (section F–G). The method was not able to identify compromised grout (section H–I).

### *Impact Echo (IE)*

IE inspection was performed on the anchorage regions of the stay cable specimens to determine the grout defects within the metal ducts embedded in these zones. The top of the deck and pylon anchorages of all four stay cable specimens were inspected at 6 in. intervals. The technique involves hitting the surface with a small impactor or impulse hammer and identifying the reflected wave energy with a displacement or accelerometer sensor mounted on the surface near the impact point. The IE tests were performed using an IE scanner test head and a PC data acquisition system. An IE velocity of 13,123 ft/s was used for the calculation of concrete thickness.

The IE testing of the anchorage regions of the stay cable specimens provides poor quality data with weak and inconsistent resonant echoes. The cross-sectional dimension of the anchorages is approximately 2 ft × 2 ft, which exceeds the range of the practical limit of the IE device. Also, IE data is clearest on plate-like structures such as walls and slabs; whereas structures with greater depth naturally produce multiple resonances due to the structure geometry, which can complicate the data analysis. As there is no conclusive information about the grout condition within the ducts in the anchorage regions, the IE inspection data from the stay cable anchorages is not presented herein.

### *Sounding*

Sounding is a simple acoustic NDE method that can be used to inspect different bridge components, including the MTEs of cable-stayed bridges. Sounding is

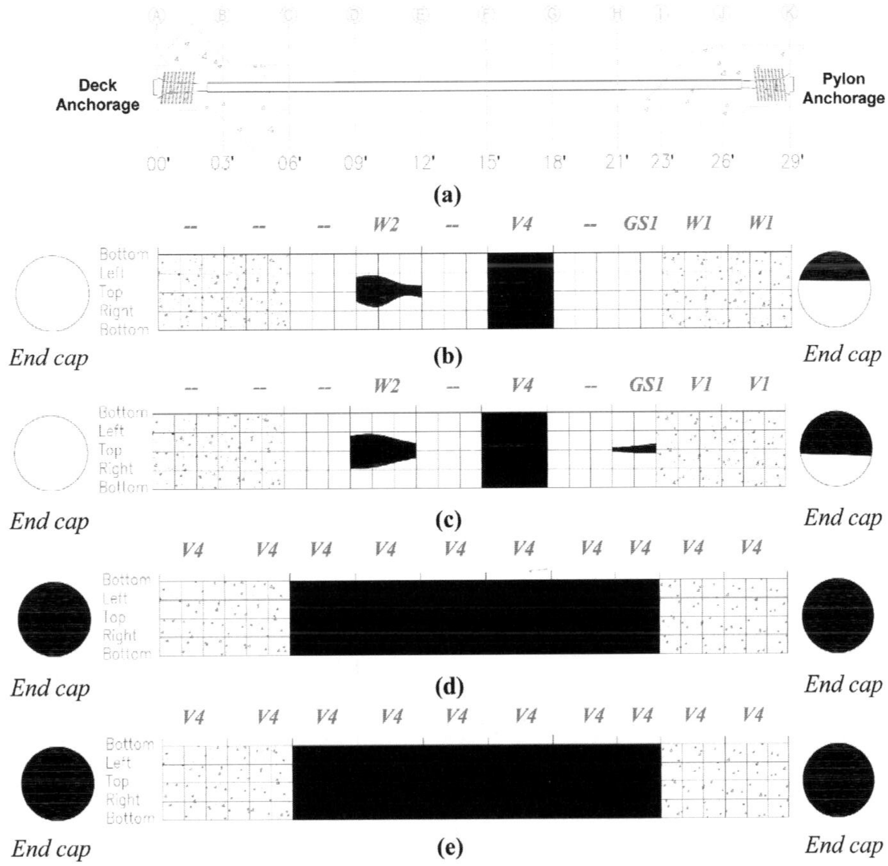


**Figure 4.** Comparison of results from sounding with defect key: (*a*) defect locations within stay cable specimens; (*b*) SC1; (*c*) SC2; (*d*) SC3; (*e*) SC4. *Note*: Regions shaded black represent low ringing sound corresponding to grout defect.

performed by tapping an impactor along the MTE of the stay cable specimen and listening for a change in the acoustic response produced by the tapping. Tapping on an area of the voided duct creates a different sound pitch, which indicates the potential presence of a voids. By recording areas of lower pitch responses corresponding to voids, a map of voided areas can be produced, and these locations can be marked for further inspection.

Inspection of the MTE along the length of the duct was conducted for all four stay cable specimens. The inspections were performed at 12 in. intervals. Sounding inspection was also conducted on the end caps of the anchorage systems by tapping the grout caps. The results from the sounding inspection were included in the void mapping sheets shown in Figure 4. The sounding maps provided accurate information on voided regions existing within the ducts. However, sounding did not differentiate between voided regions, water infiltrated regions, and regions with poor grout.

Figure 4 shows a comparison of the sounding inspection results with the defect key for each specimen. The voided and water infiltration regions along the free span of the MTEs and the end caps in the anchorage regions can be clearly identified from the sounding map. While the segregated grout defect in specimen SC1 with the metal duct could not be identified, the same defect in the HDPE duct in specimen SC2 is identifiable. This is due to the relative difficulty of interpreting the lower pitch sounds created when tapping on the metal ducts. Due to the interpretation of acoustic responses by the inspector, sounding is a subjective method and data can vary from each trial and inspector, especially for metal ducts. Sounding maps for specimens SC3 and SC4 clearly indicate ungrouted MTEs.

Sounding was very effective in determining the grout condition in the end cap. The partial and full voids in the end cap regions can also be clearly identified. However, as expected, sounding through mass concrete was not

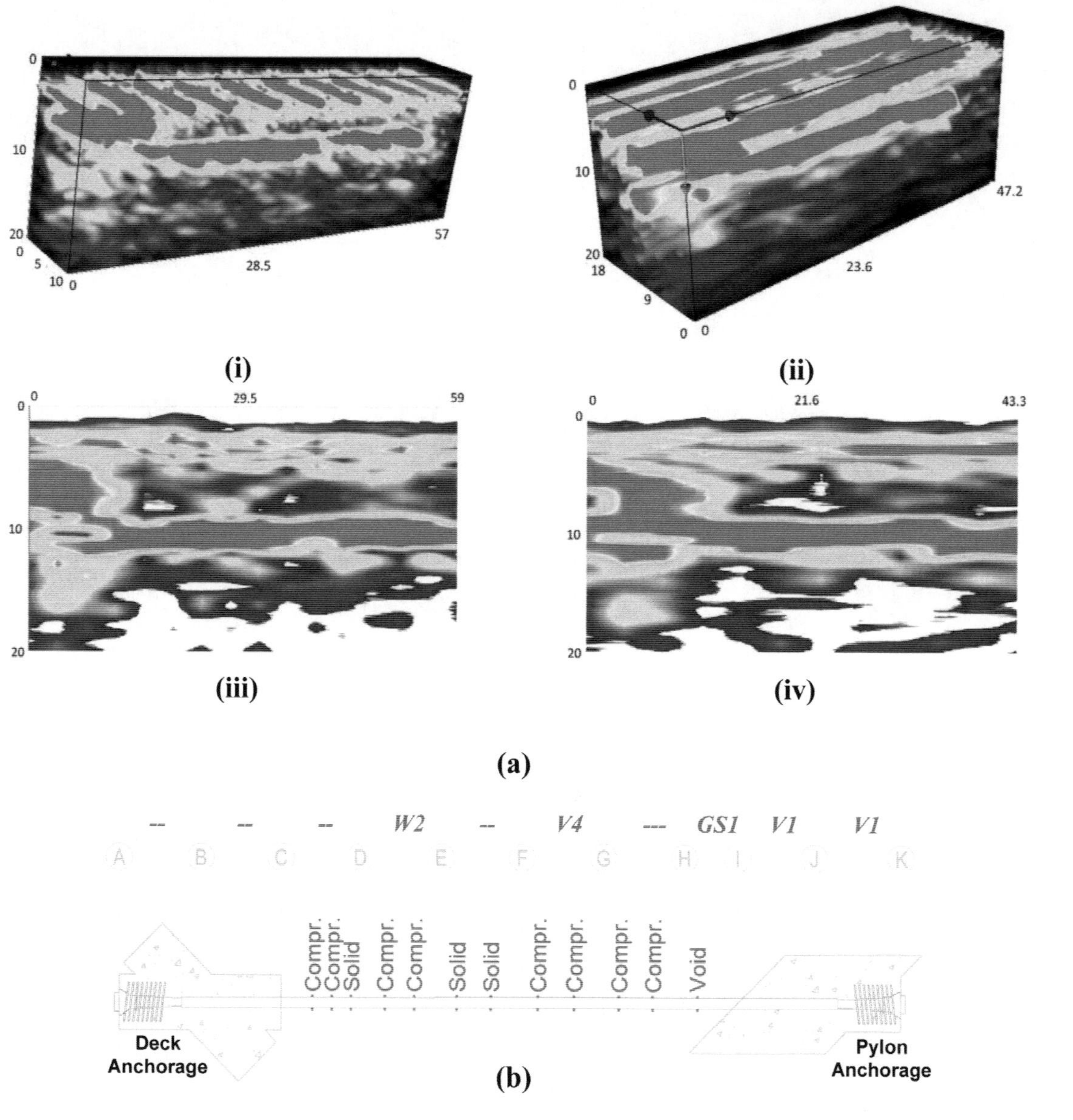


**Figure 5.** Results from UST and LFUT inspection of stay cable specimens: (*a*) UST inspection of anchorage regions of SC1 showing (*i*) 3-D map of deck anchorage parallel scan, (*ii*) 3-D map of pylon anchorage perpendicular scan, (*iii*) D-scan of deck anchorage perpendicular scan, and (*iv*) D-scan of pylon anchorage perpendicular scan (all units in inches); (*b*) comparison of results from LFUT with the defect key of SC2.
*Note*: Solid = no grout defect; Compr. = water infiltration or segregated, unhydrated or gassed grouts; Void = void defect.

able to determine any useful information about the defects in embedded ducts in the anchorage regions.

## Ultrasonic Tomography (UST)

The anchorage regions of the stay cable specimens were inspected using UST. In this technique, the emitted ultrasonic waves are reflected at material interfaces due to a change in acoustic impedance. Consequently, UST was used to identify grout defects within the metal ducts in the anchorage regions of the stay cable specimens. Considering that the top of the deck and pylon anchorages of the stay cable specimens would be accessible in an in-service cable-stayed bridge, these regions were inspected using UST. To facilitate the inspection, a 2 in. × 2 in. grid system was made on the deck and pylon anchorage regions. The A1040 MIRA was used in this investigation. The operational frequency was set to

50 kHz, and the specimen was scanned with the device oriented both parallel and perpendicular to the longitudinal axis of the MTEs.

Figure 5*a*(*i,ii*) shows a representative 3-D map created from the UST scanning of the anchorage regions of specimen SC1. Although the duct underneath the top reinforcement is visible, it is unable to conclusively decipher the grout defects within the metal ducts embedded in the anchorage regions. Figure 5*a*(*iii,iv*) presents the D-scans through the embedded metal ducts of the deck and pylon anchorages of specimen SC1. The images through the duct do not show any indication of grouted, ungrouted, or water infiltrated sections.

### *Ultrasonic Pulse Velocity (UPV)*

Inspection using UPV was performed on the deck and pylon anchorages of selected stay cable specimens to assess the grout condition within the ducts in these regions. Two 50 kHz ultrasonic transducers were used as an emitter and receiver. The testing was performed at two heights through the anchorage region, one at 12 in. from the top face, which is through the centerline of the embedded duct; and the second at 31.5 in. from the top surface, which provides a velocity baseline. Results from UPV inspection along these two locations show no apparent difference in the direct velocity, which suggests that the UPV method is not sensitive to the grout conditions within the ducts. One of the reasons that UPV is not sensitive to the defect is likely due to the small size of the defect relative to the length of the test path.

### *Low Frequency Ultrasound (LFUT)*

LFUT was used to assess the grout condition in the MTE of specimen SC2 with grouted HDPE ducts. The LFUT system is designed to generate and receive low frequency ultrasonic waves in a pitch-catch fashion, which propagates across the cross-section of the duct. The grout conditions identified by LFUT are reported to be either solid (no defects), compromised (which includes water infiltration and segregated, unhydrated, and gassed grouts), or void. The measurements were performed at intervals of 12 in. to 18 in. along the length of the ducts.

Figure 5*b* presents a comparison of actual defect locations with the results from LFUT for specimen SC2. The water infiltration (section D–E) and air void (section F–G) are reported as compromised grout (compr. in Figure 5*b*). The segment of the stay cable with segregated grout (section H–I) is reported as a void. Section E–F, which has good grout, is reported to have solid grout. However, other segments of the specimen (C–D and G–H) without any grout defects are reported to have compromised grout.

### *Visual Testing (VT)*

Currently, VT is the most common method of inspection of the MTEs of a stay cable. This entails a trained inspector monitoring the MTEs for indicators of physical changes that may be detrimental to the structure. As the grout is enclosed within opaque ducts, the true condition of the grout cannot be determined without opening the duct at the location of inspection. This means that detrimental conditions within a duct can often go undetected when inspection techniques are limited to visual inspection alone. In this investigation VT of the anchorage regions was conducted after removing the end cap. All eight end caps of stay cable specimens were removed and visually inspected, and then the photographed end cap conditions were compared with the actual defect key. Figure 6*a* shows two representative images of the end caps, which accurately identified the actual defect conditions, which are shown in parentheses.

Similarly, visual testing was also conducted through the small cut outs on the ducts that were introduced during the construction phase for placing the defects into the ducts, as shown in Figure 6*a*(*iii,iv*). Visual testing is very useful for identifying the defect condition and severity, but typically it may not be possible to detect the location of the defect at early stages.

### *Borescope*

Borescope inspection is a visual technique that can be used to inspect hard-to-reach places. Borescope inspection of the MTEs was performed using an Olympus IPLEX SX II apparatus to identify and provide visual information about the condition of the grout within the duct. Borescope inspections are relatively slow, but they provide the user with the ability to visually examine the conditions that exist within the duct. Although this method is useful for determining the severity of voids, it is essential that a void is present for the camera to physically enter the duct. Owing to this accessibility requirement, borescope inspection is referred to as a semi-destructive method. Depending on the system of stay cables, grout ports or grout caps may be temporarily opened for inserting the borescope tube. Therefore, it may be a very useful method in anchorage regions where the grout inlet/outlets can be opened for inspection and closed afterwards.

Borescope not only allows the inspector to view the grouting conditions, but also to qualitatively evaluate the severity and shape of the defects in order to recommend remediation techniques. This method does not have the ability to locate detrimental conditions by itself, so it is usually preceded by other inspection methods; usually visual testing or sounding is used to locate voids.

In this investigation, threaded entry holes were drilled into the ducts during construction and later used as entry

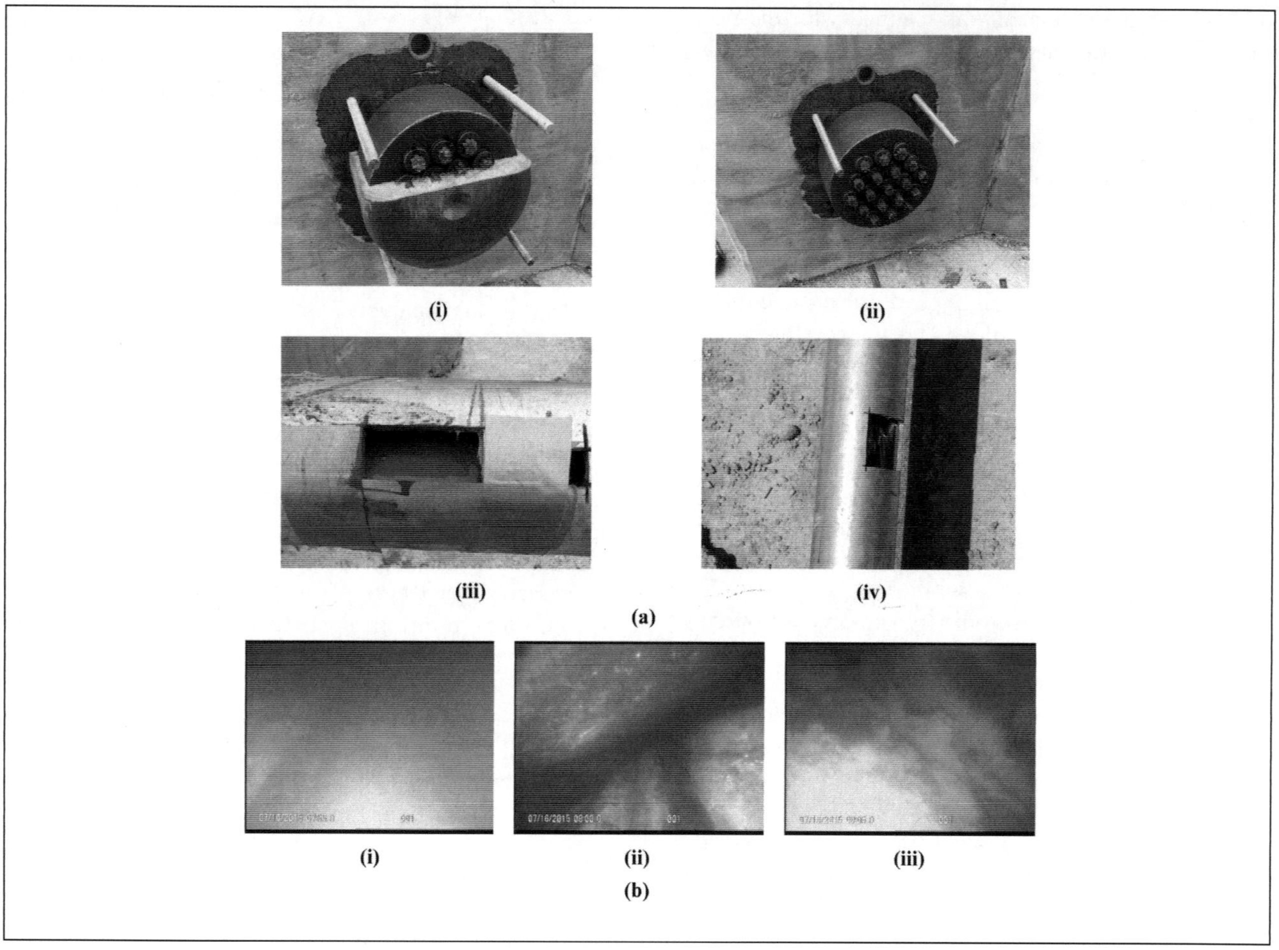


**Figure 6.** Results from visual inspection techniques: (*a*) representative end cap and open duct location photos for (*i*) SC1-pylon anchorage (W1), (*ii*) SC3-pylon anchorage (V4), (*iii*) SC1 section D–E (W2), and (*iv*) SC1 section F–G (V4); and (*b*) borescope inspection photos of SC2 (*i*) section D–E (W2), (*ii*) section F–G (V4), and (*iii*) section H–I (GS1).

ports for the borescope tube. These holes were sealed with a screw-in plug when not in use to protect the integrity of the defects. In addition to the drilled holes, grout ports were also used as entry points. Even with the presence of drilled holes, entry access was not necessarily available in each section, as fully grouted sections did not allow access of the camera even after the plug was removed.

Figure 6*b* shows representative borescope images of water infiltration, voided regions, and segregated grout conditions. The still photos appear blurry due to distortion, but video data and live operation of the device clearly showed that the camera was moving through water. Figure 6*biii* shows the segregation of grout along the top of specimen SC2 in section H–I.

## Discussion of Results

Based on the test results, Table 2 presents a summary of the capability of the various NDE techniques in locating the voids, water infiltration, and segregated grout conditions in both metal and HDPE ducts of the MTEs, and the anchorage regions of the stay cable specimens. The electromagnetic methods used for the inspection of grout defects within the MTEs of the stay cable specimens include GPR, IRT, and ECT. Among these, all the methods were able to identify the location of the voids and water infiltration defects within the HDPE ducts. However, they could not differentiate between the two defects. In addition, the methods were not applicable to metal MTEs, and could not differentiate between compromised grout conditions and intact grout. GPR and ECT were unable to provide any quantitative or qualitative information regarding the depth of the void or grout quality within the duct. Although IRT could make some estimates about the size of the defect, the accuracy associated with determining the severity of the defects is low.

All electromagnetic methods were ineffective in identifying the grout defects in the metal ducts embedded

**Table 2.** Capability of NDE Techniques to Detect Grout Conditions in MTEs and Anchorage Regions of Stay Cable Systems

| | MTEs | | Anchorage region | |
|---|---|---|---|---|
| NDE technique | HDPE duct | Metal duct | Embedded region | End cap |
| GPR | ☑V, W<br>☒GS | ☑None<br>☒V, W, GS | ☑None<br>☒V, W, GS | NI |
| IRT | ☑V, W<br>☒GS | ☑None<br>☒V, W, GS | ☑None<br>☒V, W, GS | ☑V, W<br>☒GS |
| ECT | ☑V, W<br>☒GS | NI | NI | NI |
| IE | NI | NI | ☑None<br>☒V, W, GS | NI |
| Sounding | ☑V, W, GS<br>☒None | ☑V, W<br>☒GS | NI | ☑V, W<br>☒GS |
| UST | NI | NI | ☑None<br>☒V, W, GS | NI |
| UPV | NI | NI | ☑None<br>☒V, W, GS | NI |
| LFUT | ☑V, W, GS<br>☒None | ☑None<br>☒V, WS, GS | NI | NI |
| VT* | ☑V, W, GS<br>☒None | ☑V, W, GS†<br>☒None | NI | ☑V, W, GS†<br>☒None |
| Borescope* | ☑V, W, GS<br>☒None | ☑V, W, GS†<br>☒None | ☑V, W, GS†<br>☒None | ☑V, W, GS†<br>☒None |

*Note*: GS = segregated grout; NI = not investigated; None = identifies/does not identify any defect; V = voids; W = water infiltration. Methods cannot differentiate between void and water infiltration.
☑Indicates presence of defect.
☒Indicates no defect.
*Access into ducts/end regions required.
†Defect either inaccessible or not present.

within the anchorage regions. While GPR and ECT cannot be used for the inspection of the end caps in the anchorage region, IRT could be effectively used to identify partially filled end caps. However, if the end cap is completely void or filled with grout, then it is not possible to identify the actual condition within the end cap using IRT alone. IRT can also be used to identify the severity of the voids in the end cap regions with good accuracy.

The vibration methods used for the inspection of grout defects in the stay cable specimens include IE and sounding. IE was used for the inspection of the anchorage regions, and sounding was used for the inspection of the MTEs and the end caps of the stay cable specimen. IE did not provide any discernable information regarding the grout defects within the metal ducts embedded in the anchorage region, whereas sounding was very effective in providing information about the location of the voids and water infiltration defects within the MTEs and the end caps. However, differentiation between the voids and water infiltration defects was not possible. Sounding was also not effective in identifying grout defects within the metal ducts.

The inspection techniques involving ultrasonic waves included UST, UPV, and LFUT. UST and UPV were used for the inspection of the anchorage regions of the stay cable specimens. Neither of these methods were successful in detecting grout defects within the anchorage regions due to heavy steel reinforcement around the metal ducts. LFUT was used for the inspection of grout defects in the MTE of specimen SC2, as this was the only grouted MTE with HDPE ducts. LFUT could identify the void and water infiltration in the MTE but could not differentiate between these conditions. The method also incorrectly identified intact sections as sections with compromised conditions.

Visual testing and borescope techniques were also used for inspecting the grout conditions in the stay cable specimens. These methods are effective once the approximate locations of the grout defects are known through other inspection techniques. The two methods were able to identify the defective grout conditions in regions with sufficient accessibility. However, for both the techniques obtaining access to certain regions to carry out the inspection can be a major challenge. Borescope inspection is an ineffective tool for detecting grout defects by itself. However, it can be very useful as an evaluative tool when coupled with another NDE method that can locate defects.

## Conclusions

To evaluate the effectiveness of various NDE techniques in detecting grout defects in stay cables, four stay cable specimen mock-ups were built with known grout defects. Several NDE techniques that may be broadly classified as electromagnetic methods, vibration methods, ultrasonic methods, and visual methods were used for the inspection of grout defects in the stay cable specimens, and the following observations were made:

1. IRT, sounding, and the visual inspection techniques were able to identify the voids and water infiltration defects in the end cap regions.
2. GPR, IRT, ECT, sounding, and LFUT were capable of identifying grout defects in the MTEs with HDPE ducts.
3. Methods other than sounding, visual testing, and borescope were either ineffective or not applicable to metal ducts.
4. The NDE methods used in this investigation were capable of identifying potential vulnerable locations in the stay cable, however they could not differentiate between voids and water infiltration defects.
5. The compromised grout conditions in the MTEs were not identified by the NDE methods considered in this investigation.
6. Detecting grout defects located within metal ducts in the anchorage regions is still a challenge. This is attributed to the greater attenuation of signals due to the increased thickness of the concrete block, and the greater steel reinforcement ratio in the anchorage region.

### Acknowledgments

This project was sponsored by the American Association of State Highway and Transportation Officials, in cooperation with the Federal Highway Administration, and was conducted through the National Cooperative Highway Research Program (Grant No. NCHRP 14-28). The project team would like to thank the Senior Program Officer, Dr. Waseem Dekelbab, and all NCHRP panel members, for their input during the various phases of the project. The authors would also like to thank Mr. William R. 'Randy' Cox (ASBI) for his support for grouting certification training; and Mr. John Turner and Dr. Mike Mota (CRSI), Mr. Pete Diggs and Mr. Bryan McMurray (Gerdau Long Steel North America), Dr. Zuming Xia and Mr. John Crigler (VSL), Mr. Ray Bauer (Commercial Metals Company), Mr. Mark Huff (BASF), and Mr. Steve Koch (Sumiden Wire Products) for their in-kind contribution of material for the construction of the mock-up specimens. The authors also thank the various industry partners for conducting NDE inspection of the stay cable specimens.

### Author Contributions

The authors confirm contribution to the paper as follows: study conception and design: Dr. Stefan Hurlebaus, and Dr. Mary Beth D. Hueste; data collection: Dr. Madhu M. Karthik, Dr. Tevfik Terzioglu, and Mr. Casey Jones; analysis and interpretation of results: Dr. Madhu M. Karthik Dr. Tevfik Terzioglu, and Mr. Casey Jones; draft manuscript preparation: Dr. Madhu M. Karthik, and Dr. Tevfik Terzioglu. All authors reviewed the results and approved the final version of the manuscript.

### References

1. Watson, S. C., and D. G. Stafford. Cables in Trouble. *Civil Engineering-ASCE*, Vol. 58, No. 4, 1988, pp. 38–41.
2. Grant, A. Cables Not in Trouble. *Civil Engineering-ASCE*, Vol. 61, No. 5, 1991, pp. 61–63.
3. Saul, R., and H. Karl. Inspection and Maintenance of Cable-Stayed Bridges–German Experiences. *Proc., First International Conference on Bridge Maintenance and Management (IABMAS'02)*, CRC Press, Taylor and Francis, pp. 3–8, 2002.
4. Telang, N. M., C. A. Ligozio, and A. B. Mehrabi. *Luling Bridge—Phase II Structural Evaluation of Luling Bridge Stay Cable Array—Tasks 1, 2, and 3*. Report to the Louisiana Department of Transportation and Development, State Project 700-45, 0107, 2004.
5. Mehrabi, A. B. In-service Evaluation of Cable-Stayed Bridges, Overview of Available Methods and Findings. *Journal of Bridge Engineering*, Vol. 11, No. 6, 2006, pp. 716–724.
6. Hurlebaus, S., M. D. Hueste, M. M. Karthik, and T. Terzioglu. *NCHRP Report 14-28: Condition Assessment of Bridge Post-tensioning and Stay Cable Systems Using NDE Methods*. Transportation Research Board of the National Academies, Washington, D.C., 2017.

*The Standing Committee on Testing and Evaluation of Transportation Structures (AFF40) peer-reviewed this paper (18-04206).*

Article

TRR
JOURNAL OF THE TRANSPORTATION RESEARCH BOARD

Transportation Research Record
2018, Vol. 2672(41) 113–122

DOI: 10.1177/0361198118790835
journals.sagepub.com/home/trr
SAGE

# Comparison of Two Nondestructive Evaluation Technologies for the Condition Assessment of Bridge Decks

**Ali A. Sultan[1] and Glenn A. Washer[1]**

**Abstract**

Selection of the appropriate condition assessment tool to meet the decision-making needs for preservation, maintenance, and repair (PM&R) actions can be difficult, because different nondestructive evaluation (NDE) technologies have different performance characteristics. The objective of the research was to compare the reliability of ground-penetrating radar (GPR) and infrared thermography (IRT) for the detection of delamination in concrete bridge decks. Field tests were conducted on two in-service concrete decks constructed in the same era and having significant deterioration due to corrosion damage. Reliability of the technologies was evaluated using receiver operator characteristics (ROC) analysis, a methodology that allows for a quantitative point-by-point comparison of results from NDE technologies with ground truth. The analysis results quantified the reliability of IRT and GPR for the two candidate concrete decks. It was found that IRT exhibited a higher reliability measure as compared with GPR for these two bridge decks. The implication of the test results for implementation of these technologies is discussed. The results of the analysis provide useful data for bridge owners making decisions regarding suitable NDE technologies for the condition assessment of bridge decks.

The objective of the research was to compare the reliability of ground-penetrating radar (GPR) and infrared thermography (IRT) for the detection of areas of delamination in concrete bridge decks. This comparison was made by developing a high-resolution reliability analysis using receiver-operator characteristic (ROC) analysis, a statistical method for quantitatively assessing the reliability of diagnostic systems. The field testing consisted of implementing IRT and GPR scanning on two bridge decks constructed in 1968. Ground truth was established by a combination of multiple soundings of the deck, some selected applications of the impact echo (IE) method, and physical sampling for confirmation of the sounding results. These data were used to complete a quantitative reliability analysis. The contributions of this research are (1) the development of an ROC methodology for assessing the reliability of nondestructive evaluation (NDE) technologies for in-service bridge decks; and (2) the quantitative comparison of the performance of IRT and GPR detecting areas of subsurface delamination in two case studies assessing reinforced concrete (RC) bridge decks. The results of the analysis can provide information for decision-makers for the selection and application of GPR and IRT as part of a condition assessment strategy for highway bridge decks.

## Background

The condition assessment of concrete bridge decks is important for determining preservation, maintenance, and repair (PM&R) actions required to maintain the serviceability and safety of bridges. The RC from which the deck is formed can be affected by the corrosion of embedded reinforcing steel that manifests in concrete spalling, forming potholes in the deck that require repair to maintain the serviceability of the deck. Spalling may also occur as the result of mechanically induced stresses, construction defects, or debonding of an overlay layer from the concrete deck substrate.

The typical condition assessment method used for bridges is visual inspection, which detects the onset of damage once spalling and cracking initiates and is visually observable. Nondestructive evaluation (NDE) tools are sometimes used to detect damage before the formation of spalling when the damage is not visually observable. Sounding of the deck with a hammer or chain is sometimes part of a visual inspection; more

[1]University of Missouri-Columbia, Columbia, MO

**Corresponding Author:**
Address correspondence to Glenn A. Washer: WasherG@missouri.edu

sophisticated methods such as GPR, IE, and IRT may also be used. Historically, sounding and chain drag are the most commonly used methods implemented by bridge owners.

Bridge owners are increasingly focused toward preservation and maintenance activities to maintain bridge decks in good condition, so as to avoid significant renovation or replacement needs. In this way, bridge decks may be subject to preservation or maintenance activities based on how damaged the deck is. Decision-makers need to understand the reliability of an NDE technology and how its reliability intersects with decision-making parameters. For example, the Michigan bridge deck preservation matrix suggests different PM&R strategies based on the percentage of damaged deck area. Decks with less than 5%, 5–10%, or 10–30% damaged deck area each have different recommended actions for PM&R. To determine whether a given condition assessment procedure or technology can be useful for making PM&R decisions, the reliability of the technology should be fully understood. The capabilities of different NDE technologies need to be better described such that the selection of the appropriate technology can be made by the bridge owner with consideration of these decision parameters. In this article, we compare the capabilities of two common NDE technologies used for deck condition assessment and provide a quantitative reliability analysis of the results from two cases studies. These data can be useful for decision-makers for determining the suitability of GPR and IRT.

## *Condition Assessment Technologies*

Different NDE technologies are currently being implemented by bridge owners for the condition assessment of RC bridge decks. There are significant differences in the manner in which these different NDE technologies infer or assess the presence of damage in concrete. It is important to recognize that NDE technologies generally do not detect damage directly. Rather, damage is assessed through analysis of the response to a physical phenomenon. For example, sounding typically consists of using a hammer or rod to impact the surface of the deck, or using a chain dragged over the surface of the deck. The audible variations in tone between sound areas and areas of subsurface delamination are assessed subjectively. IE is a methodology that utilizes an impactor and receiving transducer to digitize the response of the concrete to an impact (*1*, *2*). Analysis of the response is then conducted by transforming the time series signal to the frequency domain, and assessing the variation in frequency that may result from damage (*3*). IE is generally less subjective than sounding, since quantitative data are collected and analyzed.

GPR assesses condition in a much different way. GPR systems launch an electromagnetic (EM) wave into the concrete. These EM waves are reflected by boundaries between materials of different electromagnetic properties. For example, a wave propagating in air is partially reflected when it interacts with the concrete surface of the deck. A portion of the wave is transmitted into the concrete; this wave is reflected when it interacts with the embedded reinforcing bars in the concrete, because the bars are conductive (steel). Typically the analysis of the GPR response is performed on this reflection from the top layer of reinforcing steel, although reflections from the bottom of the deck may also be used (*4*, *5*).

The EM wave amplitude and velocity when propagating through the concrete are affected by the dielectric properties of the concrete. The presence of moisture, and to some extent chlorides, affect the dielectric properties of the concrete and the changes are represented in the travel time and amplitude of the reflected signal (*6*). These data are then interpreted; areas where the reflection amplitude is reduced and/or the velocity of the EM wave is reduced are interpreted as corrosive environments (i.e., moisture and chlorides) that are potentially damaged or deteriorated areas (*7–9*).

Conventional IRT consists of interpreting thermal anomalies at the surface of the concrete. The IRT camera captures the infrared radiation emitted from the surface of an object. When there is a subsurface delamination or void, the material between the subsurface feature and the surface of the concrete has a different thermal inertia than areas where the concrete is intact. Consequently, these areas heat up more quickly when ambient temperatures are rising and/or the area is exposed to solar radiation. These areas also cool more quickly when ambient temperature is decreasing and significant solar radiation is not present, such as in the later afternoon and nighttime. Thermal anomalies can also occur if the concrete is saturated with moisture, or there is moisture present in a subsurface delamination. Subsurface damage that is dry may appear as a hot-spot during the day, while saturated areas can sometimes appear as cold spots. Debris, oil stains, discoloration of the concrete, paint, and so on, on the surface of the concrete can result in thermal anomalies of the same magnitude as the anomalies caused by the subsurface delamination. For this reason, IRT results are commonly matched with visual images such that subjective interpretation can be made when considering the source of a thermal anomaly. These anomalies are interpreted to be associated with areas of delamination, or associated with a surface feature such as debris, staining of the deck, and so forth.

Therefore, these three different NDE technologies each depend on the interpretation of physical phenomena to indirectly identify areas of damage in concrete. Sounding and IE rely on mechanical waves, GPR relies on the response to EM radiation, and IRT relies on the radiant energy associated with temperature. The question for the bridge owner is which of these approaches, which use entirely different processes to reach the same goal of detecting damage, is best suited for their condition assessment needs. Therefore, there is a need for reliability analysis that provides objective, quantitative comparison between different technologies and provides that information to decision-makers. The following section describes several methods that have been used traditionally to assess the capabilities of NDE technologies, and how the ROC analysis conducted in this research overcomes important limitations of these approaches.

## Methods of Comparing NDE Technologies

The reliability of an NDE technology is a measure of the ability of the technology to perform its intended function under a given set of operating conditions; in this research, it was assumed that the intended function of the NDE technologies was to detect subsurface delamination in RC bridge decks. Therefore, the reliability analysis compared the NDE results to the presence of subsurface delamination in two RC decks in a pointwise manner throughout the entire deck.

It can be difficult to evaluate the reliability of an NDE technology when it is used to detect damage that is not visually observable. Several different approaches have been used to assess the reliability or capability of NDE technologies for bridge decks. One method commonly used in the GPR field is to compare the total quantity of repair on a bridge to the total area of deterioration identified by GPR (*10*). In this way, location of the damage is not considered in the analysis. Some researchers have compared GPR results with half-cell potential or resistivity measurements (*8*, *11*, *12*). Such an analysis is useful for comparing one NDE result to another NDE result. This approach may not be effective for comparing NDE results to actual damage, because these technologies measure electrical properties of the RC, not physical damage.

Another method is to compare the results from one NDE technology with the results of another NDE technology that works by a different physical phenomenon. For example, comparing the results of GPR or IRT with a chain drag result (*13–19*). In this way, the EM method is compared with an acoustic method that is commonly used for deck condition assessments by bridge owners. Sometimes this comparison is made by producing a plan-view image of the deck surface to display results from the NDE technology. Areas of damage indicated in these images are compared either qualitatively (i.e., noting that the results seem "similar") or by dividing the deck area and spatially comparing the results pointwise. For example, dividing the deck into unit areas (e.g., 1 sq. ft. or 0.09 $m^2$) and then comparing the results within each unit area across the entire deck (*19*).

Some previous efforts evaluated NDE technologies for bridge decks based on cores removed from the deck (*13*, *18*, *20*, *21*). Removing a core at a single location only determines if the result is correct or incorrect at that particular point; this does not assess whether the correct spatial extent of the damage has been effectively represented by the NDE data. The number of cores is typically small, representing a very small sample relative to the total area of deck.

Both cores and spatial comparison between NDE technologies are often analyzed using some form of an *accuracy index*. An accuracy index presents the ratio of correct results to the total number of test points (i.e., samples). For example, removing 10 cores from a bridge deck to compare with NDE results at those 10 locations, or spatially comparing areas of damaged and sound concrete based on plan-view data representations. Although the accuracy index provides a quantitative measure of NDE capability, it is susceptible to bias based on the population sampled. For example, if only 10 out of 100 samples indicate damage, and the NDE technology detected no damage, then the NDE technology would have a 90% accuracy even though it did not detect any damage. For an NDE technology that is susceptible to a high number of false alarms (i.e., indicating damage where no damage exists), the greater the amount of actual damage in the bridge the more likely it is that the NDE results correspond with damage, since there would be fewer opportunities to be incorrect as the sound area of deck tends toward zero.

Another limitation of an accuracy index is that two different NDE technologies might produce the same accuracy indices, but actually perform very differently. For example, assume that 10 cores are collected and each technology shows an accuracy index of 60%, meaning the core did not agree with the NDE result on four out of the 10 cores. One technology has four false alarms, meaning the NDE technology detected a defect where none existed. The other technology had four false-negatives, meaning the technology did not detect a defect where a defect existed. Obviously, the two NDE technologies have very different performance characteristics. Depending on the risk associated with the damage, it may be much more significant to miss four defects than to call out four defects where none existed.

A limitation of all of these approaches for evaluating the capabilities of NDE technologies is that the assessment is performed based on a single analysis of the data, using a particular threshold value to discriminate

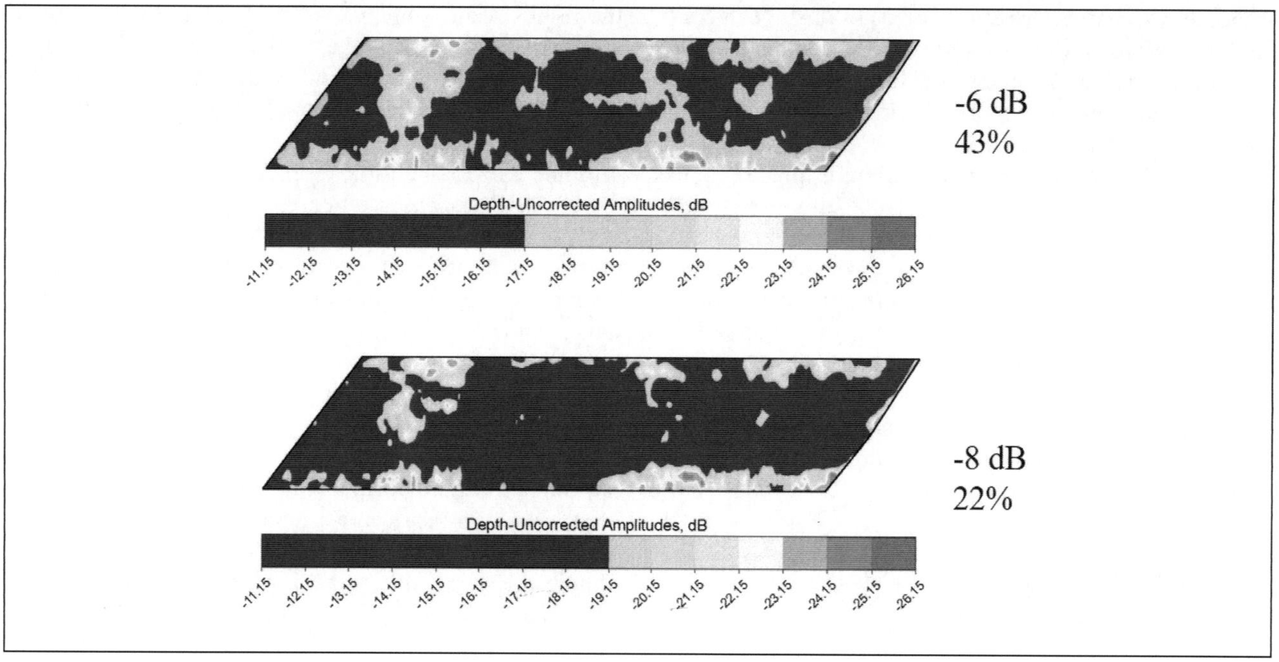


**Figure 1.** Plan-view image of bridge deck A2111 with GPR results using thresholds of −6 dB and −8 dB.

damaged areas from undamaged areas. For illustration, consider the GPR data from bridge A2111 shown in Figure 1. This figure shows GPR results with different thresholds used to identify areas of deterioration and areas of sound concrete. In these figures, the areas of the deck that do not exceed the threshold (i.e., sound areas) have been colored a uniform gray, while areas that match or exceed the threshold (i.e., deterioration) are color-coded according to the reflection amplitude. As shown in this figure, changing the threshold value from −6 dB to −8 dB changes the areas of deck identified as deteriorated by almost a factor of two (from 22% to 43%). Clearly, the selection of the proper threshold has a significant effect on the outcome of the test; a bridge owner might make a different PM&R decision if the deck was 22% deteriorated compared with 43% deteriorated. The same is true for IRT; the results depend on the threshold selected, and selecting a different threshold can results in significantly different results.

ROC analysis addresses many of the limitations of these approaches for assessing the reliability of NDE results. ROC analysis includes all of the practical thresholds to provide a more objective and comprehensive analysis of reliability. The ROC analysis is less susceptible to sampling bias, and explicitly represents the probability of detection (POD) and the probability of false alarm (POFA). In this way, the accuracy index for any practical threshold can be determined. As will be shown later, this analysis can also be used to select an optimum threshold at which an appropriate balance between false-positive and false-negative results can be achieved.

An ROC analysis considers all the possible outputs from the comparison of the ground truth and the NDE data. This includes true-positive (TP) results (both NDE and the ground truth indicated delamination), true-negative (TN) results (both NDE and the ground truth indicated no delamination), false-positive (FP) results (NDE indicated delamination, but ground truth indicated no delamination), and false-negative (FN) results (NDE indicated no delamination but the ground truth indicated delamination). For a given threshold setting, these indices can be used to calculate the POD or true-positive rate (TPR), $TPR = TP/n^+$, where $n^+$ is the total number of positive test points. The POFA or false-positive rate (FPR), $FPR = FP/n^-$, where $n^-$ is the total number of negative test points, is also determined. This process is repeated across the range of practical threshold settings. This range extends from a value at which the entire deck would be indicated as sound, such that the POFA is 0%, to a value at which the entire deck would be indicated as delaminated and the POFA is therefore 100% (i.e., all of the sound areas are indicated as delaminated).

ROC was developed by electrical and radar engineers during World War II to assess the ability of radar systems and operators to correctly identify enemy aircraft (*22*). ROC analysis has not been used widely in the field of NDE of highway bridges. The first study to demonstrate the applicability of ROC analysis for the estimation of the reliability of NDE technologies was done by Nockemann et al. in 1991 (*23*). The study examined the reliability of radiography for detection of defects in welds. Shin and Grivas (*21*) used ROC analysis to investigate the reliability of GPR for the detection of subsurface damage in a concrete bridge deck. The study used a limited number of core samples as test points for the ROC analysis. Reliability of GPR for the detection of corrosion damage was also investigated by Martino et al. using ROC analysis (*24*). In the study, GPR results were compared with half-cell potential measurements.

ROC is a statistical analysis method in which POD is plotted against POFA considering multiple threshold settings. The area under the resulting ROC curve (denoted AUC) is considered a global measure of reliability. AUC is equivalent to the probability of correctly identifying and ranking a random pair of positive and negative test points. Larger AUC values generally indicate better performance of the diagnostic system. AUC of 1.0 represents an ideal (perfect) diagnostic system. AUC of 0.5 represent a diagnostic system of no ability to differentiate positive and negative test points. Most NDE technologies are expected to have ROC curves that lie between these two extremes.

## Experimental

Two in-service bridge decks (A2111 and A2112) located along Rt 54 in Fulton, MO were evaluated by IRT and GPR. Bridge A2111 has a length of 175 ft (53 m) and a roadway width of 41 ft (12.5 m). Bridge A2112 has a length of 152 ft (46 m) and roadway width of 41 ft (12.5 m). Decks of both bridges showed signs of deterioration in the form of concrete spalling; some areas were repaired by means of asphalt or concrete patches.

IRT data were collected using a vehicle-mounted commercial thermal imaging system (ThermalStare, LLC) (*25*). Figure 2 shows the test systems used to collect data. The vehicle-mounted IRT system consisted of a 324 × 25 pixel IR camera (sensitivity <50 mK) and a visual camera, mounted at a height of approximately 10 ft (3 m), a main data collection unit, and a user interface. Data collected by the system are 14-bit values representing radiant flux. Data can be converted to temperature values, but for the purpose of this analysis conversion is not necessary and is not performed. The IRT camera collected about one lane width per pass as the system was driven down the length of the bridge. Four longitudinal passes were needed to collect data on the entire structure. The dataset of images collected by passing over the deck were then stitched together to form one image of the entire bridge deck. Vehicle speed during data collection was approximately 6–8 mph (9.6–13 km/h). The weather conditions when testing bridge A2111 were partly cloudy to cloudy, and data were collected between the hours of 10 a.m. and 2 p.m. For bridge A2112, weather conditions were sunny and data were collected between 9:30 a.m. and 11:30 a.m.

GPR data were collected using a commercial data acquisition controller device and a 1.6 GHz ground-coupled antenna (GSSI, Inc.). GPR testing was performed by making scans at 2 ft (0.6 m) intervals in the traffic direction of the bridge (perpendicular to the uppermost reinforcement bars) (*26*). The parameters used for collecting data included collecting 24 scans per foot (79 scans per meter) longitudinally along the bridge deck (*27*).

Once the required data were collected and transferred to a PC, the individual 2D scans were combined into a single 3D file. Data in the 3D file were then processed, considering time-zero correction, noise reduction, and migration. Migration was performed to collapse the exact position of the steel bars and provide accurate determination of the reflection amplitude, two-way travel time, and depth associated with each embedded steel bar. RADAN

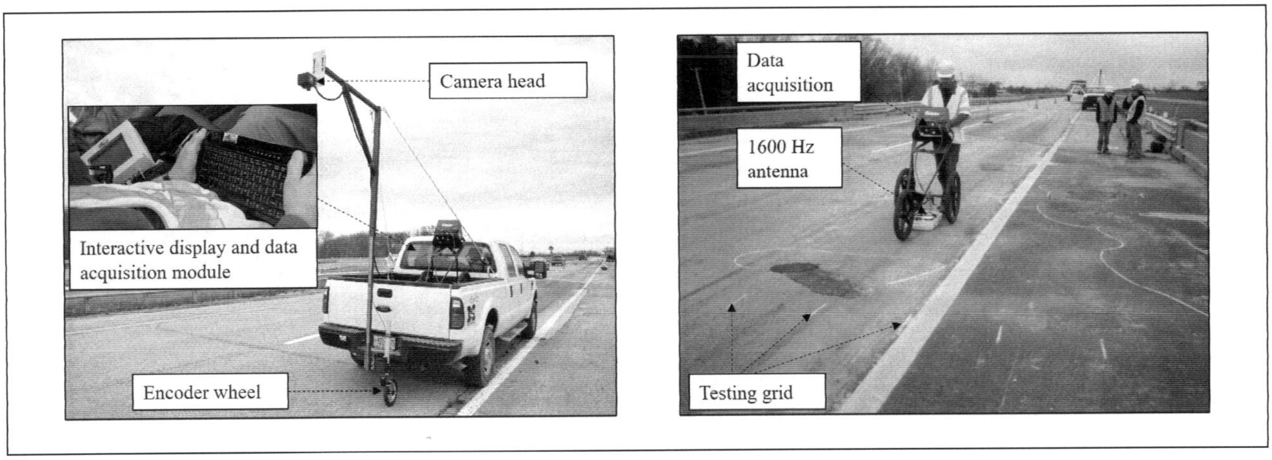


**Figure 2.** Photograph of NDE systems showing vehicle-mount IRT (*left*) and GPR (*right*).

7 software (GSSI, Inc.) was used to process the data from both case studies. The deterioration mapping function was used to extract the reflection amplitudes of steel bars. Datasets were then transferred to a third-party software for analysis and to produce plan-view color-coded images representing the GPR results. Depth correction intended to address signal losses and gains caused by depth variations of steel reinforcement was completed as recommended by Barnes et al. (*28*).

Ground truth was established to accurately represent the location and extent of areas of delamination. Hammer sounding results from three different individuals working in collaboration was used to draw the outlines of the delaminated areas across the entire deck; these results were transferred to a plan-view diagram of the deck showing the location and extent of the areas of delamination. It is rational to use sounding as a measure of ground truth, since this is the primary method implemented by bridge owners and therefore forms a meaningful comparison. The IE method was used on some selected areas in each deck to confirm the results of sounding as they apply to the spatial extent of the delamination. Physical sampling was conducted by drilling a 1.27 cm (0.5 inch) hole into the deck, and using a borescope to observe the subsurface cracking that formed the delamination to confirm the sounding results. Six physical samples were taken on each bridge. Both the IE testing and physical sampling confirmed the location and extent of areas of delamination found by sounding; this provided confidence that the sounding was identifying areas accurately.

### *Analysis of Data*

Results from the GPR and IRT scanning were compared with the established ground truth data and the results in relation to detection of areas of delamination were analyzed. GPR data were compared at "test points" that were defined by the spatial position of each reinforcing bar along a GPR scan line, where the reflection amplitude of the GPR signal was analyzed. If the position of the reinforcing bar was within an area identified as delaminated in the ground truth data, the test point was *positive*. If the ground truth data indicated that the position of the bar was not within an area identified as delaminated, the test point was *negative*. The ground truth at each test point was then compared with the GPR result at the same test point for a given threshold value.

For the IRT data, the IRT images produced by the camera covered the extent of the surface of the bridge. Therefore, comparison is done on a pixel-by-pixel basis between the IR image and a plan-view drawing produced of the ground truth data. This process involved overlaying the image produced by the IRT system and the diagram produced showing the ground truth, and evaluating each pixel in the IRT image. A threshold value for the radiant energy was used to classify the image pixels as either sound pixels or delaminated pixels. Then, each pixel was compared to the ground truth data at that position. According to this step, a single pair of POD and POFA indices can be determined by counting the total number of TPs and the total number of FPs (i.e., sound pixels that were incorrectly classified as positives) for each threshold.

As was discussed earlier, a different threshold value might result in a different pair of POD and POFA, and this consequently means a different reliability level. Therefore, it is necessary to investigate the reliability of NDE technologies with multiple threshold settings and then summarize the reliability with a comprehensive index regardless of the threshold settings. Fortunately, this can be achieved by ROC analysis where POD and POFA can be estimated with all practical threshold settings, and the desired reliability index can be defined by the area under the resulting ROC curve (*29*, *30*).

## Results

Figure 3 shows the IRT and GPR results from bridge A2111. The areas of delamination identified by the ground truth survey are encircled. Figure 3*a* shows the IRT results in plan-view. In this figure, the magnitude of the radiant flux is color-coded; generally, areas shown in red are areas of delamination indicated at a certain threshold value. Figure 3*b* shows the resulting GPR data in plan-view with the amplitude of reinforcing bar reflections color-coded. The reinforcing bar reflections have been depth-corrected. Figure 4 shows the IRT and GPR results for bridge A2112.

These images can be assessed qualitatively to identify areas of delamination based on the colors displayed, which can be subjective. In contrast, the ROC analysis objectively compares the quantitative measurements produced by the NDE technology to quantitative threshold values, without subjective assessment.

### *ROC Analysis*

Figure 5 shows the ROC results for bridges A2111 and A2112. As mentioned earlier, the ROC curve is produced by implementing a series of different threshold settings and calculating the POD and POFA for each setting. Considering Figure 5*a*, which shows the ROC curve for bridge A2111, the threshold setting at the origin is set such that no data exceed this threshold. In this case, no damage is detected and both the POD and POFA are equal to zero. In the upper-right corner of the graph, the threshold setting is such that all data points exceed the

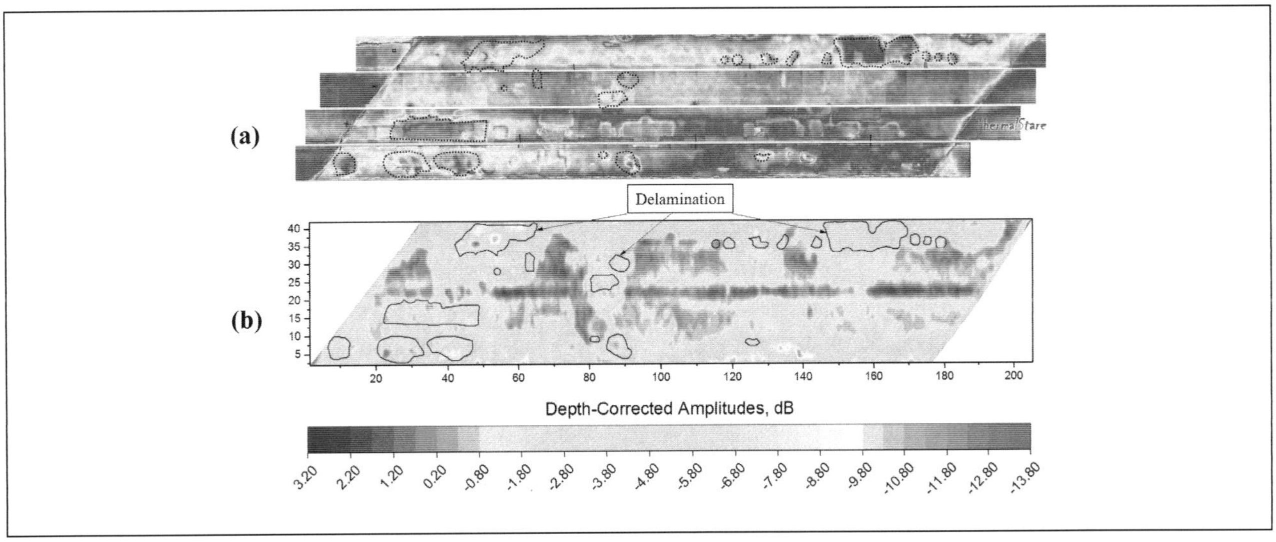


**Figure 3.** Images of bridge A2111 with areas of delamination shown with black outlines: (*a*) IRT; (*b*) GPR.

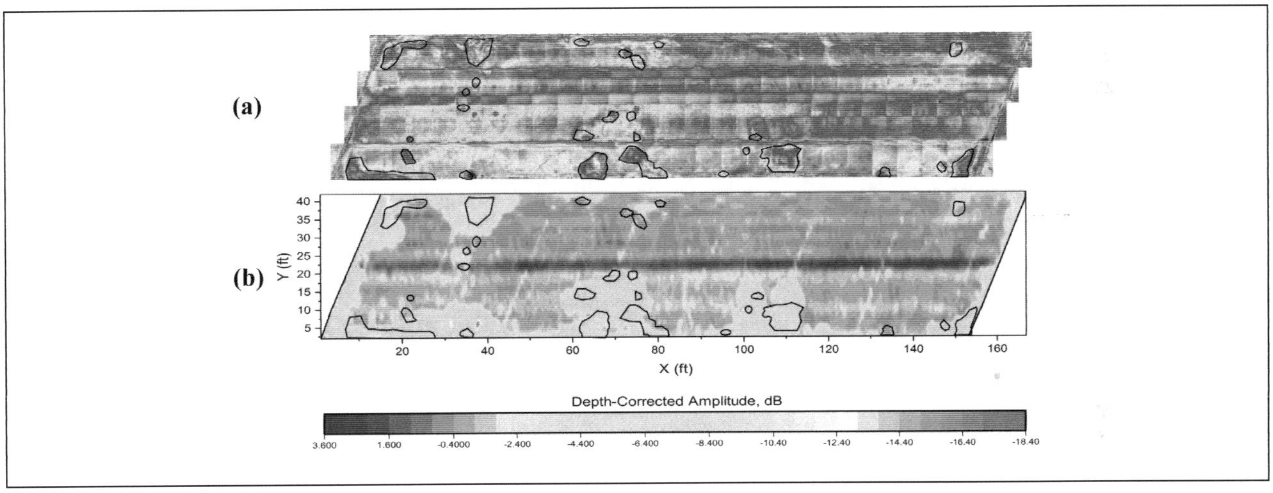


**Figure 4.** Images of bridge A2112 with areas of delamination shown with black outlines: (*a*) IRT; (*b*) GPR.

threshold. In this case, the entire surface of the deck is identified as damaged, and the POD and POFA are each 100%. Between these two points, the threshold has been advanced incrementally, with the POD and POFA calculated at each threshold. The datum for each threshold setting was joined with a line to form a curve that describes the performance of the technology across all practical threshold settings, and the area under this curve represents the reliability measure, AUC. As described previously, an ideal diagnostic system with 100% POD at all thresholds would form a vertical line from (0,0) to (0,1) on the ROC curve, and have an AUC of 1.0. The dashed line in Figure 5 (reference line) represents the lower bound of a diagnostic system with no ability to differentiate positive and negative test points, with an AUC of 0.5.

The ROC curve can be used to determine an optimum threshold setting based on the test results. An optimum threshold could be considered at the datum lying closest to the coordinate (0,1) on the graph, shown with a circle around the datum in the figure. At this threshold setting, IRT results indicate a POD of ~75% and a POFA of ~30%. For GPR, the POD would be ~62% and the POFA would be ~37%. For situations in which it is desirable to detect all of the damage, and false calls are acceptable, threshold values resulting in data to the right of these points will result in a higher POD and POFA. If it is desirable to limit the number of false calls, a threshold resulting in a point to the left can be selected, with the associated consequence of increasing the number of FNs (i.e., not detecting all of the damage).

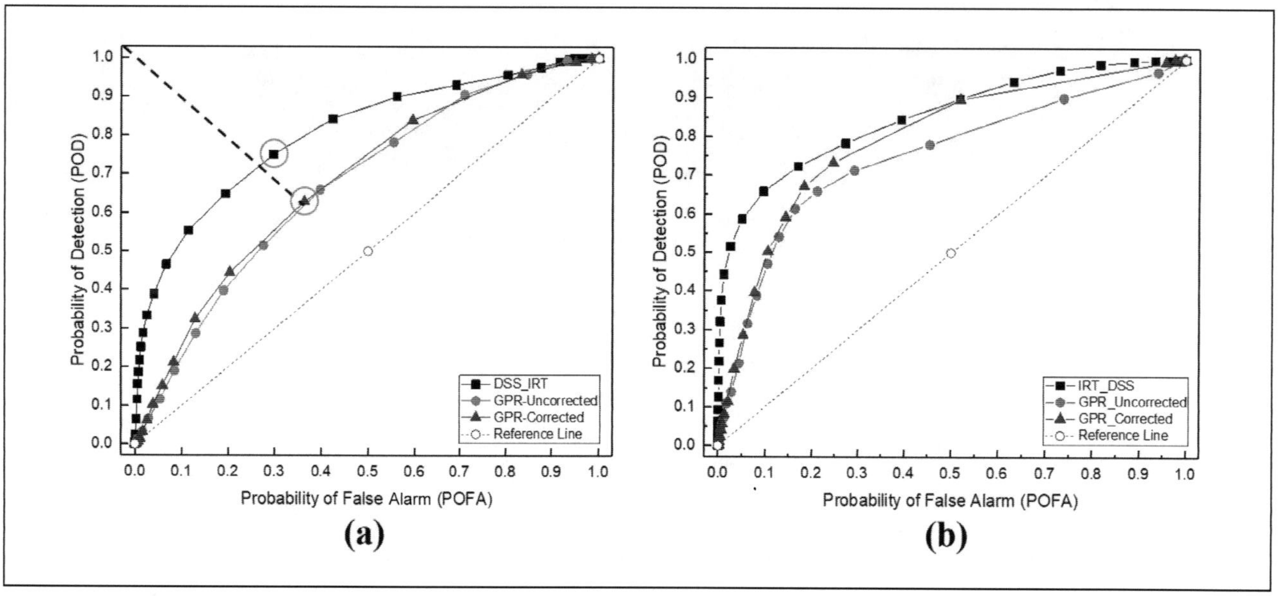


**Figure 5.** ROC curves for bridges (*a*) A2111 and (*b*) A2112.

For test bridge A2111, a dataset of 8,602 GPR test points was evaluated. The ROC curve was constructed considering threshold values for the normalized amplitude of the reflected waves within the range of −11 dB to −27 dB. The AUC was found to be 0.67 when the GPR data were not depth-corrected. Depth-correcting the GPR data resulted in an AUC of 0.68, only slightly higher.

A total of 7,296,000 pixels were considered for the IRT analysis. Obviously, the number of test points available for assessing IRT is much greater than GPR, because the GPR data are only actually captured at 24 inch (0.6 m) intervals across the bridge width, and at each reinforcing steel bar along the length of the bridge. The IRT image captures the entire surface of the bridge. The area under the ROC curve was increased relative to the GPR, with an AUC of 0.80.

These data indicate that for a randomly selected test point, considering all threshold values, the likelihood of a correct decision (delaminated or not delaminated) for IRT scanning was 80%. For the depth-corrected GPR, the likelihood that a randomly selected test point was correctly decided was 68%.

A total of 7,533 GPR test points was considered for bridge A2112. Figure 5*b* shows the resulting ROC curve. The ROC curve was constructed considering threshold values for the normalized amplitude of the reflected waves within the range of −11 dB to −26 dB. The AUC was found to be 0.75. Depth-correcting the GPR data resulted in an AUC of 0.78. The IRT results for bridge A2112 were derived from 6,688,000 points (pixels) in the image. It was found that the AUC was 0.85.

It is notable in Figure 5 that for both bridge decks, the results from GPR typically lie to the right of IRT, indicating that the POFA is larger for GPR at any given threshold as compared with IRT. These data are consistent with the notion that GPR is detecting a corrosive environment which may later result is corrosion damage (i.e., delamination). It is notable that at one point in Figure 5*b*, when the POD is 0.9 and the POFA is ˜0.52, the performance of IRT and depth-corrected GPR were the same for POD and POFA.

## Conclusions

The objective of the research was to compare the reliability of GPR and IRT for the detection of areas of delamination in concrete bridge decks. This comparison was made by developing a high-resolution reliability analysis using ROC. The testing shows that IRT has a larger AUC value for both bridge decks, indicating that IRT has a higher reliability measure than GPR. These data indicate that overall, regardless of the chosen threshold value, IRT is more accurate for the detection of delamination than is GPR in almost all cases except one, as noted above.

The data indicate that GPR commonly has a higher POFA as compared with IRT. In practical terms, this means that GPR is overestimating the area of delamination, resulting in false calls (FPs). This may be consistent with the idea that GPR is detecting corrosive environments that extend beyond the limits of the existing delamination. This is an important consideration for implementing GPR for the purpose of PM&R decisions.

If the desired decision-making data are to identify the spatial location and total quantity of repairs or patches for *existing* damage, then IRT may be a better choice as compared with GPR. If the desired decision-making data relate to the overall corrosion damage or potential deterioration in the deck, for purposes such as prioritizing one bridge over another, then GPR may be a suitable choice. This carries the assumption that the deterioration in the deck is caused by corrosion.

The contributions of this research are (1) the development of an ROC methodology for assessing the reliability of NDE technologies for in-service bridge decks; and (2) the quantitative comparison of the performance of IRT and GPR for detecting areas of subsurface delamination in two case studies assessing RC bridge decks. The results of the analysis provide information from two cases studies for decision-makers selecting suitable condition assessment strategies for highway bridge decks.

## Author Contributions

The authors confirm contribution to the paper as follows: study conception and design: AAS 30%, GAW 70%; data collection: 70%, 30%; analysis and interpretation of results: AAS 50%, GAW 50%; draft manuscript preparation: AAS 50%, GAW 50%. All authors reviewed the results and approved the final manuscript.

## References

1. Sansalone, M. Impact-Echo: The Complete Story. ACI Structural Journal, Vol. 94, No. 6, 1997, pp. 777–786.
2. Carino, N. J. Impact Echo: The Fundamentals. *International Symposium on Non-Destructive Testing in Civil Engineering*, Berlin, Germany, 2015.
3. Sansalone, M. J., and W. B. Streett. *Impact Echo: Nondestructive Evaluation of Concrete and Masonry*. Bullbrier Press, Jersey Shore, PA, 1997.
4. *Standard Test Method for Evaluating Asphalt-Covered Concrete Bridge Decks Using Ground Penetrating Radar*. ASTM International, West Conshohocken, PA, 2015.
5. Maser, K. R. Integration of Ground Penetrating Radar and Infrared Thermography for Bridge Deck Condition Evaluation. *Proc., 7th International Symposium on Nondestructive Testing in Civil Engineering*, Nantes, France, 2009.
6. Senin, S. F., and R. Hamid. Ground Penetrating Radar Wave Attenuation Models for Estimation of Moisture and Chloride Content in Concrete Slab. *Construction and Building Materials*, Vol. 106, 2016, pp. 659–669 .
7. Wamweya, A. *Application of Ground Penetrating Radar (GPR) for Bridge Deck Condition Assessment using a 1.5 GHz Ground-Coupled Antenna*. Thesis. Geological Engineering, Missouri University of Science and Technology, 2009
8. Maser, K., N. Martino, J. Doughty, and R. Birken. Understanding and Detecting Bridge Deck Deterioration with Ground-Penetrating Radar. *Transportation Research Record: Journal of the Transportation Research Board*, 2012. 2313: 116–123.
9. Maser, K. R., and W. M. K. Roddis. Principles of Thermography and Radar for Bridge Deck Assessment. *Journal of Transportation Engineering*, Vol. 116, No. 5, 1990. pp. 583–601.
10. Maser, K., A. Carmichael, B. Johnson, P. Keranen, B. Holdhusen, and D. Green. Network Level Bridge Deck Condition Assessment in Minnesota. *Structural Materials Technology*, New York, NY, 2012.
11. Gucunski, N., S. Nazarian, A. Imani, and H. Azari. Performance of NDT Technologies in Detection and Characterization of Reinforced Concrete Deck Deterioration. *Proc., ASCE's 2014 Geo-Congress: Geo-Characterization and Modeling for Sustainability*, Atlanta, GA, 2014, pp. 2436–2449.
12. Martino, N., K. Maser, R. Birken, and M. Wang. Determining Ground Penetrating Radar Amplitude Thresholds for the Corrosion State of Reinforced Concrete Bridge Decks. *Journal of Environmental and Engineering Geophysics*, Vol. 19, No. 3, 2014, pp. 175–181.
13. Yehia, S., O. Abudayyeh, I. Abdel-Qader, and A. Zalt. Ground-Penetrating Radar, Chain Drag, and Ground Truth: Correlation of Bridge Deck Assessment Data. *Transportation Research Record: Journal of the Transportation Research Board*, 2008. 2044: 39–50.
14. Cheng, C.-C., T.-M. Cheng, and C.-H. Chiang. Defect Detection of Concrete Structures Using Both Infrared Thermography and Elastic Waves. *Automation in Construction*, Vol. 18, No. 1, 2008, pp. 87–92.
15. Kee, S.-H., T. Oh, J. S. Popovics, R. W. Arndt, and J. Zhu. Nondestructive Bridge Deck Testing with Air-Coupled Impact-Echo and Infrared Thermography. *Journal of Bridge Engineering*, Vol. 17, No. 6, 2012. pp. 928–939.
16. Oh, T., S.-H. Kee, R. W. Arndt, J. S. Popovics, and J. Zhu. Comparison of NDT Methods for Assessment of a Concrete Bridge Deck. *Journal of Engineering Mechanics*, Vol. 139, No. 3, 2013, pp. 305–314.
17. Khan, F., and I. Bartoli. Detection of Delamination in Concrete Slabs Combining Infrared Thermography and Impact Echo Techniques: A Comparative Experimental Study. *Presented at Structural Health Monitoring and Inspection of Advanced Materials, Aerospace, and Civil Infrastructure*, San Diego, CA, 2015. https://doi.org/10.1117/12.20840962015
18. Gucunski, N., and National Research Council. *Nondestructive Testing to Identify Concrete Bridge Deck Deterioration*. Transportation Research Board, Washington, D.C., 2013.
19. Barnes, C.L., and J.-F. Trottier. Effectiveness of Ground Penetrating Radar in Predicting Deck Repair Quantities. *Journal of Infrastructure Systems*, Vol. 10, No. 2, 2004. pp. 69–76.
20. Cardimona, S., B. Willeford, J. Wenzlick, and N. Anderson. *Bridge Deck Condition Studies in Missouri Utilizing Ground Penetrating Radar*. No. RDT01-012. Missouri Department of Transportation Research, Development and Technology, Jefferson City, MO, 2001.
21. Shin, H., and D. A. Grivas. How Accurate is Ground-Penetrating Radar for Bridge Deck Condition Assessment?

*Transportation Research Record, Journal of the Transportation Research Board*, 2003. 1845: 139–147.
22. Brown, C. D., and H. T. Davis. Receiver Operating Characteristics Curves and Related Decision Measures: A Tutorial. *Chemometrics and Intelligent Laboratory Systems*, Vol. 80, No. 1, 2006, pp. 24–38.
23. Nockemann, C., H. Heidt, and N. Thomsen. Reliability in NDT: ROC Study of Radiographic Weld Inspections. *NDT & E International*, Vol. 24, No. 5, 1991, pp. 235–245.
24. Martino, N., K. Maser, R. Birken, and M. Wang. Determining Ground Penetrating Radar Amplitude Thresholds for the Corrosion State of Reinforced Concrete Bridge Decks. *Journal of Environmental and Engineering Geophysics*, Vol. 19, No. 3, 2014, pp. 175–181.
25. Washer, G., J. Dawson, P. Ruiz-Fabian, A. Sultan, M. Trial, and P. A. Fuchs. *Field Testing of Hand-Held Infrared Thermography*. Phase II TPF-5(247) Final Report. Missouri Department of Transportation, Jefferson City, MO, 2016.
26. Hooks, J., and J. Weidner. *Long-Term Bridge Performance Program Protocols, Version 1*. Federal Highway Administration, McLean, VA, 2016.
27. Varnavina, A. V., A. K. Khamzin, E. V. Torgashov, L. H. Sneed, B. T. Goodwin, and N. L. Anderson. Data Acquisition and Processing Parameters for Concrete Bridge Deck Condition Assessment Using Ground-Coupled Ground Penetrating Radar: Some Considerations. *Journal of Applied Geophysics*, Vol. 114, 2015, pp. 123–133.
28. Barnes, C. L., J.-F. Trottier, and D. Forgeron. Improved Concrete Bridge Deck Evaluation Using GPR by Accounting for Signal Depth–Amplitude Effects. *NDT & E International*, Vol. 41, No. 6, 2008, pp. 427–433.
29. Candy, J., and E. Breitfeller. *Receiver Operating Characteristic (ROC) Curves: An Analysis Tool for Detection Performance*. Lawrence Livermore National Laboratory (LLNL), Livermore, CA, 2013.
30. Colak, E., F. Mutlu, C. Bal, S. Oner, K. Ozdamar, B. Gok, and Y. Cavusoglu. Comparison of Semiparametric, Parametric, and Nonparametric ROC Analysis for Continuous Diagnostic Tests Using a Simulation Study and Acute Coronary Syndrome Data. *Computational and Mathematical Methods in Medicine*, Vol. 2012, 2012, p. 7.

*The Standing Committee on Testing and Evaluation of Transportation Structures (AFF40) peer-reviewed this paper (18-04337).*

Article

TRR
JOURNAL OF THE TRANSPORTATION RESEARCH BOARD

Transportation Research Record
2018, Vol. 2672(41) 123–131

DOI: 10.1177/0361198118781145
journals.sagepub.com/home/trr
SAGE

# Stress Measurement in Seven-Wire Strands using Higher Order Guided Ultrasonic Wave Modes

**Brennan Dubuc[1], Arvin Ebrahimkhanlou[1], and Salvatore Salamone[1]**

**Abstract**

This paper investigates the use of higher order longitudinal guided modes for stress measurement within individual wires of a steel strand. The effect of stress on the phase velocity of higher order modes is studied using an approximate theory, which does not require the solution of dispersion curves. To validate the proposed approach, a prestressing bed was designed to apply a tensile load to a strand up to 25% ultimate tensile strength while recording guided wave signals. Guided waves were excited within individual wires of a strand, and the stress sensitivity of their phase velocity was used for stress measurement. Stress measurement was performed with higher order modes using the approximate theory with parameters for a steel of similar carbon content (Hecla 17), as well as with calibrated parameters. Using the Hecla 17 parameters, roughly 15% mismatch in stress was observed, whereas roughly 5% error was observed using calibrated parameters. Stress measurement was also performed using the fundamental mode, in order to compare the accuracy of higher order modes with the mode used previously in the literature. The greater stability of higher order modes across mode and frequency yielded significantly increased stress measurement accuracy, using both Hecla 17 and calibrated parameters.

Seven-wire strands act as critical load bearing elements in civil infrastructure, including cable-stayed bridges, prestressed girders, and other prestressing/post-tensioning applications. The level of load (and therefore stress) in a strand in any of these applications has significant importance for the safety of the structure. Knowledge of the load carried by a strand is therefore valuable information in any nondestructive testing or structural health monitoring effort. In the last 20 years, guided ultrasonic waves have emerged as a potential nondestructive technology for probing the level stress in strands (*1–3*). This has been motivated by their sensitivity to the stress distribution throughout a strand (and not merely on the surface), as well as their complementary nondestructive use for passive and active damage detection.

One of the first experimental studies for the effect of stress on guided waves in a strand was performed by Kwun et al. (*4*). Magnestrictive sensors were used in their study, which excite guided waves throughout the strand cross-section as a whole. Another early study was performed by Chen et al., in which the fundamental longitudinal guided mode L(0,1) was excited within individual wires of a strand (*5*). In addition, an approximate theory based on the unstressed dispersion equation of a rod was proposed to predict the effect of stress. Lanza di Scalea et al. studied guided modes excited within both individual wires and the whole strand, in addition to increasing the accuracy of the approximate theory proposed by Chen et al. (*6*). Chaki and Bourse studied the L(0,1) mode within individual wires of a strand, while using experimental calibrations to perform stress measurement on small-scale prestressed concrete specimens (*7*). More recently, Nucera and Lanza di Scalea correlated nonlinear guided wave behavior with the stress level in prestressed strands (*8*).

In studies on the fundamental L(0,1) mode, a nonlinear stress dependence has been found at lower stress levels ($<35\%$ ultimate tensile strength [UTS]), which diverges from the linearity predicted by rod theory (*6, 7*). The divergent and nonlinear behavior of this mode can therefore cause difficulty for stress measurement applications, especially at lower stress levels. This is important for quantifying prestress losses, which may reach lower stress levels, and are a major concern for structural

[1]Smart Structures Research Group, Department of Civil, Architectural and Environmental Engineering, University of Texas at Austin, Austin, TX

**Corresponding Author:**
Address correspondence to Brennan Dubuc: bdubuc@utexas.edu

performance. Furthermore, the approximate rod theory applied in the literature can yield errors in theoretical predictions of longitudinal modes.

To address these difficulties, the present paper studies higher order longitudinal modes L(0,$n$), $n > 5$ for stress measurement within individual wires. An approximate theory designed specifically for higher order modes was used, which yields an explicit equation for the effect of stress in a rod. In this paper, the term rod is used to refer to a solid circular cylinder with free boundaries, whereas wire is used to refer to the cylinders composing a strand. To validate both the theory and stress measurement, a custom-designed prestressing bed was used to record higher order and fundamental modes in a stressed strand. An approximation for the experimental measurement of stress using higher order modes was also used, which relies only on explicit equations.

This paper is organized as follows: first, the approximate theory for higher order modes in a stressed rod is presented. The guided wave experiments on a stressed strand are then described, as well as the approximation for experimental measurements. Theoretical and experimental results are then presented for both the higher order and fundamental modes, and the stress measurement accuracy for the two types of modes is demonstrated. Finally, concluding remarks and suggestions for future work are outlined.

## Theoretical Formulation

This section presents the approximate theory for higher order modes in an axially stressed rod. In addition, the stress state within individual wires of a seven-wire strand is presented, in which other stress components are shown to be negligible compared with the axial stress.

### *Approximate Theory for Stressed Rod*

The stress state of individual wires within a tensioned strand is relatively complex (*9*). In addition to axial stress, the peripheral wires are subjected to stresses from moment and twisting. Although these additional stresses exist, it has been shown that they are negligible with respect to the axial stress (*6*). Each wire within the seven-wire strand is therefore modeled under pure axial stress. Furthermore, the waveguide of an individual wire is modeled as a rod with free boundaries for higher order modes, as there are frequency regions where they have relatively little energy leakage into a bordering medium (*10*).

Concerning higher order modes, the waveguide under consideration is an isotropic elastic rod of diameter $d$ and density $\rho$ subjected to an axial stress $\sigma$. A schematic of

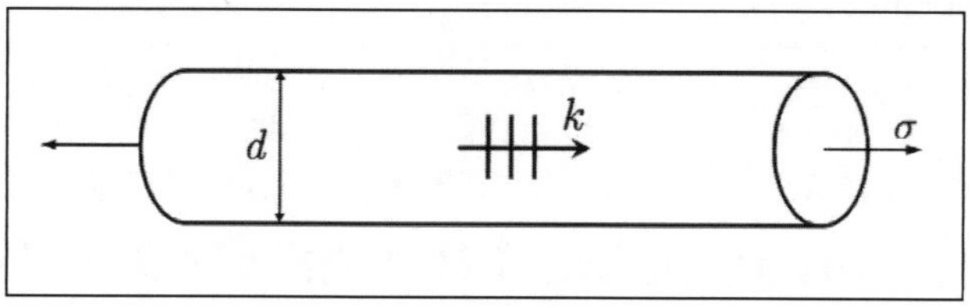


**Figure 1.** Schematic of isotropic elastic rod of diameter $d$ under axial stress $\sigma$, with wavenumber $k$ of longitudinal guided wave shown.

the stressed rod can be seen in Figure 1. For small deformations, the axial strain $e$ due to stress may be expressed

$$e = \frac{\lambda + \mu}{\mu(3\lambda + 2\mu)}\sigma \tag{1}$$

where $\lambda, \mu$ are the Lame constants of the rod.

The effect of stress on ultrasonic wave propagation (referred to as the acoustoelastic effect [*11*]) produces a change in phase velocity $\Delta c_p$ relative to an unstressed velocity $c_p^0$. From here onward, a superscript 0 is used to indicate that a velocity corresponds to the unstressed state. The stress-induced velocity change is due to both the deformation of the medium and its elastic behavior at varying stress levels (*12*). For relatively stiff materials, such as metals, the elastic behavior must be described by a stress–strain relation up to the second order in strain (*13*). This implies that the strain energy function of the medium must be expanded up to the third order. For an isotropic medium, the strain energy function may be expressed in terms of the Lame $\lambda, \mu$ and Murnaghan $l$, $m$, $n$ elastic constants (*14*),

$$U = \tfrac{1}{2}(\lambda + 2\mu)I_1^2 - 2\mu I_2 + \tfrac{1}{3}(l + 2m)I_1^3 - 2mI_1I_2 + nI_3 \tag{2}$$

where $I_1, I_2, I_3$ are the invariants of the Lagrangian strain tensor. The Murnaghan constants are referred to as third order elastic constants. Although a second order stress–strain relation must be assumed at the outset of the theory, the analysis for any waveguide is then linearized in strain on the assumption of small deformations (*13*). This linearization has been verified experimentally, allowing the use of the linear relation from Equation 1 to determine the induced strain (*13*, *15*).

Suhubi has solved for the dispersion equation of longitudinal modes in an axially stressed rod composed of an isotropic elastic medium with an arbitrary strain energy function (*16*). Longitudinal modes (both higher order and fundamental) may then be solved for by inserting the strain energy function of Equation 2 into this dispersion equation and retaining only first order terms in strain. However, the dispersion equation does not explicitly relate phase velocity $c_p$ to angular frequency $\omega$, and numerical methods must therefore be used to calculate the dispersion curves in $\omega$-$c_p$ space. As this is not a trivial task, an alternative to the rod theory is studied in this paper (for higher order modes in particular), based entirely on explicit equations.

The approximate theory of this paper relies on the fact that there are regions in the higher order mode dispersion curves where the unstressed phase velocity $c_p^0$ of each mode plateaus across frequency at the bulk longitudinal velocity, $c_l^0 = [(\lambda + 2\mu)/\rho]^{1/2}$ (*10, 17*). These are the same regions where the modes have relatively little energy leakage into a bordering medium, supporting the rod waveguide approximation for individual wires (*10*). Since the unstressed phase velocity tends to the bulk longitudinal velocity $c_l^0$, the effect of stress on the bulk velocity is used to approximate the effect of stress on higher order modes in the plateau region.

The phase velocity $c_l$ of a bulk longitudinal wave propagating along a principal stress in an isotropic medium may be expressed

$$c_l = c_l^0 + \kappa_t \sigma \tag{3}$$

where $\kappa_t$ is referred to as an acoustoelastic constant, governing the linear dependence between phase velocity and stress. A subscript $t$ is used to indicate that this is a theoretical acoustoelastic constant. The parameter $\kappa_t$ is solely determined by the properties of the medium, and may be expressed (*18*)

$$\kappa_t = \frac{(\lambda + 2l)\mu + (2\lambda + 6\mu + 4m)(\lambda + \mu)}{2(3\lambda + 2\mu)\mu[(\lambda + 2\mu)\rho]^{1/2}} \tag{4}$$

The bulk phase velocity change is then

$$\Delta c_l = \kappa_t \sigma \tag{5}$$

This may then be used as an explicit equation that approximates the phase velocity change of higher order modes,

$$\Delta c_p \big|_{\mathrm{p}} \approx \kappa_t \sigma \tag{6}$$

where the notation $|_{\mathrm{p}}$ indicates that it is evaluated at the plateau region.

## Experiments

This section describes the experiments conducted on a stressed strand. The tests are used to study higher order and fundamental mode propagation within individual wires, and to subsequently assess stress measurement accuracy using the two mode types.

### *Test Setup and Procedure*

A prestressing bed was designed to apply tensile axial load to a seven-wire strand, which can be seen in Figure 2. The frame was designed with the capability of installing ultrasonic transducers at a number of locations on

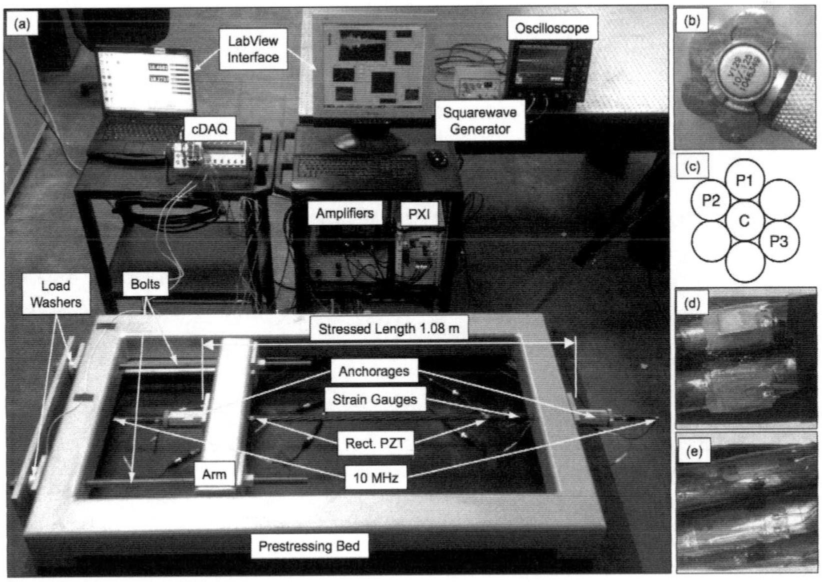


**Figure 2.** (*a*) Custom-designed prestressing bed experimental setup and data acquisition system, (*b*) 10 MHz transducers used to excite higher order modes, (*c*) Schematic of strand cross-section, with core (C) and peripheral (P1,P2,P3) wires indicated, (*d*) Rectangular piezoelectric lead zirconate titanate (PZTs) used to excite fundamental mode, and (e) Strain gauges installed on peripheral wires.

various wires in a strand. In the frame design, the ends of a strand are fixed with two anchorages between the base of the frame and a moveable central arm. Tension may then be applied by adjusting the arm's position using two bolts connecting the frame and the arm.

The tested strand had core and peripheral wire diameters of $d_c$ = 5.22 and $d_p$ = 5.08 mm, respectively, and a lay angle of $\beta = 7.9°$. The nominal diameter of the strand was 15.2 mm (0.6 in.). The length of the strand under stress (i.e., within the anchorage zone) was 1.08 m (3.54 ft), which was defined as the distance between the base of each anchorage. The total load was recorded with two load washers (Interface LW25100-20K) placed around the bolts. The axial strain within individual wires was recorded by placing a strain gauge (Micro Measurements C2A-06-250LW-350) on the surface of each of the six peripheral wires, oriented along the axis. Load washer and strain gauge readings were recorded with a National Instruments data acquisition system (DAQ), and monitored using a LabView software interface.

Before guided wave measurements were taken, the strand was first loaded and unloaded in order to develop stable contact with the anchorages. Unstressed reference waveforms were then recorded while the frame was unloaded. Load was then applied and held constant at incremental steps while guided wave measurements were taken. The maximum applied load was 98 kN ($\approx$22 kips), equal to roughly 25% of the UTS. Due to some variability in the strain readings between different peripheral wires, the values of the six strain gauges were averaged, resulting in a single peripheral wire axial strain $e_p$. As the strain in the core wire was not measured directly, the peripheral wire strain was converted to core wire strain (*9*),

$$e_c = e_p/\cos^2(\beta) \tag{7}$$

The axial stresses in the core and peripheral wires, $\sigma_c$ and $\sigma_p$, were then recorded by converting the axial strain to axial stress through an inversion of Equation 1,

$$\sigma_{p,c} = \frac{\mu(3\lambda + 2\mu)}{\lambda + \mu} e_{p,c} \tag{8}$$

Higher order longitudinal modes were excited within individual wires using a 10 MHz transducer (Olympus V129-RM) bonded to each end of the wire's cross-section with hot glue, as seen in Figure 2*b*. An Olympus 5077PR squarewave pulser/receiver was used to generate excitation waveforms for higher order modes. Waveforms were recorded on a LeCroy HDO6104 oscilloscope, sampled at 250 MHz, and averaged 500 times. The higher order mode signals excited by the squarewave pulser/receiver had a wide range of frequency content. Therefore, in order to study one particular mode at a time, the signals were processed through a zero-phase-shift Butterworth bandpass filter.

To avoid the effect of anchorages on the fundamental mode, it was excited with rectangular piezoelectric lead zirconate titanate (PZT) patches installed within the anchorage zone. The PZTs were bonded on the surface of the P1 and P2 wires using epoxy, and oriented along the wire axis, as seen in Figure 2*d*. The fundamental mode was excited at different frequencies using 5.5 cycle Hamming windowed tonebursts centered at 1.17 and 1.59 MHz-mm. Waveform excitation and reception was performed with a National Instruments PXI data acquisition system with arbitrary waveform generator (PXI-5412) and oscilloscope (PXIe-5105) cards. The waveforms were sampled at 20 MHz and averaged 250 times. No filtering was necessary for the fundamental mode, since it was excited with a narrowband toneburst.

### Phase Velocity Change Measurement

The phase velocity change at a given stress was computed from the time change $\Delta t$ for a point of constant phase within a waveform. Two examples of $\Delta t$ measurements may be seen in Figure 3 for a higher order and fundamental mode. The L(0,9) waveform is shown in its plateau region at 29.9 MHz-mm for two values of stress in Figure 3*a*, while the L(0,1) waveform is shown at 1.17 MHz-mm in Figure 3*b*.

As the time change was small with respect to the arrival time, it was converted to a phase velocity change using

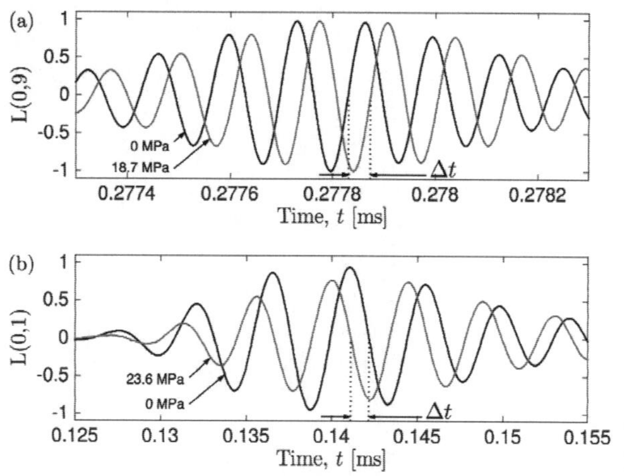


**Figure 3.** (*a*) L(0,9) and (*b*) L(0,1) waveform at two values of stress, showing time changes used for phase velocity change measurement. Waveform amplitudes each normalized against their peak value.

$$\Delta c_p = -[(c_p^0)^2/l]\Delta t \tag{9}$$

where $l$ is the length of propagation under stress. The unstressed phase velocity $c_p^0$ at a given frequency for the mode under consideration was therefore computed and used to convert the time change to phase velocity change. However, since the unstressed phase velocity for higher order modes in the plateau region asymptotes to the unstressed bulk velocity $c_l^0 = [(\lambda + 2\mu)/\rho]^{1/2}$, Equation 9 may be simplified into an explicit equation for measuring $\Delta c_p|_p$,

$$\Delta c_p|_p \approx -[(c_l^0)^2/l]\Delta t \tag{10}$$

## Results

In this section, analytical results for the effect of stress on guided wave phase velocity in a rod are first presented. These are based on both rod theory and the proposed approximate theory. Experimental results are then presented and compared with the theory for both the higher order and fundamental modes propagating within individual wires of the strand. Stress measurements using phase velocity changes are then made using theoretical and experimentally calibrated acoustoelastic constants.

### *Analytical Results*

Dispersion curves as a function of the frequency-thickness product $fd = \omega d/2\pi$ for the phase velocity change in a Hecla 17 steel rod are shown in Figure 4. The

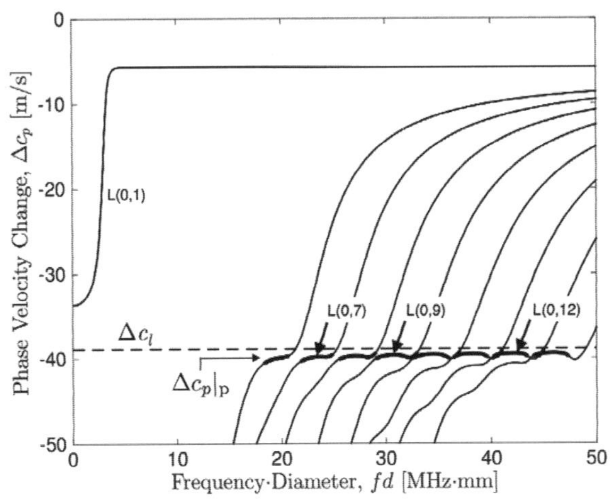


**Figure 4.** Phase velocity change for higher order and fundamental modes due to 400 MPa axial stress in Hecla 17 steel rod (solid line). $\Delta c_p|_p$ regions bolded, with their approximate theory from $\Delta c_l$ overlain (dashed line).

**Table 1.** Density (kg/m$^3$) and Elastic Constants (GPa) for Hecla 17 Steel

| $\rho$ | $\lambda$ | $\mu$ | $l$ | $m$ | $n$ |
|---|---|---|---|---|---|
| 7,825 | 110.5 | 82.0 | −328 | −595 | −668 |

curves correspond to the higher order and fundamental modes under $\sigma = 400$ MPa axial stress. The material properties of Hecla 17 steel are listed in Table 1 (*19*). This is selected as a representative steel due to the unavailability of third order elastic constants $l$, $m$, $n$ for the strand steel. Third order elastic constants have been shown to be sensitive to carbon content (*20*), and therefore Hecla 17 is selected due to its similar carbon content of 0.6%. The curves were computed from the rod theory of Suhubi, using the strain energy function from Equation 2 (*16*). The plateau regions $\Delta c_p|_p$ of higher order modes are indicated by bold lines, which are defined where the variation of $\Delta c_p$ is within $\pm 1\%$. The explicit solution from the approximate theory, based on the bulk velocity change $\Delta c_l$, is overlain as a dashed line, and is constant across frequency. It may be seen that the approximate theory describes the stable frequency and mode dependence across the plateau regions with roughly 2% error.

### *Higher Order Mode Experiments*

To validate the application of the approximate theory for higher order modes, the consistency of the phase velocity change $\Delta c_p|_p$ across different modes and frequencies is demonstrated in Figure 5. Here, the experimental velocity change in the P1 wire is plotted for the L(0,7) mode at 22.6 MHz-mm and for the L(0,12) mode at both 41.7 and 40.0 MHz-mm. Figure 5, *a* and *b* correspond to the L(0,7) and L(0,12) mode at the center of their respective plateau regions, whereas Figure 5*c* corresponds to L(0,12) at the low frequency limit of its plateau region. The plateau regions of these modes may be seen as the bold lines in Figure 4. The phase velocity change from the approximate theory in Equation 6 is overlain in each plot, using material properties for Hecla 17 steel.

The approximate theory predicts a constant phase velocity change across frequency, which is validated by the fact that the experimental velocity change is nearly identical between the three plots in Figure 5. Linear regression based on least squares was used to calculate an experimentally calibrated acoustoelastic constant $\kappa_e$ for each mode and frequency, which can be seen in the same figure. Since the constants for each mode and frequency were nearly identical, an average value was taken, yielding $\kappa_e = -0.085$ (m/s)/MPa. The slope of the theoretical $\Delta c_p|_p$ was found to be $\kappa_t = -0.097$ (m/s)/MPa. The slight mismatch between the theoretical and experimental

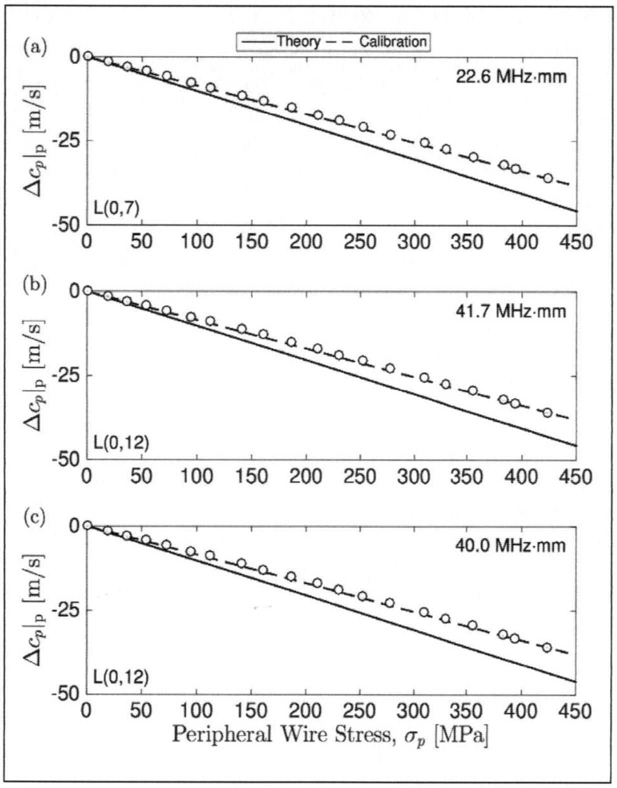


**Figure 5.** Experimental phase velocity change in P1 wire for various higher order modes and frequency-thicknesses in the plateau region. Linear regression for experimental points overlain as dashed line, and approximate theoretical change for a Hecla 17 rod overlain as solid line.

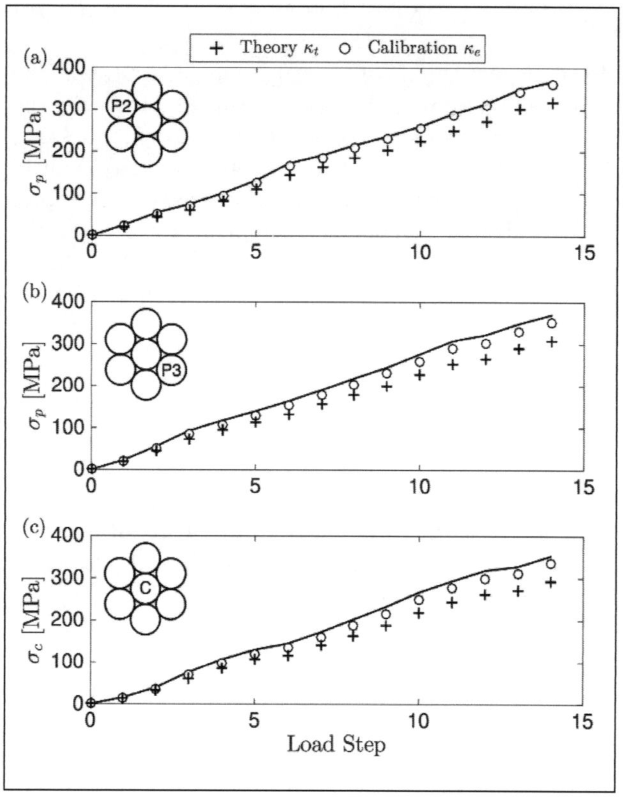


**Figure 6.** Stress measurement in various wires (P2,P3,C) using L(0,9) higher order mode at 29.9 MHz-mm. Measurements shown using theoretical and calibrated acoustoelastic constants, with calibration from P1 wire.

slopes ($\approx 15\%$) was expected due to the unavailability of third order elastic constants for the strand. It was found that the theoretical slope varied around 10% when testing the theory for different lower carbon content steels (0.2% to 0.4%, material properties obtained from [*19*]). This suggests that the Hecla 17 elastic constants are slightly different from those of the strand.

*Stress Measurement.* Stress measurement using data for the L(0,9) mode at 29.9 MHz-mm in wires P2, P3, and C is shown in Figure 6. The plateau region of this mode may be seen in Figure 4, centered at 29.9 MHz-mm. The stress recorded through strain gauge measurements from Equation 8 is plotted at each load step. For each wire, the measured phase velocity change in the plateau region $\Delta c_p|_\text{p}$ is used to solve the inverse problem of computing stress through Equation 6. Stress measurement is therefore performed according to

$$\sigma = \alpha \Delta c_p|_\text{p} \tag{11}$$

where $\alpha$ equals either $1/\kappa_t$ or $1/\kappa_e$, depending on whether the theoretical $\kappa_t$ or experimentally calibrated $\kappa_e$ acoustoelastic constant is used. Two sets of stress measurement data points are plotted in Figure 6 according to $\alpha = 1/\kappa_t$ (crosses) and $\alpha = 1/\kappa_e$ (circles).

Since $\kappa_t$ was not determined from the exact type of steel composing the strand, the stress measurement using this value has a slight degree of mismatch ($\approx 15\%$) compared with the recorded stress. The stress measurement using $\kappa_e$, however, estimates the recorded stress in each wire to within 5% error. It may be noted that $\kappa_e$ was calibrated from different modes and frequencies, as well as from a different wire, compared with those used for stress measurement. The theoretical stability across mode and frequency is therefore demonstrated by the consistent stress measurement in each wire. Furthermore, the accurate stress measurement in the core wire, using the peripheral wire calibration, suggests that contact with anchorages has a negligible influence on the phase velocity change.

## Fundamental Mode Experiments

The experimental relation between phase velocity change $\Delta c_p$ and stress for the fundamental mode L(0,1) is shown in Figure 7 at two *fd* values for the P1 wire. The

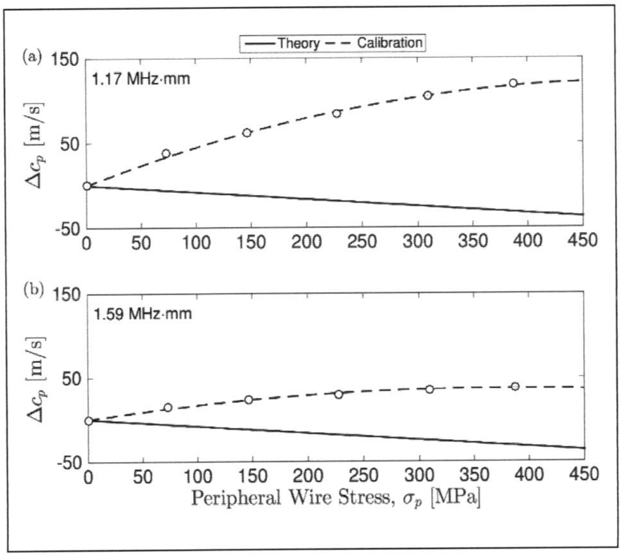


**Figure 7.** Experimental phase velocity change in P1 wire for fundamental mode at two frequency-thicknesses. Quadratic regression for experimental points overlain (dashed line), as well as approximate theoretical change for a Hecla 17 rod (solid line).

theoretical $\Delta c_p$ obtained from the rod theory, shown in Figure 4, is also overlain. The mode from rod theory has relatively little dispersion at these frequency-thicknesses. This results in the comparable phase velocity changes predicted by theory in the two plots.

In contrast to the theory, the experimental $\Delta c_p$ has the opposite sign, is nonlinear, and has strong frequency dependence. Due to nonlinearity, quadratic regression based on least squares was used to fit an equation between $\Delta c_p$ and stress at each *fd*,

$$\Delta c_p = \kappa_1 \sigma + \kappa_2 \sigma^2 \tag{12}$$

The values of $\kappa_1, \kappa_2$ found through least squares at each *fd* are shown in Table 2. Calibration curves using the quadratic fit are overlain in Figure 7, which provide a good description of the nonlinear behavior.

The phase velocity change as a function of stress has been shown to be nonlinear in the literature (*6*, *7*). The nonlinearity was shown to be strongest at lower stresses ($<35\%$ UTS), which is within the range of stress studied here. In the frequency range studied in this paper, the fundamental mode has been shown to have a high degree of energy leakage between adjacent wires (*21*). This weakens the idealization of modeling each wire as a rod waveguide with free boundaries, resulting in the significant discrepancy observed between the experimental points and the rod theory in Figure 7.

**Table 2.** Calibration Parameters ([m/s]/MPa) for Fundamental Mode at Each Frequency-Thickness (MHz-mm)

| *fd* = 1.17 | *fd* = 1.59 |
|---|---|
| $\kappa_1 = 4.93 \times 10^{-1}$<br>$\kappa_2 = -4.95 \times 10^{-4}$ | $\kappa_1 = 1.93 \times 10^{-1}$<br>$\kappa_2 = -2.50 \times 10^{-4}$ |

*Stress Measurement.* Stress measurement was performed in the P2 wire using the experimental calibration from the P1 wire, which can be seen in Figure 8. Due to the significant divergence from rod theory, only stress measurement predictions based on the calibration from Equation 12 are shown in this figure. The calibrated quadratic fit is used to solve the inverse problem, where phase velocity change is converted to stress,

$$\sigma = \frac{(\kappa_1^2 + 4\kappa_2 \Delta c_p)^{1/2} - \kappa_1}{2\kappa_2} \tag{13}$$

Compared with the stress measurement results using higher order modes, there are significantly higher errors when using the fundamental mode. The strong frequency dependence of the phase velocity change (see Figure 7) results in less reliable conversions to stress, and errors up to 15% at 1.17 MHz-mm and 40% at 1.59 MHz-mm were found. In addition to its relatively strong frequency dependence, it has been noted that the experimental $\Delta c_p$ is not a monotonic function of stress (*6*, *7*). This is typically observed beyond 25% UTS, where $\Delta c_p$ reaches a maximum and then decreases, gradually becoming negative. This yields problematic ambiguities when solving the inverse problem of converting phase velocity change to stress, since multiple stress values result in the same

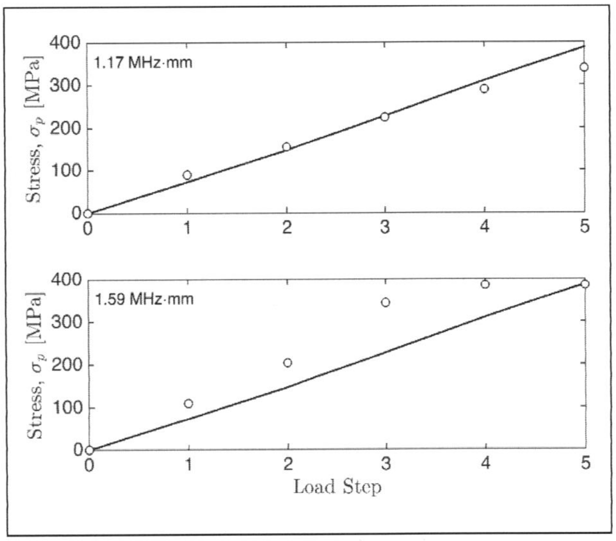


**Figure 8.** Stress measurement in P2 wire at two frequency-thicknesses using calibrated fundamental mode acoustoelastic constants $\kappa_1$, $\kappa_2$ from P1 wire.

$\Delta c_p$ value. This is in contrast to the higher order modes, in which the conversion from phase velocity change to stress is unique.

## Conclusion

The effect of stress on the phase velocity of higher order modes propagating within individual wires of seven-wire strands was studied for the purpose of stress measurement. The fundamental mode was also studied, which allowed for the accuracy of stress measurement to be put into context.

An approximate theory based on acoustoelastic bulk waves was used to predict the effect of stress on higher order modes in a stressed rod through an explicit equation. The approximate theory described the rod theory for a Hecla 17 steel rod to within 2% error, without the need to solve for dispersion curves. Stress measurement was performed using a custom-designed experimental setup for both loading a seven-wire strand and collecting guided wave signals. With higher order modes, stress measurements for various wires were accurate to within 15% using the approximate theory (with material properties for a steel of similar carbon content), and to within 5% by calibrating the theory. The modes, frequencies, and wire used for calibration were different from those used for stress measurement, demonstrating the stability of the higher order mode stress dependence. With the fundamental mode, measurements were accurate to within 40% based on calibration alone. The significant divergence of this mode from the rod theory made calibration the only option for stress measurement. A summary of the benefits of higher order modes for stress measurement includes: 1) ability to use approximate theory; 2) greater stability across mode and frequency; 3) unique conversion from phase velocity to stress measurement; and 4) negligible interaction with anchorages.

Since the required third order elastic constants for the strand steel were not available in the literature, those for a steel with similar carbon content (Hecla 17) were used for theoretical predictions. The slight mismatch ($\approx 15\%$) between the theoretical and experimental phase velocity of higher order modes suggests that the constants of the strand were slightly different from those for Hecla 17 steel. Although a steel with similar carbon content may be used as a representative material, future work should determine the elastic constants of strand steel.

In addition, the excitation of higher order modes in this paper was performed with access to the cross-section of the wires. However, this type of excitation may not be appropriate in some settings. Therefore, future work should also investigate other potential forms of excitation. These may include sensors based on magnetic induction or fiber optics, for instance, which may be able to be instrumented along the length of a strand. Although higher order modes have the least attenuation among all modes in embedded strands (*10*), sensing strategies should consider the issue of attenuation from sources like grout and bends. These results would also benefit from future work on measuring stress change due to real damage processes, such as corrosion, which may also increase attenuation.

### Acknowledgments

This work was supported by the Texas Department of Transportation (TxDOT), project #16-337.

### Author Contributions

The authors confirm contribution to the paper as follows: study conception and design: BD, AE, SS; data collection: BD, AE; analysis and interpretation of results: BD, AE, SS; draft manuscript preparation: BD, AE, SS. All authors reviewed the results and approved the final version of the manuscript.

### References

1. Salamone, S., I. Bartoli, R. Phillips, C. Nucera, and F. Lanza di Scalea. Health Monitoring of Prestressing Tendons in Post-Tensioned Concrete Bridges. *Transportation Research Record: Journal of the Transportation Research Board*, 2011. 2220: 21–27.
2. Bartoli, I., S. Salamone, R. Phillips, and F. Lanza di Scalea. Use of Interwire Ultrasonic Leakage to Quantify Loss of Prestress in Multiwire Tendons. *Journal of Engineering Mechanics*, Vol. 137, No. 5, 2011, pp. 324–334.
3. Bartoli, I., G. Castellazzi, A. Marzani, and S. Salamone. Prediction of Stress Waves Propagation in Progressively Loaded Seven Wire Strands. In M. Tomizuka, C. Yun, J. Lynch (Eds.), *Sensors and Smart Structures Technologies for Civil, Mechanical, and Aerospace Systems 2012*, Vol. 8345, International Society for Optics and Photonics, San Diego, CA, 2012, p. 834505.
4. Kwun, H., K. A. Bartels, and J. J. Hanley. Effects of Tensile Loading on the Properties of Elastic-Wave Propagation in a Strand. *The Journal of the Acoustical Society of America*, Vol. 103, No. 6, 1998, pp. 3370–3375.
5. Chen, H.-L., Y. He, and H. V. GangaRao. Measurement of Prestress Force in the Rods of Stressed Timber Bridges Using Stress Waves. *Materials Evaluation*, Vol. 56, No. 8, 1998, pp. 977–981.
6. Lanza di Scalea, F., P. Rizzo, and F. Seible. Stress Measurement and Defect Detection in Steel Strands by Guided Stress Waves. *Journal of Materials in Civil Engineering*, Vol. 15, No. 3, 2003, pp. 219–227.
7. Chaki, S., and G. Bourse. Guided Ultrasonic Waves for Non-Destructive Monitoring of the Stress Levels in Prestressed Steel Strands. *Ultrasonics*, Vol. 49, No. 2, 2009, pp. 162–171.
8. Nucera, C., and F. Lanza di Scalea. Monitoring Load Levels in Multi-Wire Strands by Nonlinear Ultrasonic Waves.

*Structural Health Monitoring*, Vol. 10, No. 6, 2011, pp. 617–629.
9. Machida, S., and A. J. Durelli. Response of a Strand to Axial and Torsional Displacements. *Journal of Mechanical Engineering Science*, Vol. 15, No. 4, 1973, pp. 241–251.
10. Pavlakovic, B. N., M. J. S. Lowe, and P. Cawley. High-Frequency Low-Loss Ultrasonic Modes in Imbedded Bars. *Journal of Applied Mechanics*, Vol. 68, No. 1, 2001, pp. 67–75.
11. Toupin, R. A., and B. Bernstein. Sound Waves in Deformed Perfectly Elastic Materials. Acoustoelastic Effect. *The Journal of the Acoustical Society of America*, Vol. 33, No. 2, 1961, pp. 216–225.
12. Thurston, R. N., and K. Brugger. Third-Order Elastic Constants and the Velocity of Small Amplitude Elastic Waves in Homogeneously Stressed Media. *Physical Review*, Vol. 133, No. 6A, 1964, pp. A1604–A1610.
13. Hughes, D. S., and J. L. Kelly. Second-Order Elastic Deformation of Solids. *Physical Review*, Vol. 92, No. 5, 1953, pp. 1145–1149.
14. Murnaghan, F. D. *Finite Deformation of an Elastic Solid.* John Wiley & Sons, New York, 1951.
15. Egle, D. M., and D. E. Bray. Measurement of Acoustoelastic and Third-Order Elastic Constants for Rail Steel. *The Journal of the Acoustical Society of America*, Vol. 60, No. 3, 1976, pp. 741–744.
16. Suhubi, E. S. Small Longitudinal Vibration of an Initially Stretched Circular Cylinder. *International Journal of Engineering Science*, Vol. 2, No. 5, 1965, pp. 509–517.
17. Meeker, T. R., and A. H. Meitzler. Guided Wave Propagation in Elongated Cylinders and Plates. In W. Mason, R. Thurston (Eds.), *Physical Acoustics*, Academic Press, New York, 1964, pp. 111–167.
18. Pao, Y.-H., and U. Gamer. Acoustoelastic Waves in Orthotropic Media. *The Journal of the Acoustical Society of America*, Vol. 77, No. 3, 1985, pp. 806–812.
19. Smith, R. T., R. Stern, and R. W. B. Stephens. Third-Order Elastic Moduli of Polycrystalline Metals from Ultrasonic Velocity Measurements. *The Journal of the Acoustical Society of America*, Vol. 40, No. 5, 1966, pp. 1002–1008.
20. Heyman, J. S., S. G. Allison, and K. Salama. Influence of Carbon Content on Higher-Order Ultrasonic Properties in Steels. In *Ultrasonics Sumposium*, Atlanta, Ga., IEEE, New York, 1983, pp. 991–994.
21. Treyssede, F. Investigation of the Interwire Energy Transfer of Elastic Guided Waves Inside Prestressed Cables. *The Journal of the Acoustical Society of America*, Vol. 140, No. 1, 2016, pp. 498–509.

*The Standing Committee on Testing and Evaluation of Transportation Structures (AFF40) peer-reviewed this paper (18-00817).*

*Article*

TRR
JOURNAL OF THE TRANSPORTATION RESEARCH BOARD

Transportation Research Record
2018, Vol. 2672(41) 132–144


DOI: 10.1177/0361198118773920
journals.sagepub.com/home/trr
SAGE

# Nondestructive Testing of Steel Corrosion in Prestressed Concrete Structures using the Magnetic Flux Leakage System

**Sadegh Shams[1], Al Ghorbanpoor[2], Shibin Lin[3], and Hoda Azari[4]**

**Abstract**

The Magnetic Flux Leakage (MFL) method can be nondestructively used to disclose the location and extent of corrosion or fracture in prestressed strands in concrete structures. In this study, parameters with the greatest effect on the performance of the MFL system are investigated using numerical simulations. The MFL system under study is based on two permanent magnets to magnetize embedded strands and Hall-effect sensors to detect normal magnetic flux leakage. The system is assessed using magnetostatic and transient numerical analysis to effectively simulate the MFL system. Results have been verified by laboratory and field experiments. Both normal and axial Hall-effect sensors are modeled in simulations to better identify magnetic signals at the corrosion zone. The sensor lift-off and the magnetic field masking by lateral reinforcements on nearby pitting corrosion are addressed as two main drawbacks of using MFL systems to detect corrosion in prestressed concrete structures. To provide more details about the flux leakage interference between the pitting corrosion and lateral reinforcements in prestressed concrete structures, linear and oriented magnets/sensor arrays are proposed and analyzed numerically.

Magnetic flux leakage (MFL) systems are effective nondestructive evaluation means used in the transportation and energy industries. Kusenberger et al. proposed the MFL system for assessing the steel components of prestressed concrete structures (*1*, *2*). The basic idea of MFL inspection is that the magnetic field emitted in a metallic object is distorted near the cross-section loss. The magnetic field distortion can be significant and detectable when the ferromagnetic materials are locally magnetized to full or near saturation. The amplitude of a flux leakage signal is generally proportional to the magnetization level and increases at cross section losses. The magnetic sources can be either permanent or electro magnets. The magnets are moved parallel to ferromagnetic strands to detect any leakage emerging from cross-sectional loss. As concrete is essentially a non-magnetic material with a relative magnetic permeability of unity, it has negligible influence on the magnetic measurements. However, the application of MFL systems in prestressed concrete structures has not been well developed because of inaccessible steel reinforcements and operations on heterogeneous concrete structures. The MFL results can be significantly affected by the distance between the magnetizer/sensors and the object of interest known as sensor lift-off. Also, signals originating in desired objects can be interfered with by signals from nearby secondary ferromagnetic objects. In the current MFL system, two permanent magnets are used as the magnetizer to subject a prestressed concrete structure to the magnetic field (*3*, *4*). The magnets/sensors assembly travels linearly along the length of the structure while maintaining a constant distance from the concrete surface. When the magnetic field reaches the corroded

[1]Engineering Software Consultants Inc., Turner-Fairbank Highway Research Center, McLean, VA
[2]College of Engineering & Applied Science, University of Wisconsin-Milwaukee, Milwaukee, WI
[3]Center for Advanced Infrastructure and Transportation, Rutgers University, Piscataway, NJ
[4]Turner-Fairbank Highway Research Center, Federal Highway Administration, McLean, VA

**Corresponding Author:**
Address correspondence to Shibin Lin: shibinlin@gmail.com

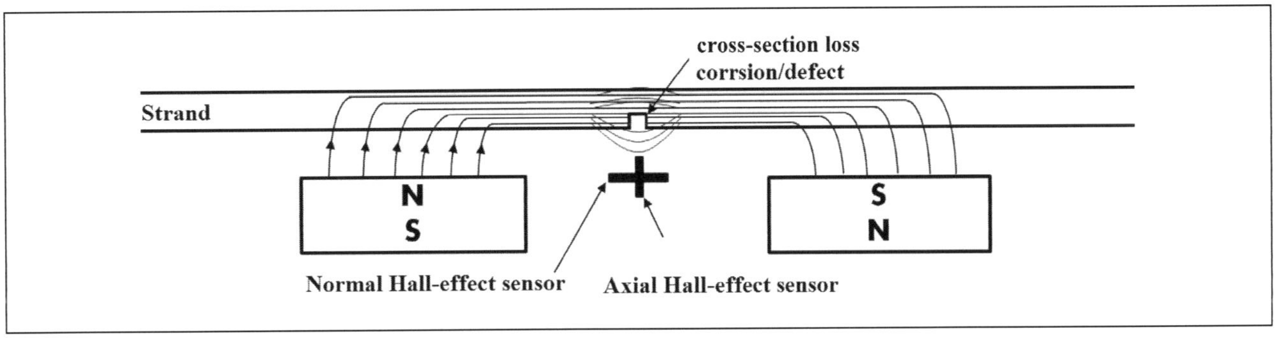


**Figure 1.** Schematic view of the magnetic field in a defective strand.

section of the strand (Figure 1), the magnetic field can be distorted, or it can be stated that corrosion creates flux leakage. Figure 1 shows the schematic view of the MFL system used in this study.

In this application of the MFL system, the direction of the applied magnetic field is aligned with the longitudinal strands embedded in concrete. Hall-effect sensors are attached to a moving magnetizer that passes over the defects to measure the flux leakage field generated by defects. Hall-effect sensors are mounted axially or normally with respect to the direction of the magnetic field, producing a voltage output that changes proportionally with the magnetic field. The magnetic field can be influenced by several parameters, including the distance between magnets/sensors and the strand (sensor lift-off), the size and the number of strands, secondary ferromagnetic elements in the structure, and size of defects. These parameters are addressed in this paper using numerical analysis of the MFL system. The magnetic field masking by lateral reinforcements on the nearby pitting corrosion is demonstrated to make the corrosion detection difficult. An oriented magnet/sensor array is proposed to overcome this drawback. Among available modelling approaches, finite element analysis has been successfully used in the simulation of the MFL system.

## Experimental Evaluation of the MFL System

The Federal Highway Administration (FHWA) sponsored a study that was performed at the University of Wisconsin-Milwaukee (UWM) to develop a more effective and field worthy magnetic-based NDE system for the detection of steel corrosion in prestressed concrete I-girder bridges (*4*). Figure 2*a* shows the developed MFL system operating on prestressed concrete I-girders. The existing MFL system consists of two strong permanent magnets and ten Hall-effect sensors in two layers. The sensors are positioned in an 8-inch-wide area at the mid distance between the two magnets. As the MFL system travels along a concrete structure, an encoder device records the traveled distance and recorded data is synchronized with its respective position. During each test scan, the data acquisition system records flux leakage in the normal direction from all sensors. Test results are shown in the form of graphs of the normalized voltage amplitude and the travel distance. Figure 2, *b–d* show results from field tests on a prestressed concrete bridge box girder using the MFL system. The tests were conducted as a part of a study led by the Lehigh University's ATLSS Research Center (*5*). Figure 2*b* shows the signal trend when no corrosion existed in strands, indicating peaks related to the stirrups.

Figure 2, *c* and *d* show the effect of different defect sizes on the output signals of Hall-effect sensors. Note the peak signals of the stirrups are disturbed when corrosions are nearby. Results were verified by removing the concrete and inspecting the strands. The MFL system has demonstrated its effectiveness during several nondestructive evaluation tests in laboratories and field tests of prestressed concrete girders and slabs (*6*). The system can disclose both cross-sectional loss and corrosion zones in prestressed strands.

## Simulation of the MFL System using Finite Element Method

The simulation of the MFL system can be used to optimize the shape, orientation, and size of the magnets, system layouts, and field sensor selection. ANSYS-Maxwell is used to simulate magnetic field distributions from the MFL system (*7*). Field variations are investigated under different influencing parameters, defect size, and strand layouts. Using the Maxwell-3D module, the magnetostatic Maxwell's equation of the MFL system is solved for a finite region of space with appropriate boundary conditions at each time step. The equation is then integrated to prepare a transient solution. Initial conditions of the system, such as system velocity, sensor definition, and eddy current effects are specified to obtain an

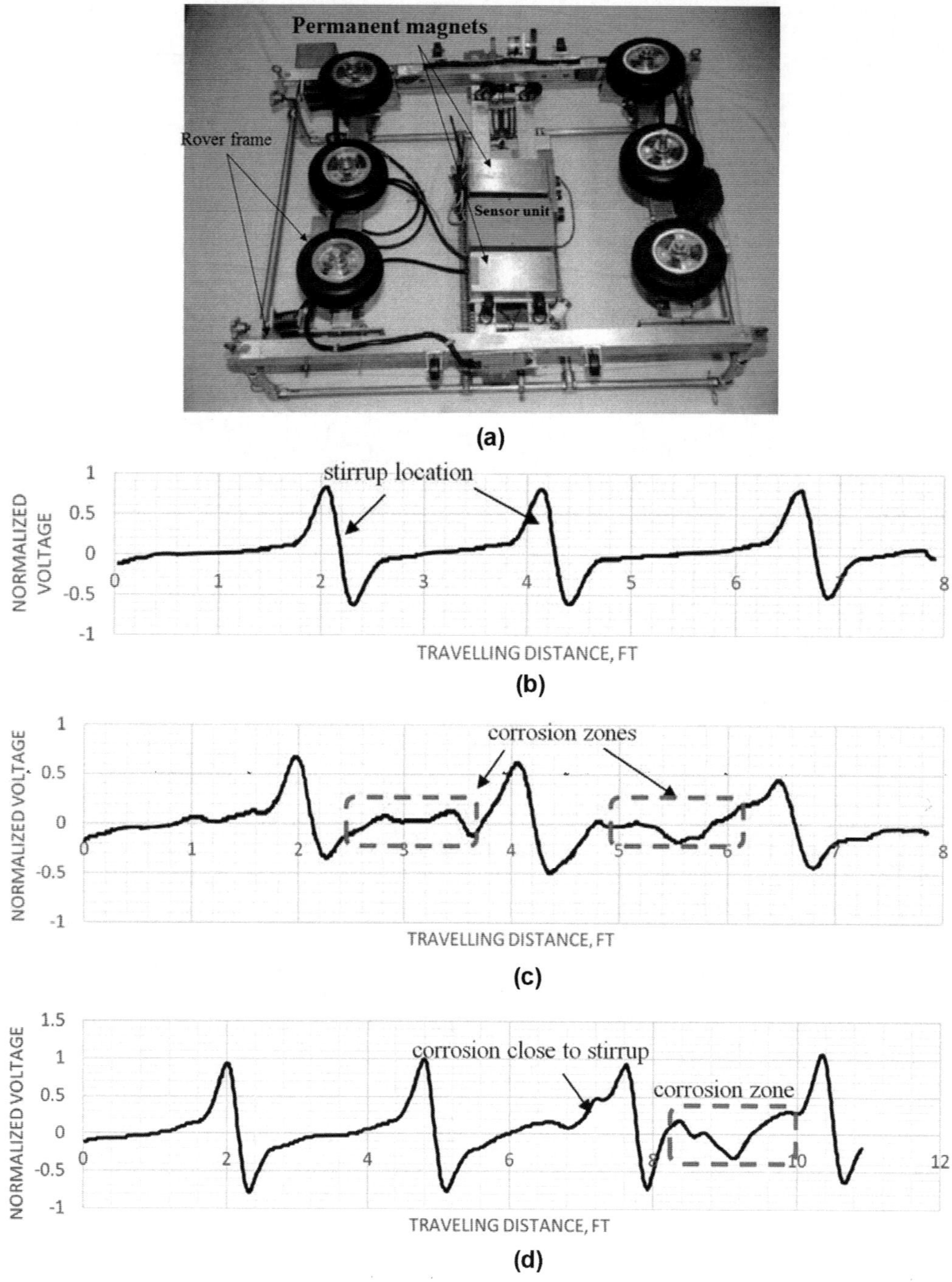


**Figure 2.** The MFL system assembly and field outputs, (*a*) the system prototype, (*b*) showing no leakage except at stirrup locations, (*c*) showing leakage at corrosion zones far from stirrup locations and (*d*) showing leakage at corrosion zones at vicinity of stirrup locations.

accurate magnetic field distribution. The transient analysis of the system is very sensitive to mesh, which affects both the numerical stability and computational efficiency of the solution. The Maxwell's equation for magnetostatics, given for the flux density B and the net field H, is

$$\nabla . B = 0, \nabla \times H = 0 \tag{1}$$

The magnetic scalar potential $\psi$ is $H = -\nabla\psi$. The governing equation is written as the Poisson's equation:

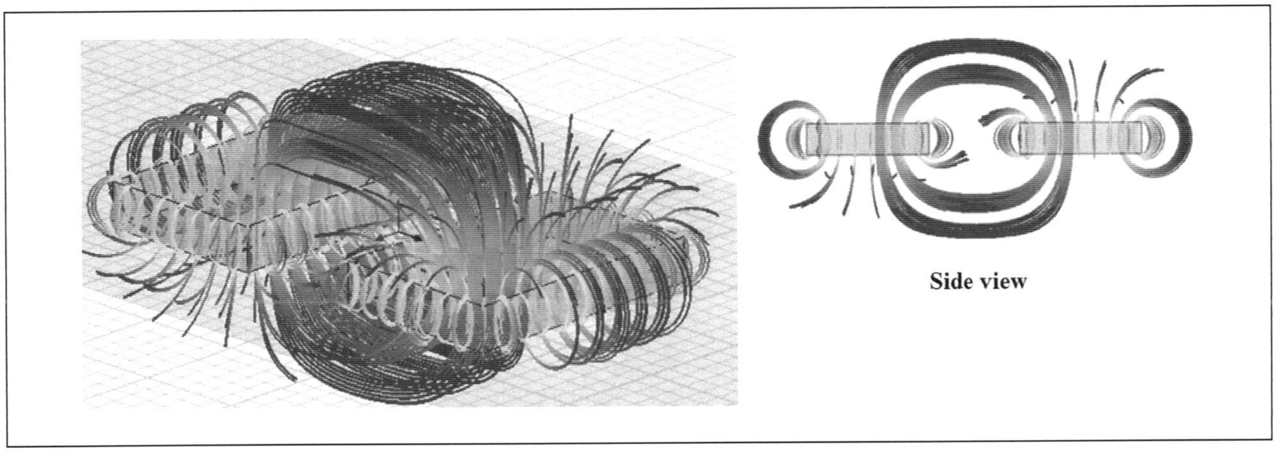


**Figure 3.** Distribution of the magnetic field for the designed permanent magnet system configuration.

$$\nabla.\mu\nabla\psi = -\nabla.B_{rem} \quad (2)$$

where $B_{rem}$ is the remnant magnetization of the material and $\mu$ is the material's relative magnetic permeability constant. To obtain the relevant set of algebraic equations, the geometry of the problem is discretized using an adaptive meshing procedure. This results in tetrahedron 3D elements for the problem with less than 0.5% error energy to convergence. Solids and air are automatically meshed by the software. Several numerical modules, such as magnetostatic and transient analyses involving eddy currents effects, are integrated to accurately simulate the system's behavior. Before the transient simulation, an optimization analysis was performed to minimize the magnet size while maintaining the magnetic saturation of the object of interest. Among many properties of permanent magnets that are taken into consideration, the magnetization curve has the most influential effect on the MFL applications. The magnetization power of a permanent magnet can be found from its B–H hysteresis curve. For this study, the initial design started with a magnetostatic analysis of two 12 in × 8 in × 2 in magnets made of neodymium grade NdFe35 material. The magnets' properties included $\mu = 1.099$, coercivity $H_c = -890000A/m$, and remanence $B_r = 1.23$ T. The magnetic flow resulting from the two-magnet configuration is shown in Figure 3. To examine the field circulation emitted in the air, the depiction in the figure does not account for the any nearby ferromagnetic object.

As is shown in Figure 3, the field circulates well with uniform flux lines between the two magnets. To determine the effect of the magnets geometry and layout on the resulting flux density, magnetostatic simulations were parametrically performed on magnets with a range of widths, thicknesses, and layouts. The magnetic flux density can be measured around the magnets by integrating the average magnitude of the flux density over a virtual surface. Flux density values were measured at locations changing from the surface of the magnets to 3 inches above the surface. The results of the parametric study are illustrated in Figure 4 where, for example, M8L6t2 is referred to two magnets with 6 inches distance to each other and each has 8 inches width and 2 inches thickness.

As shown in Figure 4*a*, it was found that a pair of 4-in-wide magnets produced flux density values of up to about 20% lower than that by the 8-in-wide magnets, depending on the height from the surface of the point where the measurement was made. Also worthy of note is that the average magnetic flux density at the magnets' surface is slightly higher for the 4-in-wide magnets because of the presence of the focused magnetic field. Figure 4*b* shows results for a set of 4-in-wide magnets with thickness values of 1.0 in, 1.5 in, and 2.0 in. It is shown that a changing magnet thickness can influence the resulting flux density more drastically than can the width of the magnets. Figure 4*c* illustrates the effect of changing the clear distance between the magnets from 5 in to 7 in. The effect of changing the clear distance between the magnets is not significant within the range used in the study. A conclusion from Figure 4 is that, within assessed configuration limits, the magnetic flux density can be influenced more by the thickness of the magnet than the magnet width and the distance between the magnets. In a subsequent analysis, a prestressed strand was located 2.5 in above the top surface of the magnets (Figure 5). The initial configuration considers a strand with a diameter of 0.5 in and two 8 in × 4 in × 2 in magnets. A cross-sectional loss of 5%, representing severe pitted corrosion, is modeled in the strand. The magnetic field distribution in Figure 5 permits investigation of any disturbance in the flux lines at the vicinity of the defect.

As can be seen in Figure 5, the strand at 2.29 T is highly saturated, and consequently the flux lines are

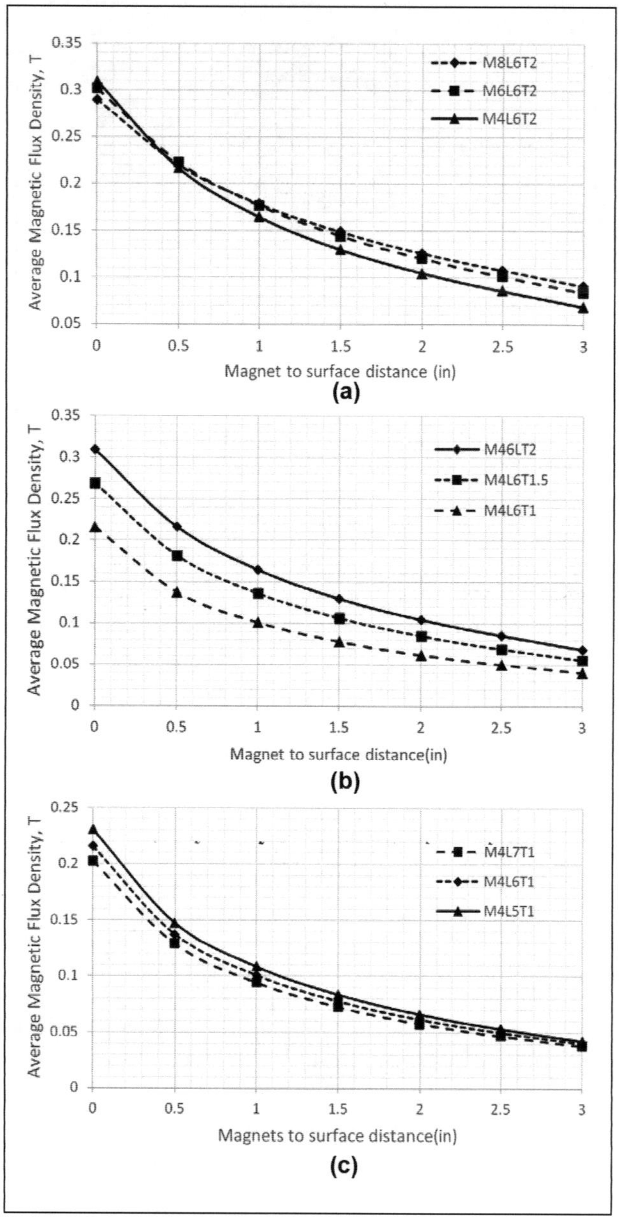


**Figure 4.** (*a*) Average magnetic flux density in surfaces changed from $0 \leq Z \leq 3$in, (*b*) average magnetic flux density of magnet with different thicknesses, and (*c*) average magnetic flux density for magnet while distance between magnets are changed from 5 in to 7 in.

disturbed at the vicinity of the defect. The magnet's overall dimensions are later reduced to 8 in × 4 in × 1.5 in to reduce the system's overall weight. With the new dimensions, the magnetic flux density in the strand was decreased to 2.037 T, which is about 10% less than the saturation level achieved with the larger magnets. As the strand is still saturated, this reduction is not significant to mask the magnetic flux leakage at the defect zone. In developing transient magnetic fields, time-varying and moving permanent magnets are combined with linear materials. Then the induced fields are associated with the motion in discrete time steps. The mechanical transient approach is based on defining velocity and the system's mass. A vacuum band/space is defined along the path of the magnets and a low constant speed of 5 in/sec was used for the motion of the magnets within the vacuum band. Figure 6 shows the geometrical MFL setup for the system.

As shown in Figure 1, virtual surfaces were modeled to represent transverse and axial Hall sensors at mid-distance between the magnets. Figure 7 shows the magnetic flux density changes during the movement of the magnets along the strand. The eddy current effect was considered in the transient analysis. The magnetic flux density remains approximately zero, except at the vicinity of the cross-sectional loss (consistent with magnetic theory). The normal magnetic flux density fluctuates the most at the edges of the defect. It is nearly zero at the center of the defect (Figure 7*a*). The magnetic signal for the axial flux density sensor is shown in Figure 7*b*. The signal offset at the beginning and at the end of the plot in Figure 7*a* is related to the surface magnetization and eddy current effects.

Note that the duration of signal peaks are not matched with the defect width, however the magnetic flux leakage peak duration is proportional to the distance of the defect to the sensor (the sensor lift-off), which is 2.5 in in this case. The signal trends have been verified by experiments (Figure 2) and the theory explaining the normal magnetic flux density (NMFD) leakage is expected to be diminished at the center of the defect zone and to be maximized there through the axial magnetic flux density (AMFD) leakage measurement. The magnitudes of the leakage across the defect length are 3.34 G in the normal and 3.0 G in the axial directions for the 5% cross-sectional loss in the strand.

## Investigation of Impacting Parameters

Several analyses were performed by applying changes to the overall system configuration and the analysis process. These analyses include defect size effects and the effect of multiple strands associated with lateral reinforcements (stirrups).

### *Defect Size Effect*

The effect of width and depth of the defect were investigated. The model has five different percentages of cross section loss by changing the depth of defect (as sketched in Figure 5) with a constant width of 0.2 in. Figure 8 illustrates the results of normal (Figure 8*a*) and axial

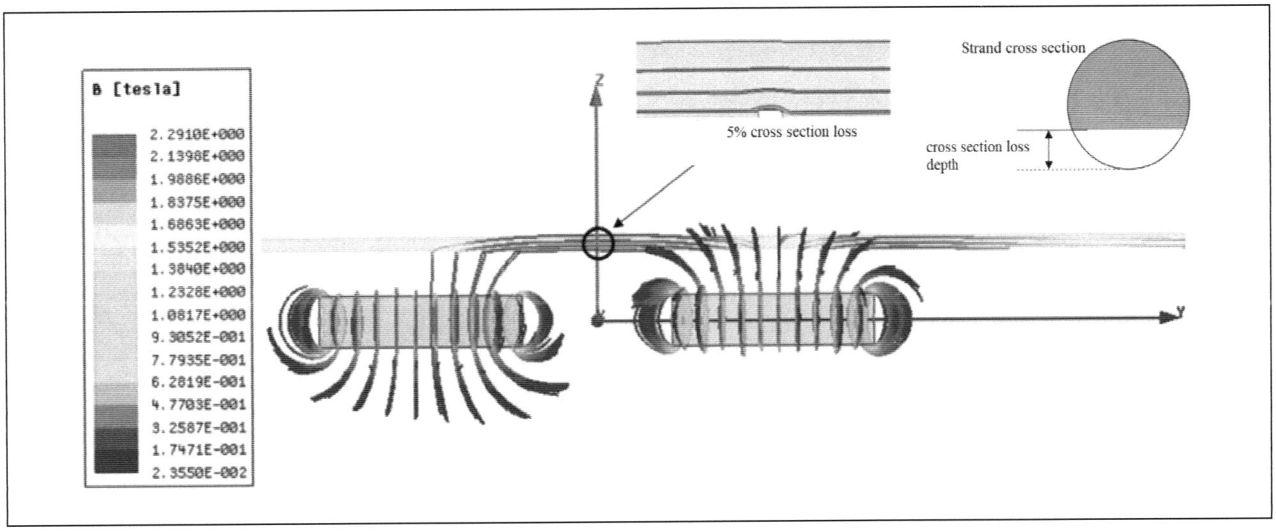


**Figure 5.** The distribution of the magnetic field and consequent disturbance of the flux lines at the defect zone.

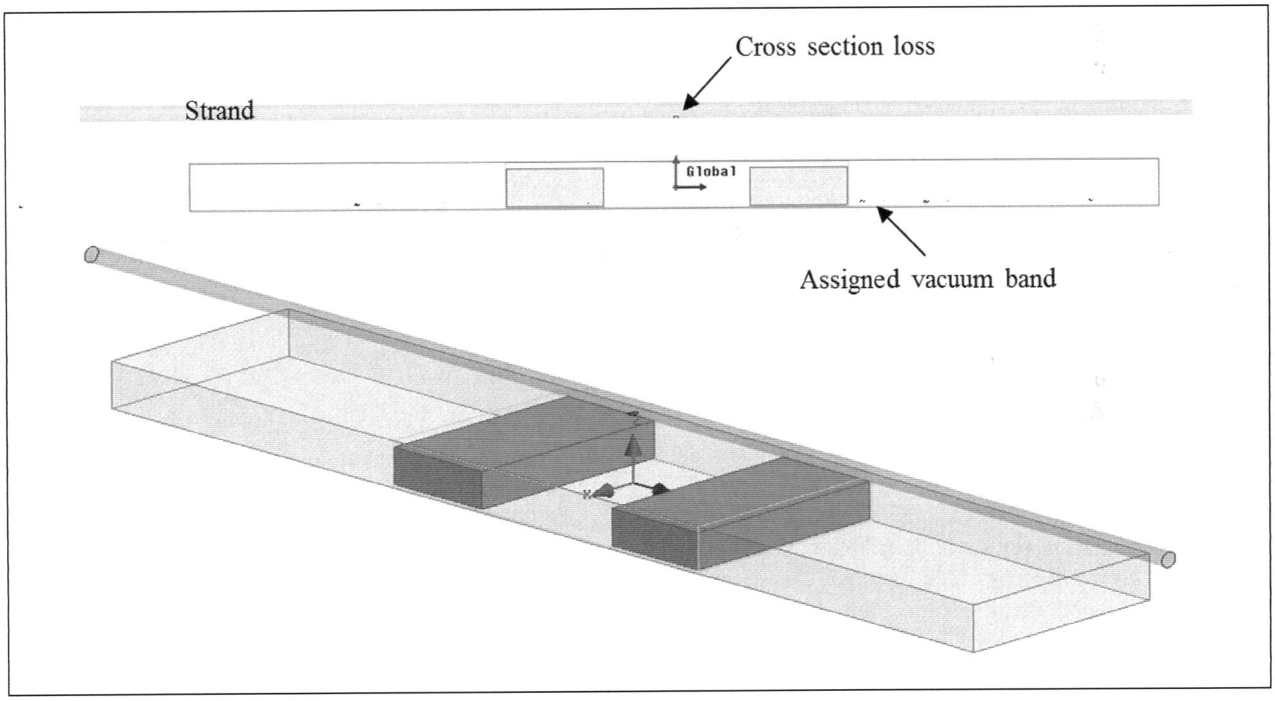


**Figure 6.** Initial setup showing the travel direction of the system.

(Figure 8*b*) magnetic flux density changes along the strand. The defect depth significantly impacted the peak to peak values of the flux density in both Hall effect sensor directions.

The width of the defect has been changed from 0.1 in to 0.5 in, with the 0.5-in defect being the furthest to the right. The cross-section loss of all created defects remains at 5%. The normal and axial flux densities are shown in Figure 8*c* and *d*, respectively. The density shows distinctive changes in the amplitude of flux for different width sizes. However, the overall amplitude remains in the order of 5% of cross-section loss that was measured for the defect depth. Results demonstrate that the defect width signals are smaller in comparison to the defect depth peaks associated with magnetic flux leakage signal offset (Figure 8*c*).

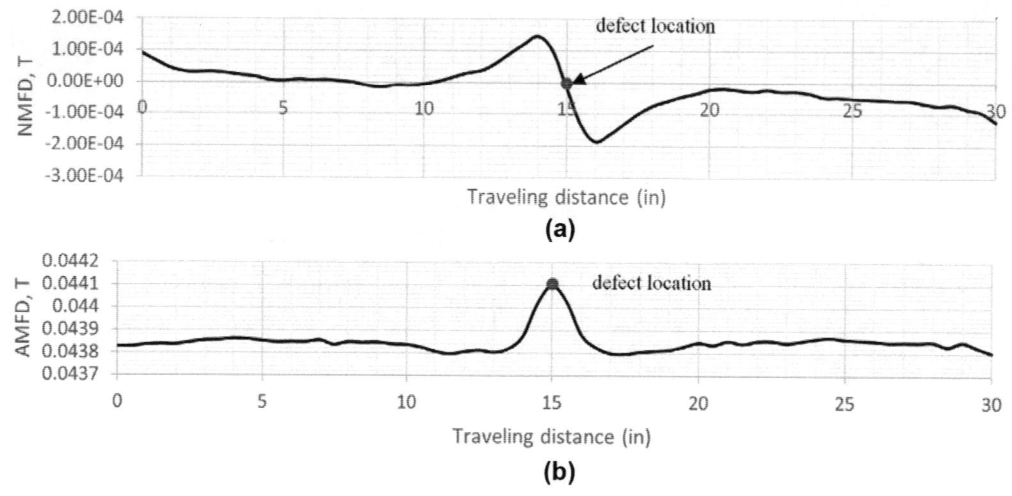


**Figure 7.** (*a*) Normal and (*b*) axial magnetic flux density signals showing distinctive leakage at the defect.

### *Lateral Reinforcements (Stirrups) Effect*

The ferromagnetic components embedded in concrete structures may include longitudinal steel reinforcements, adjacent steel reinforcements, stirrups (lateral reinforcements), tie wires, and steel chairs. Normally, stirrups are present more frequently in prestressed concrete members and have the most impact on the magnetic flux signal. In the simulation shown in Figure 9, two stirrups are included in the finite-element model containing five longitudinal strands each with 5/8 in diameter. Figure 9*a* shows the system layout and the resulting magnetic field for two 0.5-in diameter stirrups placed 8 inches from the defect's center.

Figure 9*b* and *c* show the normal and axial magnetic flux density changes from transient analysis, respectively. A significant result shown by the analysis is that the leakage signal induced by the defect can be shielded by the strong leakage emerging from the stirrups. Consequently, detection of magnetic field leakage emitted from defects can be difficult when corrosion or defects are located nearby. If stirrups are located closer to the sensor than the defect, the magnetic field leakage from the defect might be completely masked. In the analysis, the NMFD at the stirrup zone is 41.9 G, but it is 4.7 G for the cross-section loss. For the same layout, the AMFD is 17.1 G and 2.0 G for the stirrup and the defect, respectively.

The contribution of the stirrups in obtaining magnetic flux values is significant. To separate the effects of multiple strands and stirrups, an additional analysis was performed in which 5 strands and no stirrups were included in the model. The NMFD was measured at approximately 1.45 G, which is smaller than the value obtained from just one strand in the model. This analysis suggests that the saturation of the strand is decreased to lower levels leading to the lower flux magnitude (peak to peak measurements) at the defect zone. On the other hand, although stirrups can mask the magnetic signals from defects, they can also produce a data offset level in the acquired signals. This offset level reduces the peaks of the magnetic flux signals at the defect zone. In general, a lower magnetization level might result from adding more strands to the model (Figure 9*c*).

## Sensor Array

As it was demonstrated in the previous section, the magnetic flux leakage from the corrosion can be masked with stronger signals of secondary metallic objects such as stirrups embedded in concrete structures. Signal processing techniques are only helpful for defects located far from the stirrups. The oriented sensor array is proposed in this section to improve the MFL system to overcome this drawback. The effects of a linear and planar oriented magnetic field are investigated. Several numerical simulations are conducted on whether the planar rotation of the magnet/sensor unit around a fixed normal axis could provide useful information to differentiate between MFL signals from defects and lateral reinforcements. As mentioned earlier, the dominant MFL signals produced by adjacent lateral reinforcements make the interpretation of the recorded data difficult. The finite-element model in this study includes magnet rotation increments of 10°. The velocity of the system is set to 5 in/s as a slow and constant velocity of the probe is advantageous to reach the maximum magnetization of the strand. In practice, velocities in the range of 3–15 in/s are suitable. The

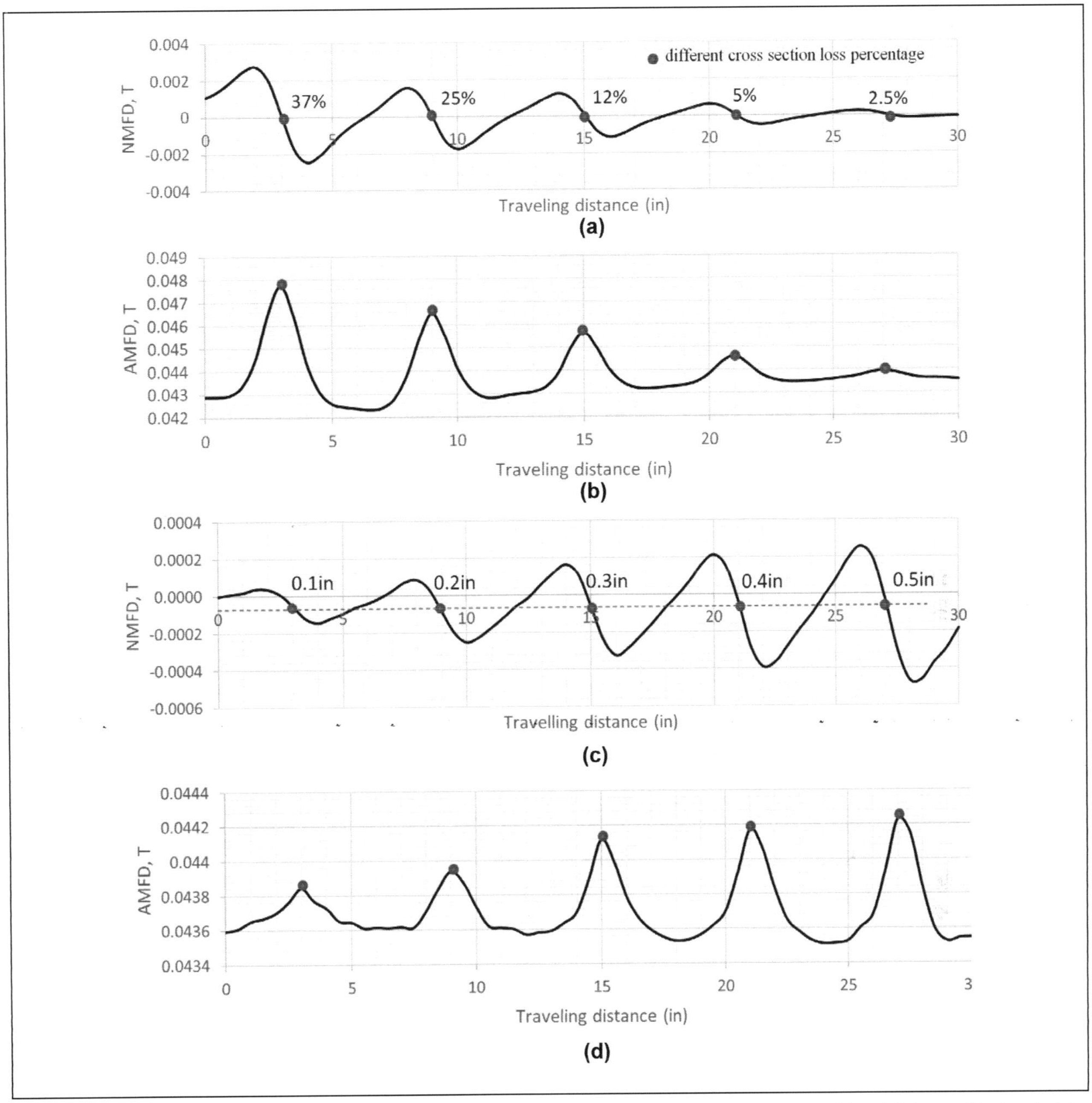


**Figure 8.** Normal and axial magnetic flux densities showing leakage at different defect depths in (*a*) and (*b*) and defect width sizes in (*c*) and (*d*).

magnet size used in the study is 4 in × 8 in × 1.5 in. The models contain a sensor array of 30 axial and normal Hall-effect sensors aligned at the mid-distance between two magnets (Figure 10). In the reference analysis (original system) with linear sensor array, no rotation was considered. However, in the next section, the sensor array is oriented (the top view of the layout is shown in Figure 10) to examine the possibility of separating signal peaks resulting from the corrosion and the stirrup.

The normal and axial magnetic flux density data obtained from the linear sensor array are shown in Figure 11. The magnitudes of the flux density changes are gradually reduced as the resultant sensor distance from the strand has been increased (sensor lift-off effect). Thus, for sensor 8, the maximum flux density change can be detected with higher flux values, but sensors 1 and 15 show lower magnitudes. In the non-oriented model, sensors are symmetric with respect to sensor 8 located in the

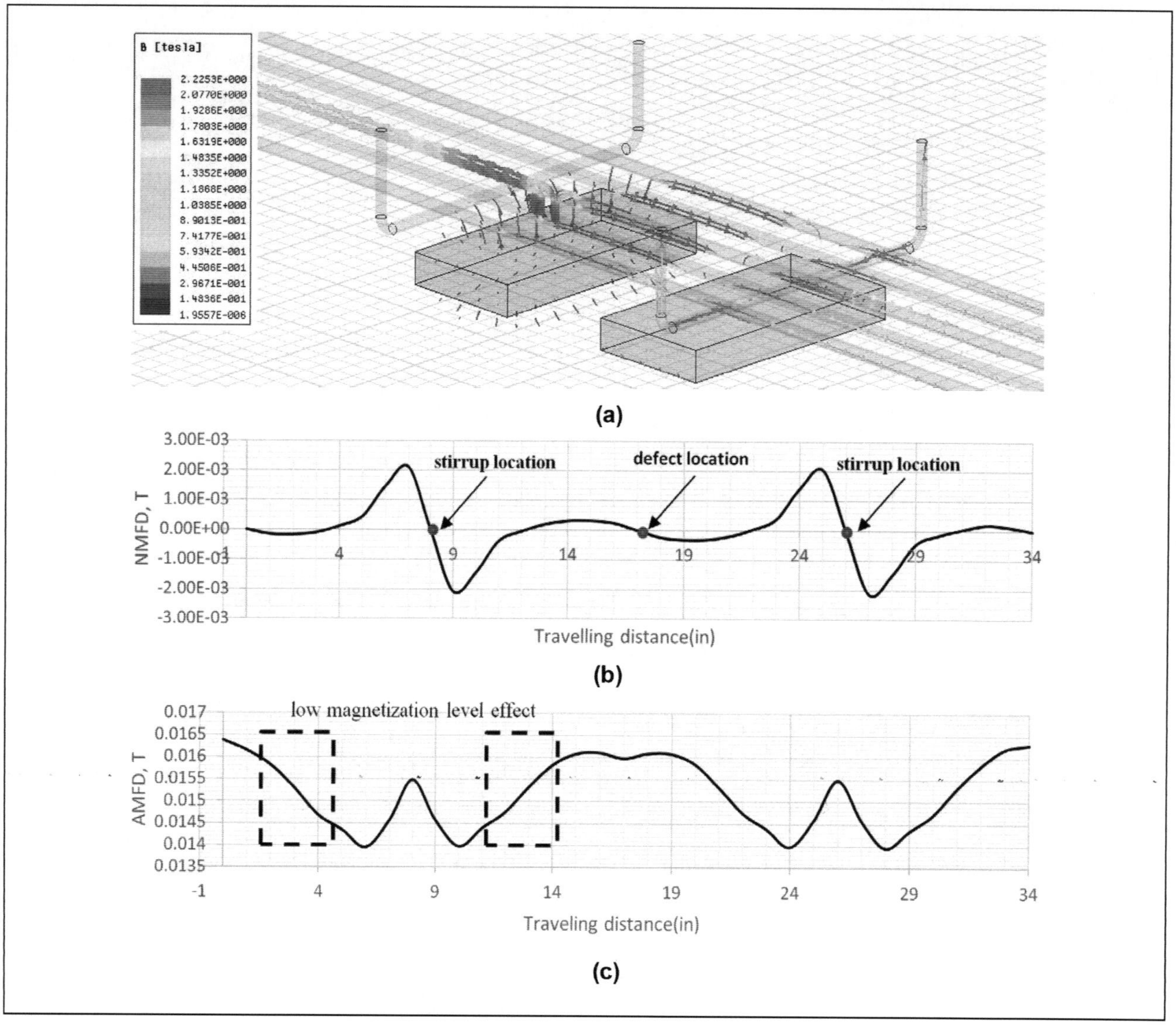


**Figure 9.** (*a*) Simulation model and magnetic field distribution, (*b*) influence of stirrups on the normal magnetic flux density change, and (*c*) influence of stirrups on the axial magnetic flux density changes.

plane of symmetry. Accordingly, signals from sensors 8 to 1 and signals from 8 to 15 are identical. In Figure 11, the change at the corrosion zone for both normal and axial signals cannot be identified as the magnetic flux leakage is mostly dominated by the stirrup field.

There are some slope changes observed between the stirrup zone and defect zone. To investigate these zones, the derivatives of curves are extracted. The derivatives of the magnetic flux density curves obtained from sensors 1 and 8 in the normal and axial directions are illustrated in Figure 12.

A differentiation of the original outputs shows more signal disturbances at the corrosion defect zone. Thus, investigation of the derivatives of the signals can be beneficial.

## Oriented Sensor Array

The effect of the oriented sensor array unit on defect detection of the MFL system is considered in this section. In each movement along the strand, each sensor is positioned in different projections to the magnetic field, resulting in different magnetic flux leakage trends. With this rotation, the stirrup should be able to pass a higher magnetic field, causing less change in the magnetic flux produced by the stirrup. Whereas the magnetic flux leakage emerging from the corrosion can be increased as the cross-section reduction is increased by $cos^{-1}(10°)$. Figure 13 shows the signals extracted from two sensors. The magnetic flux leakage around the defect zone shows higher disturbances than shown in the original case.

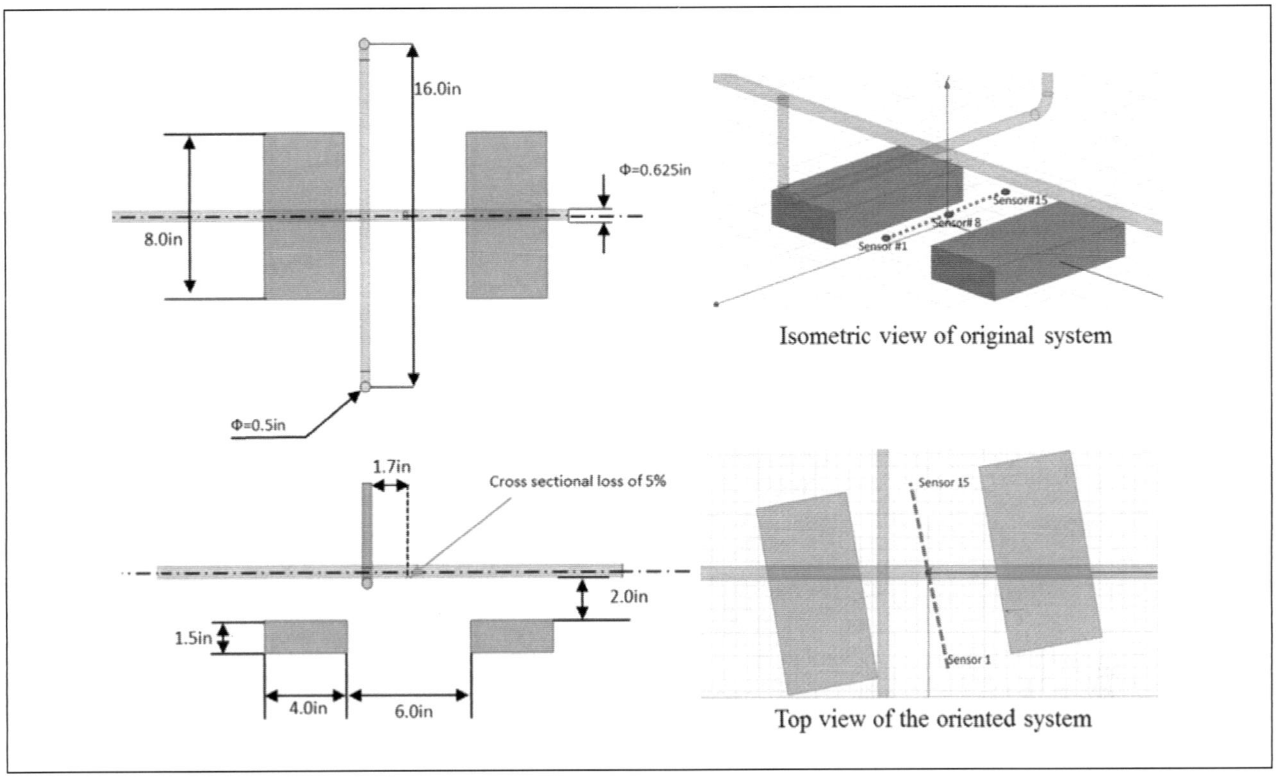


**Figure 10.** Linear and oriented sensor array models with 30 axial and normal Hall-effect sensors array.

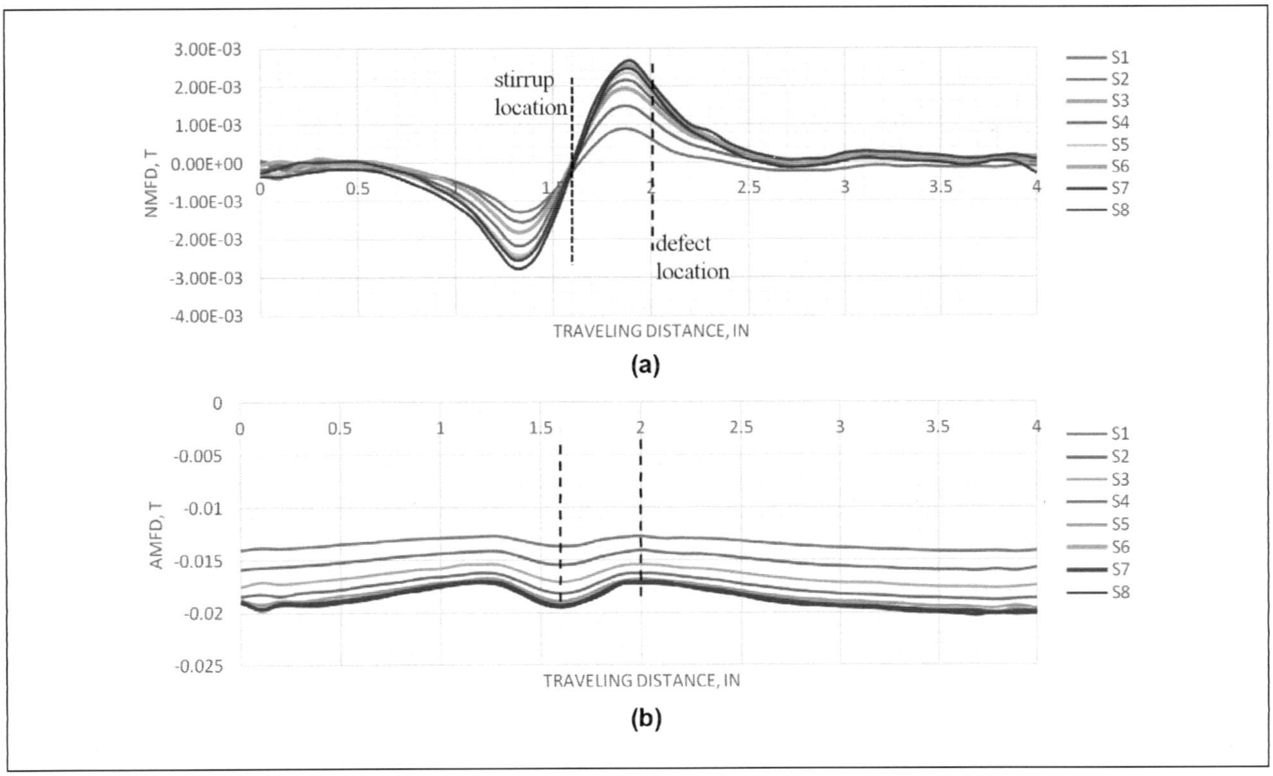


**Figure 11.** Normal (*a*) and axial (*b*) magnetic flux density changes during a system travel extracted from the sensor array.

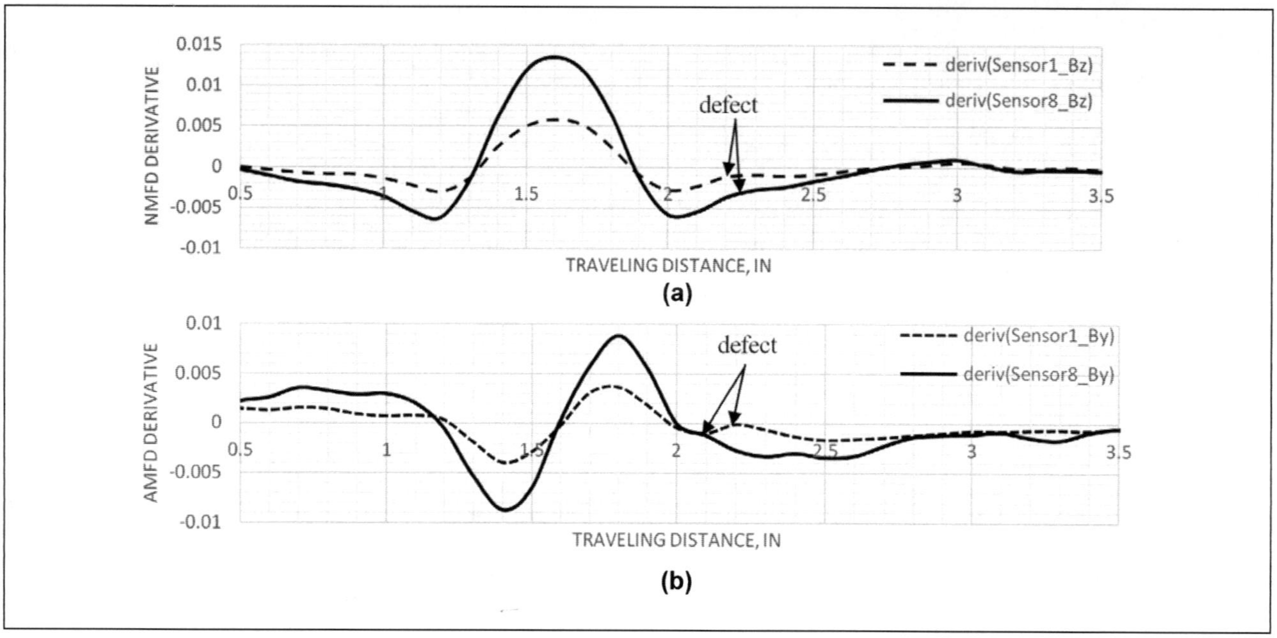


**Figure 12.** Differentials of the normal (*a*) and axial (*b*) magnetic flux densities.

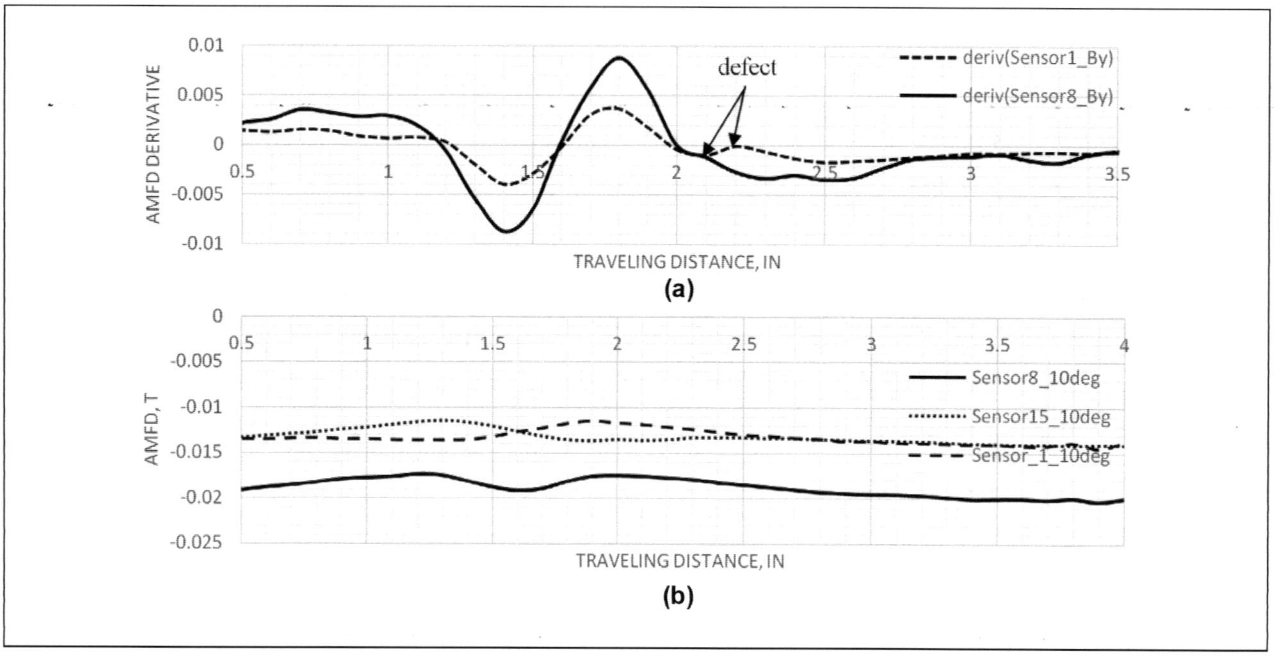


**Figure 13.** (*a*) Normal and (*b*) axial magnetic flux density change during system traveling with 10° orientation.

As expected, the amplitude of magnetic flux from the stirrup is reduced and the location of the slope change derived by the corrosion shows a higher magnetic flux leakage. To investigate the changes in the rate of the magnetic flux densities, the derivatives of the signals have been extracted (Figure 14). The signals show magnifications in flux densities around the corrosion zones.

Figure 14 shows very distinctive changes in the rate of the flux density at the location of the corrosion, especially for sensors 1 and 15 in comparison with linear sensor array results. Concerning the axial sensor in Figure 14*b*,

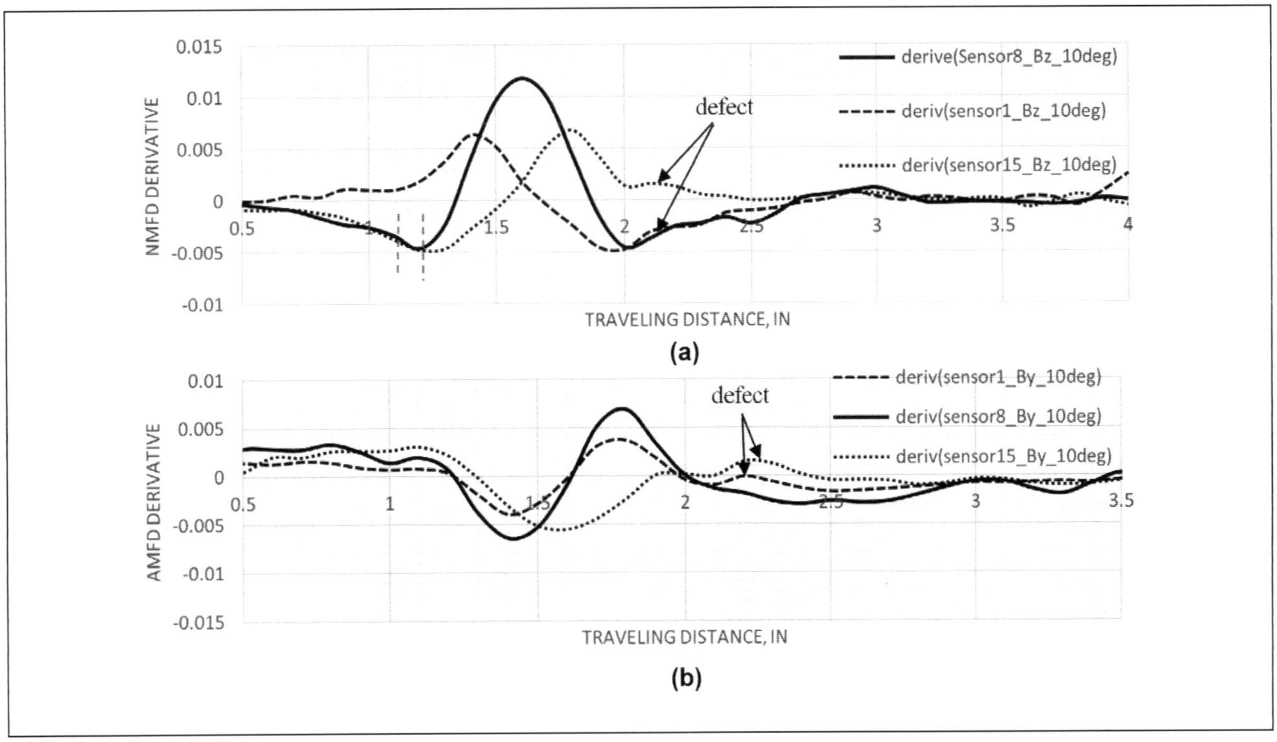


**Figure 14.** Differentials of the normal (*a*) and axial (*b*) magnetic flux densities of 10° orientation.

sensors 1 and 15 show a complete separation of the corrosion zone from the stirrup flux density leakage. Separated signals help to filter out all the stirrups' signals through the data acquisition process.

## Conclusions

In this paper, comprehensive numerical magnetostatic and transient analyses were performed to simulate a MFL system and to verify its application for detecting corrosion in prestressed concrete structures. Magnetostatic simulations used to optimize the MFL system layout including the size of the magnets and distances between the magnets. It is verified that the thickness of the permanent magnet is the most significant parameter influencing the flux density in the current system and axial Hall-effect sensors can emphasize the magnetic flux leakage. Several magnetic transient analyses have been performed to address the effect of defect size, lift-off sensor and magnetic interference from secondary ferromagnetic materials embedded in a concrete structure. A sensor array with an oriented magnetic field has been investigated through conducting appropriate numerical simulations. The oriented magnetic field shows an improved distinct change in the masked signals for the corrosion.

### Author Contributions

The authors confirm contribution to the paper as follows: study conception and design: Sadegh Shams, Al Ghorbanpoor; Simulations: Sadegh Shams; data collection: Shibin Lin, Hoda Azari; analysis and interpretation of results: Al Ghorbanpoor, Sadegh Shams, Shibin Lin; draft manuscript preparation: Sadegh Shams, Hoda Azari, Shibin Lin. All authors reviewed the results and approved the final version of the manuscript.

### References

1. Kusenberger, F. N., and J. R. Barton. *Detection of Flaws in Reinforcement Steels in Prestressed Concrete Bridges.* Research Report, Federal Highway Administration, Washington, D.C., 1981.
2. Ghorbanpoor, A., G. R. Steber, and S. L. Shew. *Evaluation of Steel in Concrete Bridges: The Magnetic Field Disturbance (MFD) System.* Research Report, Federal Highway Administration, Washington, D.C., 1991.
3. Ghorbanpoor, A., and S. Shi. Assessment of Corrosion of Steel in Concrete Structures by Magnetic Based NDE Techniques. *Proc., American Society for Testing and Materials STP* 1276, 1996.
4. Ghorbanpoor, A., R. Borchelt, M. Edwards, and E. Abdelsalam. *Magnetic-Based NDE of Prestressed and Post-Tensioned Concrete Members - The MFL System.* Research Report, Federal Highway Administration, Washington, D.C., 2000.

5. Jones, L., S. Pessiki, C. Naito, and I. Hodgson. *Inspection Methods & Techniques to Determine Non-visible Corrosion of Pre-stressing Strands in Concrete Bridge Components.* Research Report, Pennsylvania Department of Transportation, Harrisburg, Pennsylvania, 2010
6. Sawade, G., and H. J. Krause. *Non-Destructive Evaluation of Reinforced Concrete Structures Non-Destructive Testing Methods.* Volume 2 in Woodhead Publishing Series in Civil and Structural Engineering, Woodhead Publishing, Cambridge, UK, 2010, pp. 215–242.
7. ANSYS® Academic Research *Electromagnetics Suite, Release 17.1.0.*

*The Standing Committee on Testing and Evaluation of Transportation Structures (AFF40) peer-reviewed this paper (18-05070).*

Article

TRR
JOURNAL OF THE TRANSPORTATION RESEARCH BOARD

Transportation Research Record
2018, Vol. 2672(41) 145–155

DOI: 10.1177/0361198118780711
journals.sagepub.com/home/trr
SAGE

# Reducing Thermal Reflections for Infrared Thermography Applications on Tunnel Liners with Reflective Finishes

**Dewei Meng**[1], **Shibin Lin**[1], **and Hoda Azari**[2]

**Abstract**

Infrared thermography (IRT) has been employed as a non-destructive evaluation (NDE) method to detect delaminations in bridge decks by ASTM standards since 2003. Compared to other NDE methods, IRT is fast in data acquisition and processing. In addition to applications on bridge decks, IRT can be employed to detect defects in other structures such as roadway pavements and tunnel liners. Recent IRT applications on tunnel liners with reflective finishes such as ceramic tiles show that the IRT results can be severely contaminated by infrared reflections of the vehicles and the illuminating systems in the tunnels. In this study, an algorithm is developed to mitigate the contamination caused by the thermal reflections in the IRT results. The reflection is reduced by combining thermal data from multiple thermographs. Perspective projection is employed to map thermal data to the same coordinates if the thermographs are taken with an infrared camera moving in the space. The algorithm is proved to be successful by the results from the lab experiments and field tests. The algorithm can work with IRT data collected by vehicle-mounted infrared cameras to achieve high-speed IRT for tunnel liners with reflective finishes.

Reinforced concrete is a very popular structural material because of its good performance and economic efficiency. It is broadly used in civil infrastructures such as pavements, bridges, and tunnels. Reinforced concrete takes advantage of high tensile strength of the steel reinforcement and good compressive strength of the concrete. However, reinforced concrete may experience problems caused by excessive stress and poor construction quality, such as rebar corrosion, cracking, delamination, voids, and honeycombing (*1*). These defects can cause further deteriorations and may lead to failure of the structures. It is necessary to perform structural inspections periodically to detect and evaluate these defects. Non-destructive evaluation (NDE) methods based on different physical principles are currently employed in structural inspections. Examples of popular NDE technologies are impact echo, ground penetrating radar, and infrared thermography (IRT).

IRT has been employed as an NDE method to detect delamination and debonding in concrete bridge decks since the 1980s (*2,3*). The IRT method is fast in data acquisition and data processing compared with other NDE methods (*4,5*). ASTM has published a standard test method for detecting delaminations in bridge decks (*6*). The vehicle-mounted IRT method has been used for detecting damages in bridge decks and pavements at traffic speed (*7*). IRT has been reported to be used in tunnel excavation (*8–10*). Similar to applications on bridge decks, IRT can also be applied to detect defects in tunnel liners. IRT has been employed as a fast scanning method to locate anomaly in the tunnels. Even though it is difficult for the IRT method to detect the levels and depths of the damages, it can provide targets for local in-depth testing with other NDE methods. Williams employed IRT and other NDE methods in the Moab Tunnel test, showing that IRT can detect water infiltration in the tunnel liner (*11*). Sheen made a quality evaluation of tunnel lining by using IRT to detect lower temperature on the damaged lining caused by water leakage (*12*). Kurosu proved IRT is capable of detecting the void near the surface and flaking in the subway tunnel lining (*13*). The SHRP 2 project (*14*) showed that a vehicle-mounted IRT system (*15*) can detect tile

[1]Engineering & Software Consultants, Inc., Turner-Fairbank Highway Research Center, McLean, VA
[2]Federal Highway Administration, Turner-Fairbank Highway Research Center, McLean, VA

**Corresponding Author:**
Address correspondence to Shibin Lin: shibin.lin.ctr@dot.gov

debonding, and cracks and stalactites containing moisture in the tunnels.

Infrared has a longer wavelength (0.7–14 μm) than visual light (0.4–0.7 μm) and can be reflected more easily by structural surfaces (*16*). Recent applications of IRT on tunnel liners with reflective finishes shows that thermal reflection can contaminate the IRT results and cause false positive results (*14*). Thermal reflection is partially or fully polarized and can be filtered by an infrared polarizer (*16*). However, an infrared polarizer is usually expensive, and the effectiveness depends on the angle of the incident thermal radiation. In this study, an algorithm is developed to reduce thermal reflection caused by moving and stationary infrared sources. The algorithm works with IRT data collected with regular infrared lens, so the cost could be much cheaper. The algorithm is validated by lab experiments and field tests. It is possible to integrate this algorithm with vehicle-mounted IRT to achieve high-speed tunnel scanning with high accuracy. A few factors affecting the performance of the algorithm in field applications are also discussed. The principle of the IRT method is introduced in the following section.

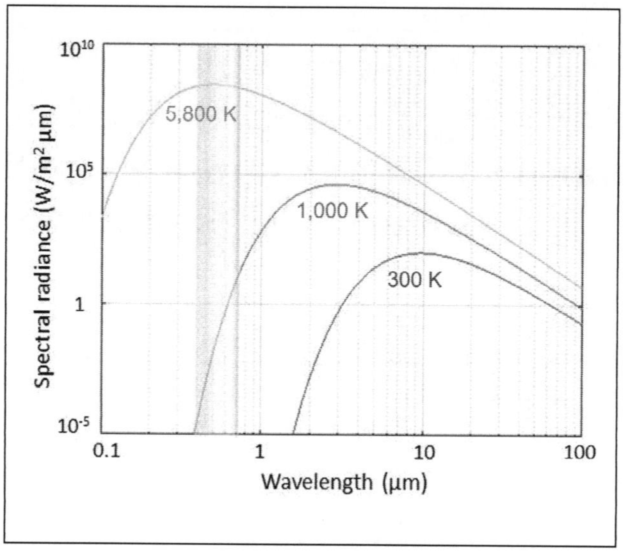


**Figure 1.** Black-body radiation spectrum under different temperatures.

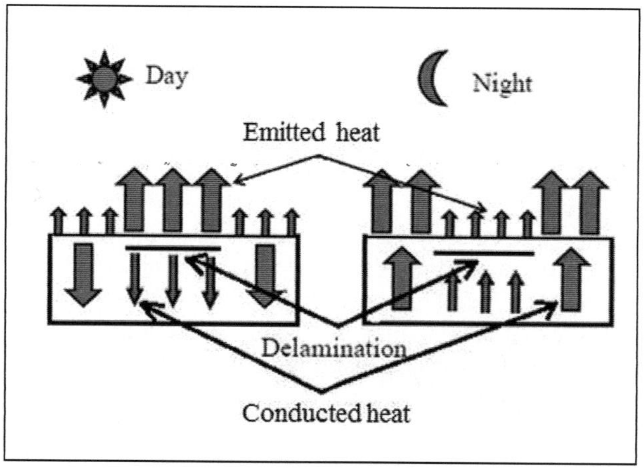


**Figure 2.** Schematic heat flow for a concrete bridge deck with a delamination.

## Principle of IRT

Heat transfers between objects by conduction, convection, and radiation. Within a solid object or between solid objects in contact with each other, heat transfers by conduction when a temperature gradient exists. Within a fluid, or between a fluid and a solid that have physical contacts, heat transfers by convection. Any object with a temperature greater than 0 K will generate thermal radiation. Between objects whose surfaces are exposed to each other, heat transfers by radiation, and no physical contact is needed. The radiation energy is related to the temperature of the object by the Stefan Boltazmann Law (Equation 1).

$$Q = \varepsilon \sigma T^4 \quad (1)$$

where

$Q$ = radiation in $W/m^2$,

$\varepsilon$ = the emissivity of the object with a range of 0–1,

$\sigma$ = the Stefan Boltzmann constant ($5.67 \times 10^{-8} Wm^{-2}K^{-4}$), and

$T$ = the surface temperature in K.

The radiation spectrum is related to the temperature of the object and can be determined by Planck's Law. Examples of radiation spectra of objects with different temperatures are shown in Figure 1. For objects with temperatures around 300 K, most of the radiation is infrared. Surface temperature can be calculated through Equation 1 by measuring the infrared radiation from an object.

Heat flows differently in the defect part compared with the intact part of a concrete structure. A schema of a concrete bridge deck with an embedded delamination is shown in Figure 2. The deck is warmed by solar radiation and hot ambient air during the daytime. For the intact part, heat flows from the surface to the deeper concrete and is accumulated in the concrete. The delamination obstructs the heat flow, resulting in a surplus of heat in the concrete above the delamination and a higher surface temperature. At night, the heat dissipates from the concrete deck to the ambient. A shortage of heat develops in the defect part as less energy is accumulated in the concrete above the delamination during the daytime. As a result, at night the defect part of the deck shows a lower

**Figure 3.** Thermograph of a concrete slab taken during the day.

temperature than the intact part. The thermograph of a concrete slab taken during the day is shown in Figure 3. The delaminated parts are shown with higher temperatures (lighter colors) compared with the intact part of the slab.

Tunnel liners are seldom exposed to solar radiation. Heat transfers from the ambient to the liners mostly by convection. Small variations in the ambient temperature make it difficult to achieve good IRT results (*17*). In addition, IRT results may be contaminated by thermal reflections of objects in the tunnel if the tunnel liner has smooth finishes. A tunnel liner with ceramic tile finishes is shown in Figure 4, *a* and *b*. Thermal reflections can be seen clearly on the liner from the thermograph in Figure 4*b*. Damage detection based on the IRT method becomes inaccurate if the collected data is contaminated by thermal reflection. It is necessary to reduce thermal reflection before drawing conclusions from IRT results.

## Reducing Thermal Reflection by Data Combination

Similar to visible light, thermal reflection from a smooth surface is partially or fully polarized depending on the angle of the incident radiation. As a result, thermal reflection can be filtered by an infrared polarizer (*16*). By filtering thermal reflection before it is captured by the photosensitive element of the infrared camera, this method is very time efficient. However, an infrared polarizer is usually expensive, and the extent of reduction depends on the angle of the incident thermal radiation.

Thermal reflection can be reduced by processing thermal data collected with a regular infrared lens. As shown in Equation 2, the total energy captured by each pixel $Q$ is composed of the thermal radiation from the structural surface $Q_{ra}$ and the thermal reflection $Q_{re}$. As the total energy $Q$ is always greater than the actual radiation $Q_{ra}$, the apparent temperature calculated via Equation 1 is also higher than the real temperature on the structural surface. Consequently, the temperature for each location on the structural surface can be approximated by the

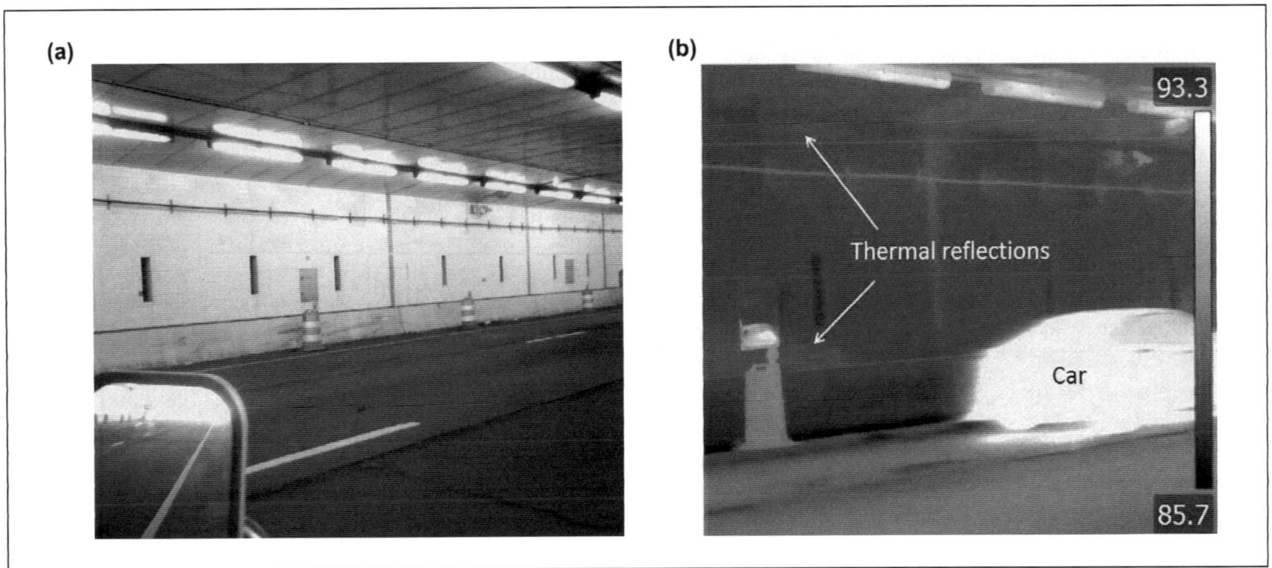


**Figure 4.** A tunnel liner with tiled finishes: (*a*) picture and (*b*) thermograph.

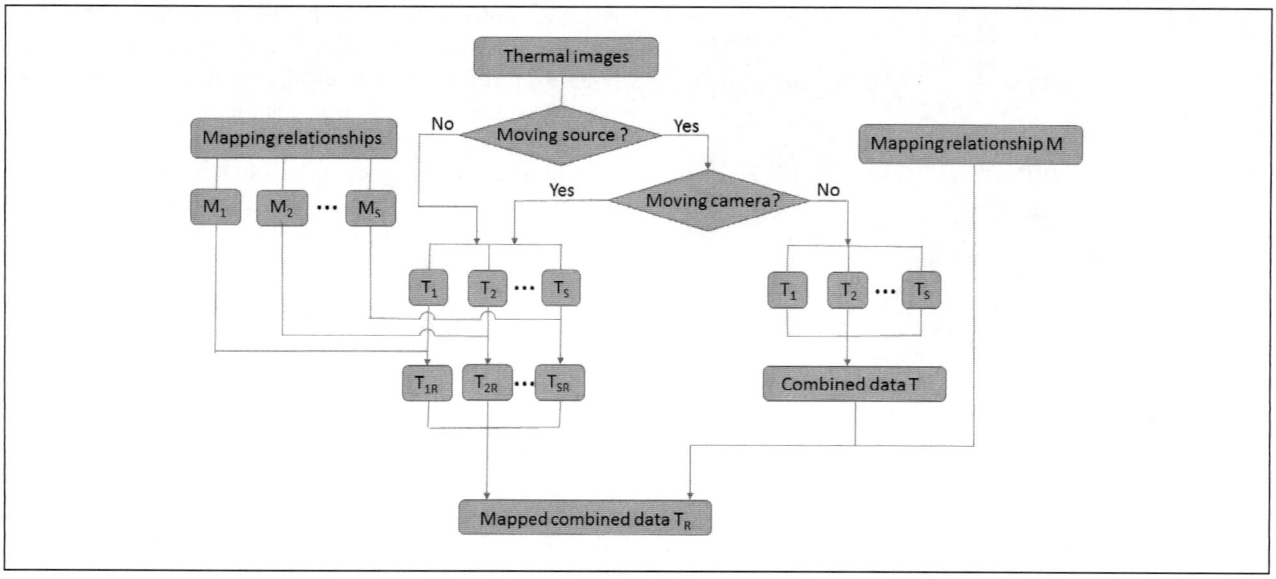


**Figure 5.** The flowchart of the algorithm for reducing thermal reflection.

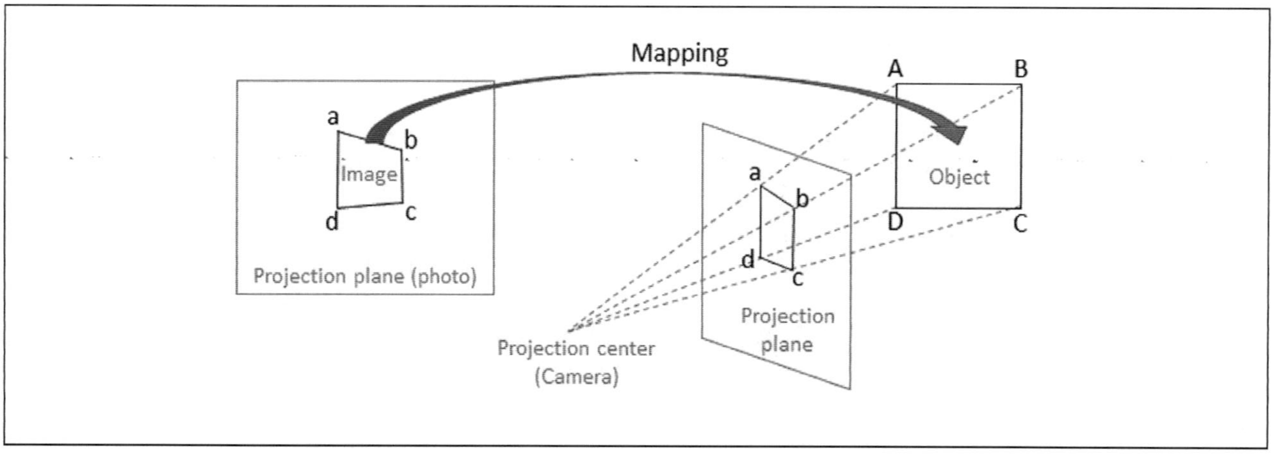


**Figure 6.** Perspective projection and the mapping relationship.

lowest apparent temperature among a series of thermographs.

$$Q = Q_{ra} + Q_{re} \tag{2}$$

An algorithm is developed to capture the lowest apparent temperature at the pixel level. The flowchart of the algorithm is shown in Figure 5. For a thermograph of $m \times n$ pixels, the thermal data is stored in an $m \times n$ matrix. A mapping relationship $M$ correlates the thermograph and the actual locations on the structural surface. The temperature data presented in the coordinates of the structural surface $T_R$ can be acquired by mapping the corresponding thermal data in the thermograph $T$ via Equation 3.

$$T \xrightarrow{M} T_R \tag{3}$$

The mapping relationship is determined by the principle of perspective projection shown in Figure 6. An area of interest is chosen as the target area on the structural surface. When the infrared camera is moving in the space, the size and shape of the target area change among the thermographs. In this case, thermal data from different thermographs should be converted to the same coordinates (the liner surface) before data combination via Equation 4.

$$T_1 \xrightarrow{M_1} T_{1R}, T_2 \xrightarrow{M_2} T_{2R}, \ldots, T_s \xrightarrow{M_s} T_{sR} \tag{4}$$

where

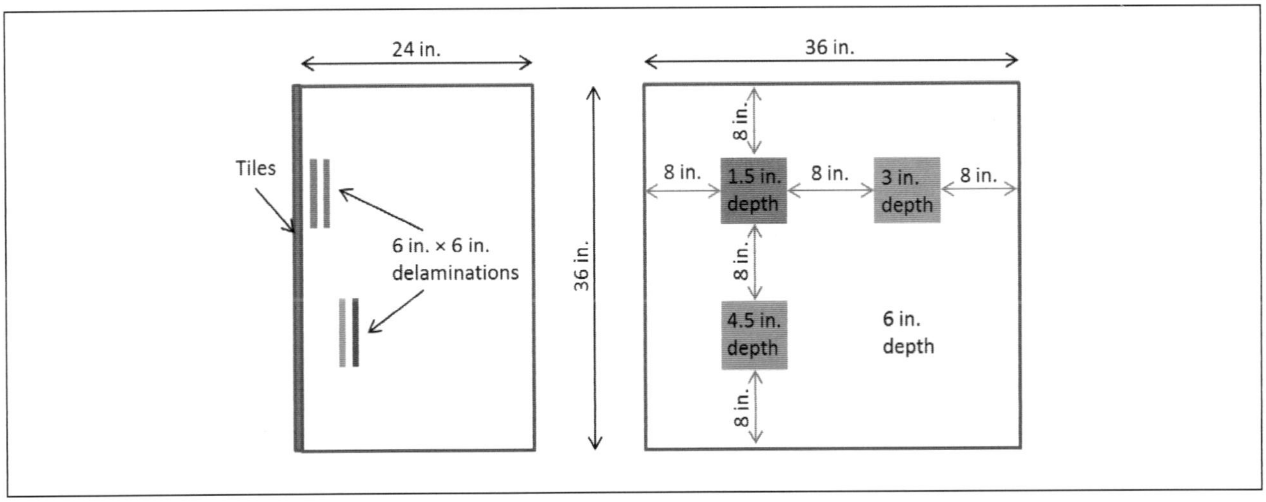


**Figure 7.** Design of the concrete specimen with artificial delaminations. Side view (*left*) and front view (*right*).

$M_1, M_2, \ldots M_s$ are the mapping relationships between the liner surface and the thermographs for a series of thermographs,

$T_1, T_2, \ldots, T_s$ are the thermal data within the target area, and

$T_{1R}, T_{2R}, \ldots T_{sR}$ are the thermal data in the target area after being mapped onto the liner surface.

The mapped data can be combined through Equation 5 by using the minimum apparent temperatures as the true surface temperature for each data point within the target area.

$$T_R = \min(T_{1R}, T_{2R}, \ldots T_{sR}) \quad (5)$$

where $T_R$ is the combined mapped data. If the infrared camera is set stationary in the tunnel, the mapping relationship for each thermograph is the same. In this case, the mapping process for a series of thermographs can be rewritten as Equation 6.

$$T_1, T_2, \ldots T_s \xrightarrow{M} T_{1R}, T_{2R}, \ldots T_{sR} \quad (6)$$

where $M$ is the mapping relationship that is the same for each thermograph. With the same mapping relationship, the algorithm can start with data combination directly, and data mapping can be omitted if presentation of data in the coordinates of the structural surface is unnecessary.

## Laboratory Experiments

Laboratory experiments were carried out to evaluate the performance of the algorithm on reducing thermal reflections caused by moving and stationary sources. A concrete block (3 Ft × 3 Ft × 2 Ft) was built with its side surface (3 Ft × 3 Ft) mounted with ceramic tiles. Four Styrofoam plates (6 in. × 6 in. × 0.5 in.) were embedded to simulate delaminations at different depths (1.5 in., 3 in., 4.5 in., and 6 in.) from the tiled surface. The schematic design of the concrete block is shown in Figure 7. Heat was applied to the tiled surface by heating lamps that were set 4 Ft away. The heating process lasted for 1 H before the heating lamps were removed from the site. The experiment started with introducing a moving thermal source and was followed by introducing a stationary thermal source.

In the first experiment, a heating lamp was introduced to simulate a moving thermal source such as headlights of a vehicle. The infrared camera was set stationary. Two thermographs were taken consecutively and are shown in Figure 8, *a* and *b*. The target area (the area within the boundary) is the same in the two thermographs. The thermal reflection may be misidentified as a damage indicator from either thermograph as it shows a higher temperature than the rest of the surface. The reflection can be recognized as a moving hot area on the tiled surface after comparing the two thermographs. As the mapping relationship for the two thermographs is the same, data combination was performed directly via Equation 5, and the result is shown in Figure 8*c*. The thermal reflection was successfully reduced, and the 1.5-in.-deep delamination was recognized from the result.

In the second experiment, the heating lamp was stationary in the space to simulate the illuminating system in a tunnel. Two thermographs were taken consecutively with the infrared camera moving parallel to the tiled surface. As shown in Figure 9, *a* and *b*, the shape of the target area and the location of the thermal reflection change in the two thermographs. The target area was mapped onto the 3 Ft × 3 Ft square, and is shown in Figure 10, *a* and *b*. Data combination was then performed, and the

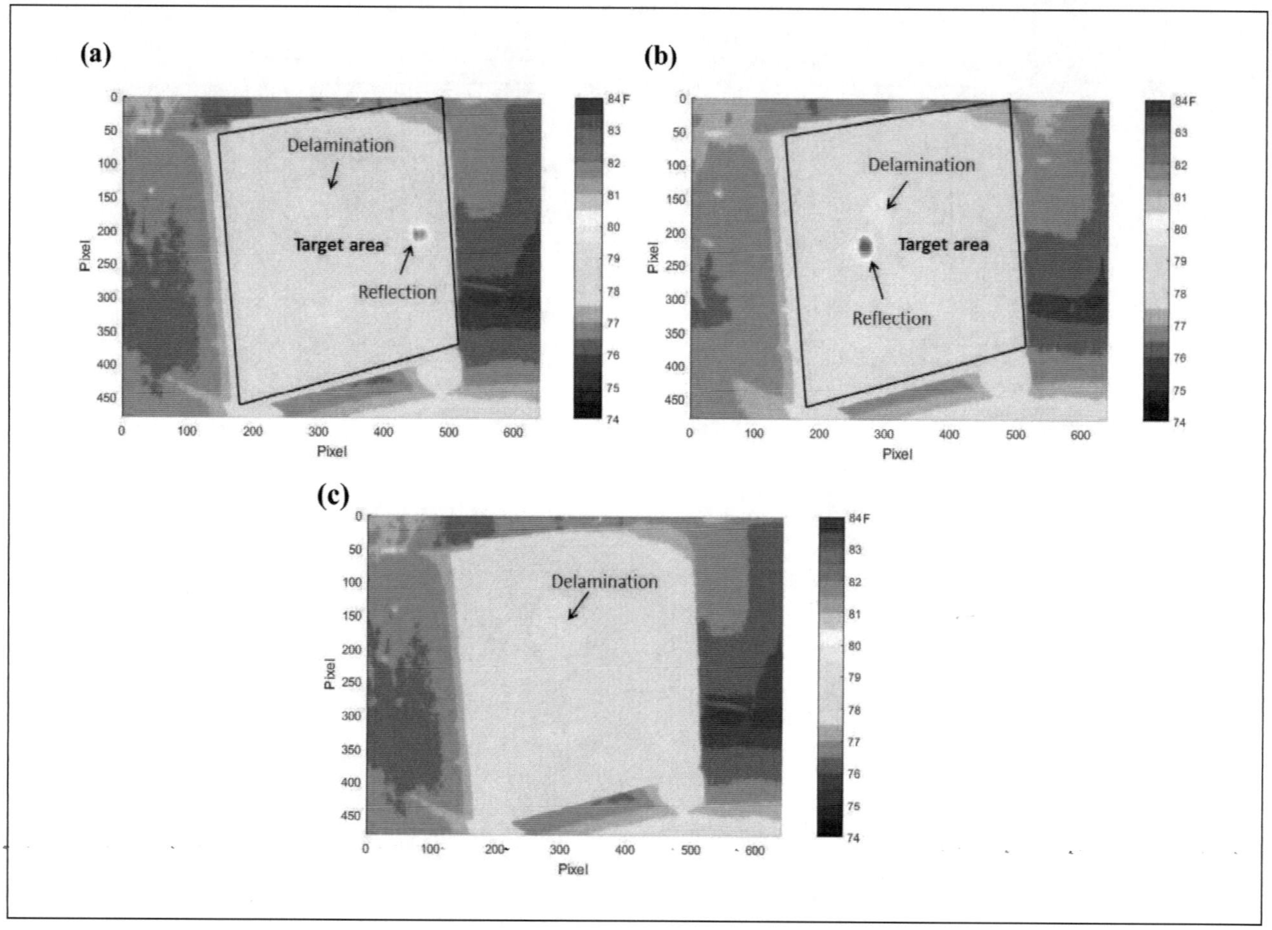


**Figure 8.** Two thermographs collected by a stationary infrared camera and reduced thermal reflection for a moving thermal source: (*a*) Thermograph 1, (*b*) Thermograph 2, and (*c*) combined data without mapping.

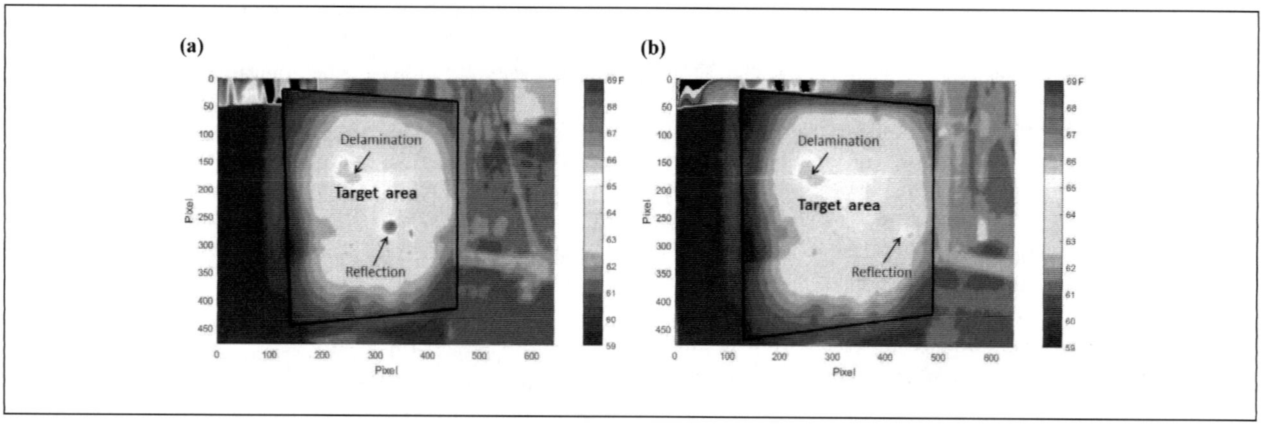


**Figure 9.** Two thermographs collected by a moving infrared camera for a stationary thermal source: (*a*) Thermograph 1 and (*b*) Thermograph 2.

result is shown in Figure 10*c*. The thermal reflection is successfully reduced after data combination, and only the 1.5-in.-deep delamination is recognized from the result.

## Field Tests

Field tests were carried out in a tunnel located on the northbound lane of Interstate 395 in Washington, D.C.

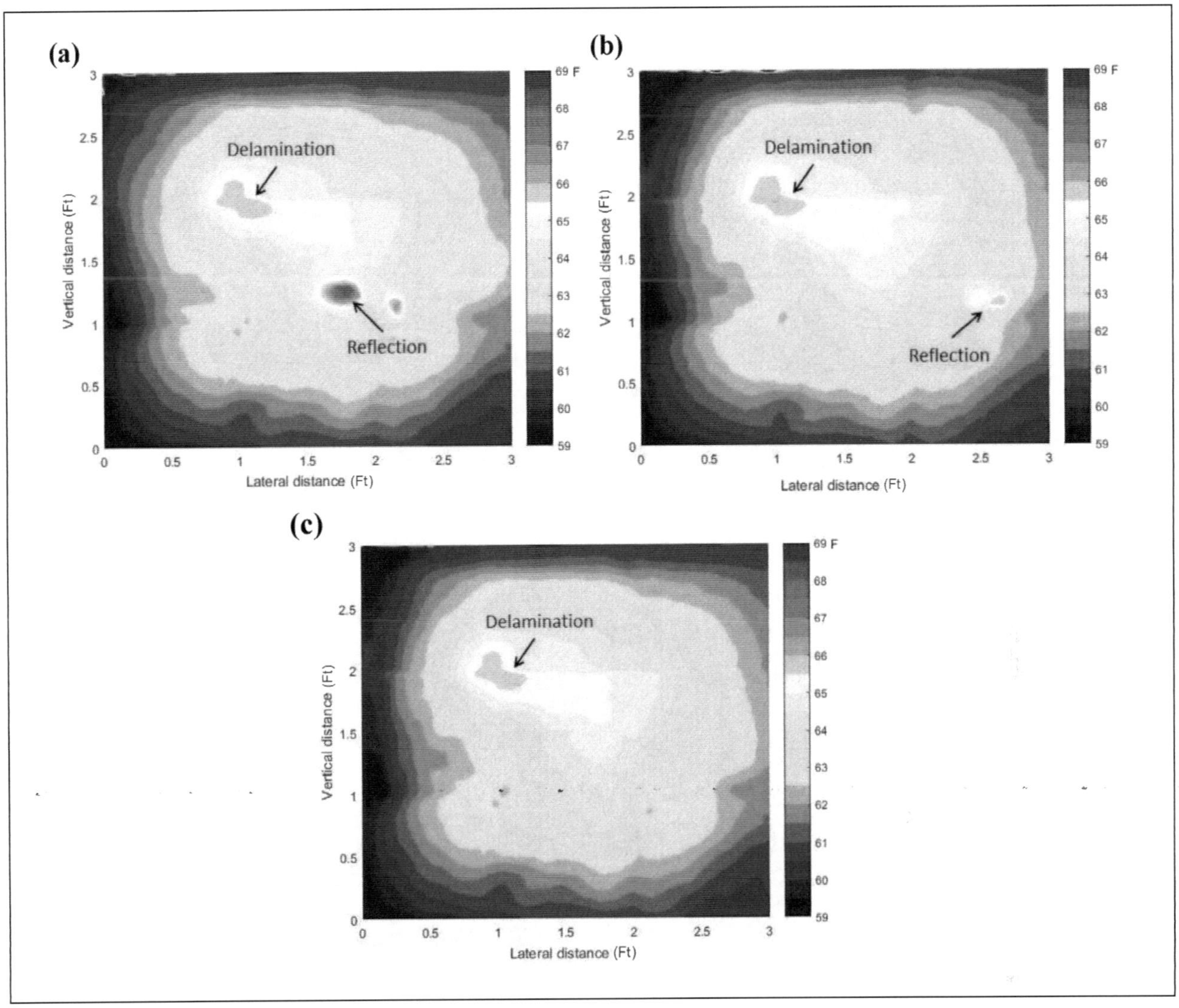


**Figure 10.** Thermal data in the target area mapped to the surface coordinates and combined thermal data after mapping: (*a*) target area in Thermograph 1, (*b*) target area in thermograph 2 and (*c*) combined thermal data.

As shown in Figure 4, the tunnel liner is covered with ceramic tiles. There is usually heavy traffic in the tunnel. Thermal reflections of the traffic and the illuminating system can cause serious contamination of the IRT results of the tunnel liner. With the infrared camera set at the height of a regular vehicle, the lower part of the liner was significantly affected by thermal reflections of moving vehicles, and the upper part of the liner was significantly affected by thermal reflections of the illuminating systems. Applications of the algorithm for the lower part and the upper part of the tunnel liner were introduced respectively.

In the first test, the infrared camera was set stationary in the hatched zone in the tunnel. Three thermographs taken consecutively are shown in Figure 11, *a, b*, and *–c*. A rectangular area from the lower part of the liner was chosen as the target area. In each thermograph, there is an area of higher temperature. When comparing the thermographs, the moving hot area was recognized as thermal reflection of a moving vehicle in the tunnel. The thermal data were combined through Equation 5 directly, and the result is shown in Figure 11*d*. Instead of being characterized in any thermograph as a potential damaged area, thermal reflection of a moving vehicle was successfully reduced after data combination.

In the second test, the infrared camera was installed on a vehicle moving in the right lane of the tunnel to reduce thermal reflection caused by the illuminating system. The target area was a 20 Ft × 4 Ft rectangular area in the upper part of the liner. Three thermographs taken consecutively for the same target area are shown in Figure 12, *a, b*, and *–c*. As the location of the infrared

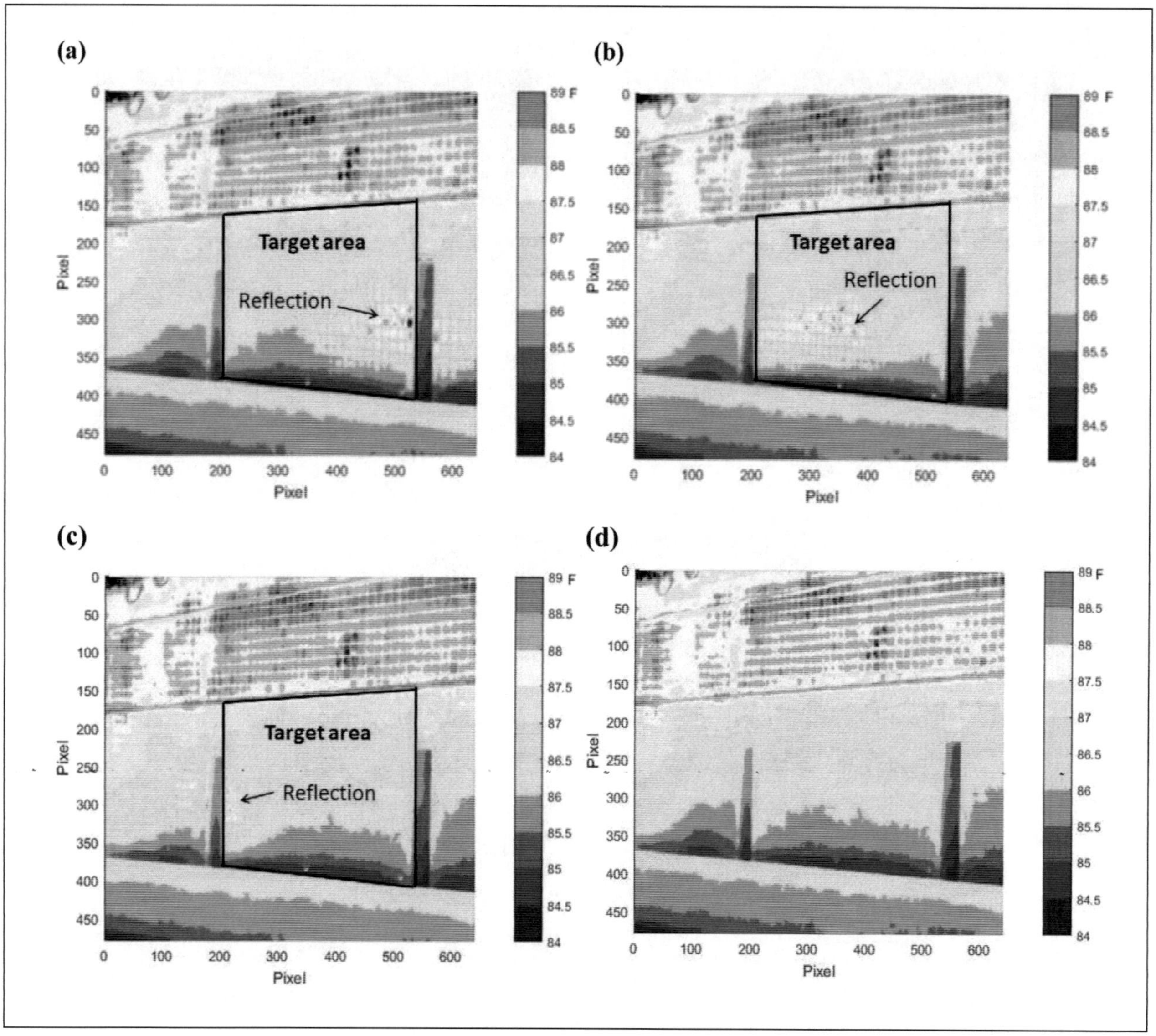


**Figure 11.** Thermographs taken consecutively by a stationary infrared camera in the tunnel, and combined data without mapping: (*a*) Thermograph 1, (*b*) Thermograph 2, (*c*) Thermograph 3, and (*d*) combined thermal data.

camera changes with the vehicle, the apparent size and shape of the target area also changes among the thermographs. The thermal data were mapped onto the rectangular target area through Equation 4, and the result is shown in Figure 13, *a, b*, and *–c*. In Figure 13, *a, b*, and–*c*, a hot area is located between 12 Ft and 15 Ft in the X direction and between 0.5 Ft and 3.5 Ft in the Y direction, with slight variations in the temperature magnitude and distribution. The mapped thermal data were then combined through Equation 5, and the result is shown in Figure 13*d*. The temperature difference between the hot area and the overall area is greatly reduced by the algorithm. The hot area is probably caused by thermal reflection of the illuminating system as the maximum temperature difference is less than 1°F.

## Discussions

The algorithm developed in this study can work with the stationary and the vehicle-mounted IRT method to detect damages in tunnel liners with reflective finishes. Based on the experiences with the field tests, a few factors should be considered for the vehicle-mounted IRT method to achieve good results. The shapes and locations of the objects change in the thermographs as the infrared camera and the vehicle on which the camera is mounted are moving in the tunnel. To locate each part of the liner accurately from the collected thermal data, the liner finishes should be divided into a series of target areas with markers that can be easily recognized from the thermographs. After data mapping and combination

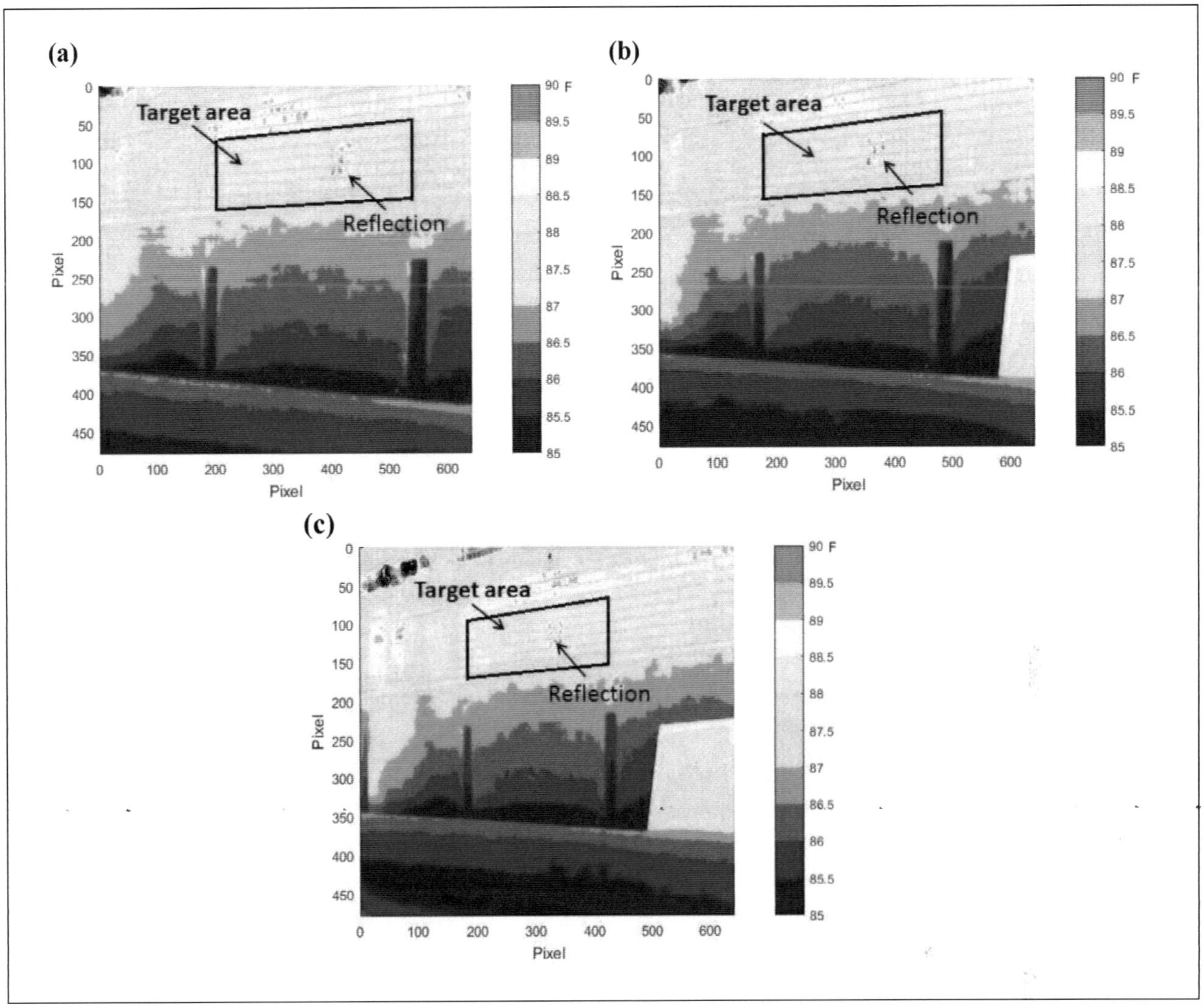


**Figure 12.** Thermographs taken consecutively by a moving infrared camera in the tunnel: (*a*) Thermograph 1, (*b*) Thermograph 2, and (*c*) Thermograph 3.

are performed for each target area, IRT data for the entire tunnel liner can be put together and presented. To acquire good results from the algorithm, the reflection needs to create an obvious displacement on the target area in the thermographs. As shown in Figure 14*a*, a larger displacement of the reflection on the target surface is equivalent to a larger angular displacement of the line connecting the infrared camera and the thermal image. Use of a wide-angle lens is recommended so that the same target area can still be captured by the infrared camera after a large angular displacement. If possible, the vehicle with the infrared camera should be moving in the lane next to the target liner so that traffic does not obstruct data acquisition. The lens of the infrared camera should be directed to the tunnel liner at an angle that prevents the reflections of the vehicle and the infrared camera from affecting the thermographs. The only disadvantage of the algorithm is that it does not work well when the stationary thermal source is very close to the reflective finishes. As shown in Figure 14*b*, the reflection of a thermal source close to the reflective surface would have a limited movement, which may cause unsatisfactory results from the algorithm.

## Conclusion

Reducing thermal reflection is necessary for IRT applications on tunnel liners with reflective finishes because thermal reflections could contaminate the damage detection results. Using an infrared polarizer is a direct but expensive method to filter thermal reflection, and the performance depends on the angle of the incident

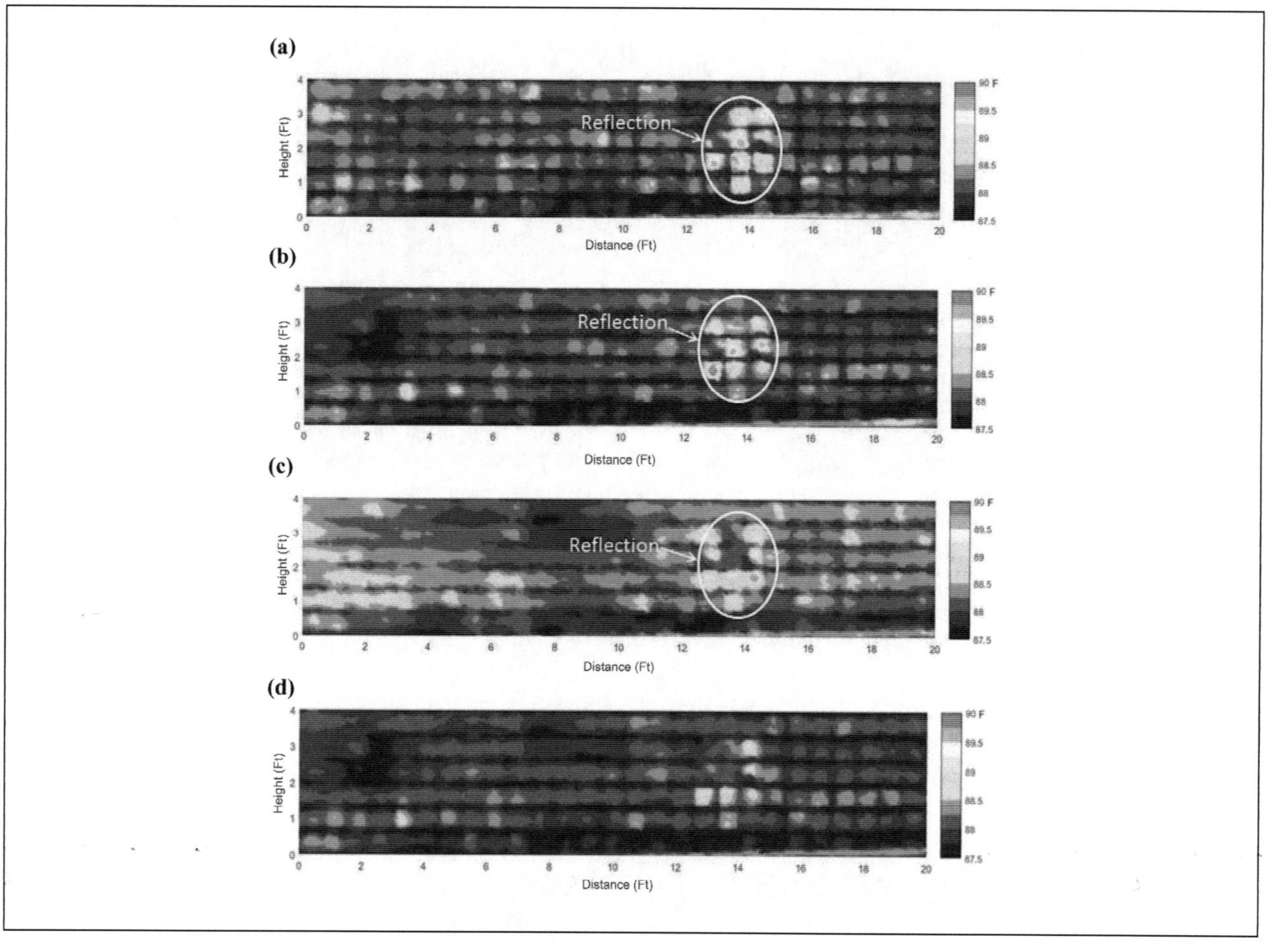


**Figure 13.** Thermal data for the target area after mapping and the combined thermal data: (*a*) data from Thermograph 1, (*b*) data from Thermograph 2, (*c*) data from Thermograph 3, and (*d*) combined thermal data.

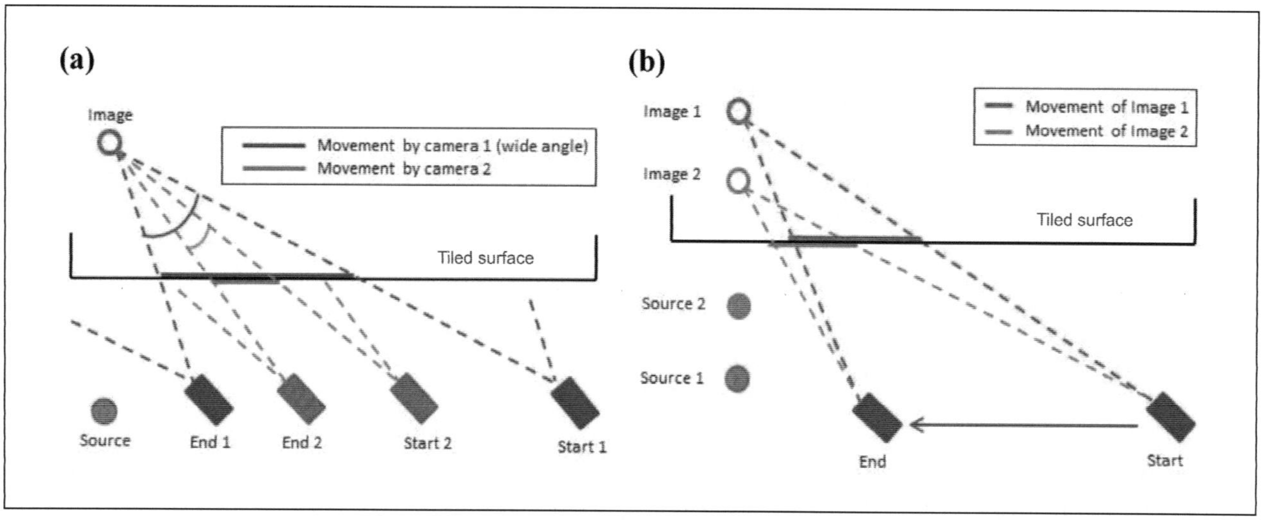


**Figure 14.** Factors affecting the performance of the algorithm: (*a*) wide-angle lens and (*b*) distance of the thermal source to the reflective surface.

thermal radiation. An algorithm was developed in this study as an alternate method to reduce thermal reflection by combining thermal data from multiple thermographs taken with a regular infrared lens. The algorithm uses the minimal apparent temperature among multiple thermographs to approximate the true value for each location on the target area. The algorithm works with the reflections caused by both moving and stationary thermal sources. With the infrared camera set stationary, the algorithm can be performed as data combination without or before mapping to save time in data processing. When the infrared camera is installed on a moving vehicle, data mapping should be performed first, followed by data combination for each target area. Both laboratory experiments and field tests showed that the algorithm worked well in reducing reflections caused by moving and stationary thermal sources. The algorithm with the high-speed vehicle-mounted IRT method could be a promising fast NDE solution for damage detection in tunnels with reflective finishes. The discussion section of this study contained a few suggestions for improving performance in field applications.

### Author Contributions

The authors confirm contribution to the paper as follows: study conception and design: Dewei Meng, Hoda Azari; data collection: Dewei Meng; analysis and interpretation of results: Dewei Meng; draft manuscript preparation: Dewei Meng. All authors reviewed the results and approved the final version of the manuscript.

### References

1. Portland Cement Association. *Concrete Slab Surface Defects: Causes, Prevention, Repair*. PCA R&D Serial No. 2155. Portland Cement Association, Skokie, Ill., 2001.
2. Clemena, G. G., and W. T. McKeel, Jr. Detection of Delamination in Bridge Decks with Infrared Thermography. *Transportation Research Record: Journal of the Transportation Research Board*, 1978. 664: 180–182.
3. Manning, D. G., and F. B. Holt. Detecting Delamination in Concrete Bridge Decks. *Concrete International*, Vol. 2, No. 11, 1980, pp. 34–41.
4. Gucunski, N., A. Imani, F. Romero, S. Nazarian, D. Yuan, H. Wiggenhauser, P. Shokouhi, A. Taffe, and D. Kutrubes. *SHRP 2 Report S2-R06A-RR-1: Nondestructive Testing to Identify Concrete Bridge Deck Deterioration*. Transportation Research Board of the National Academies, Washington, D.C., 2013.
5. Usamentiaga, R., P. Venegas, J. Guerediaga, L. Vega, J. Molleda, and F. G. Bulnes. Infrared Thermography for Temperature Measurement and Non-destructive Testing. *Sensors*, Vol. 14, No. 7, 2014, pp. 12305–12348.
6. ASTM International. *Standard Test Method for Detecting Delaminations in Bridge Decks Using Infrared Thermography*. ASTM D4788-03. ASTM International, West Conshohocken, Pa., 2013.
7. Omar, T., and M. L. Nehdi. Application of Passive Infrared Thermography for the Detection of Defects in Concrete Bridge Elements. *Proc., 2016 Conference of the Transportation Association of Canada*, Transportation Association of Canada, Toronto, Canada, 2016.
8. Barla, G., F. Antolini, and G. Gigli. 3D Laser Scanner and Thermography for Tunnel Discontinuity Mapping. *Geomechanics and Tunnelling*, Vol. 9, 2016, pp. 29–36. doi: 10.1002/geot.201500050.
9. Feng, X. *Rockburst: Mechanism, Monitoring, Warning and Mitigation*. Butterworth-Heinemann, Cambridge, Mass., 2017.
10. Zou, B., Z. Luo, J. Wang, G. Wang, and L. Hu. Infrared Thermography Analysis of Tunnel Surrounding Rock Damage under Blasting. *Journal of Engineering Blasting*, No. 4, 2016, pp. 1–6.
11. Williams, N. D. *Nondestructive Testing of Rail Tunnel Linings*. Master's thesis. Texas A&M University, 2014.
12. Sheen, Y. N. Quality Evaluation of Tunnel Lining by Using Infrared Thermography. *Journal of Chinese Corrosion Engineering*, Vol. 18, No. 3, 2004, pp. 241–248.
13. Kurosu, H. Monitoring of Subway Tunnel by Using Infrared Thermography Camera. *Journal of the Japan Society of Photogrammetry and Remote Sensing*, Vol. 55, No. 1, 2016, pp. 19–22.
14. Wimsatt, A. J. National Research Council, Transportation Research Board, and Second Strategic Highway Research Program. *SHRP 2 Report S2-R06G-RR-1: Mapping Voids, Debonding, Delaminations, Moisture, and Other Defects Behind or Within Tunnel Linings*. Transportation Research Board of the National Academies, Washington, D.C., 2014.
15. White, J., S. Hurlebaus, P. Shokouhi, A. Wittwer, and A. Wimsatt. Noncontact Techniques for Monitoring of Tunnel Linings. *Structural Monitoring and Maintenance*, Vol. 1, No. 2, 2014, pp. 197–211.
16. Henke, S., D. Karstädt, K. P. Möllmann, F. Pinno, and M. Vollmer. Identification and Suppression of Thermal Reflections in Infrared Thermal Imaging. *Inframation Proc*, Vol. 5, 2004, pp. 287–298.
17. Washer, G., R. Fenwick, S. Nelson, and R. Rumbayan. Guidelines for Thermographic Inspection of Concrete Bridge Components in Shaded Conditions. *Transportation Research Record: Journal of the Transportation Research Board*, 2013. 2360: 13–20.

*The Standing Committee on Tunnels and Underground Structures (AFF60) peer-reviewed this paper (18-00957).*

Article

TRR
JOURNAL OF THE TRANSPORTATION RESEARCH BOARD

Transportation Research Record
2018, Vol. 2672(41) 156–164

DOI: 10.1177/0361198118798458
journals.sagepub.com/home/trr
SAGE

# Blast Vulnerability Assessment of Road Tunnels with Reinforced Concrete Liners

**Fengtao Bai[1], Qi Guo[1], Kyle Root[1], Clay Naito[1], and Spencer Quiel[1]**

**Abstract**

Tunnels are a critical component of our transportation infrastructure, and unexpected damage to a tunnel can significantly and adversely impact the functionality of a transportation network. Tunnel systems are vulnerable to potential threats of intentional and accidental blast events because of their relatively unrestricted public access. These events can lead to spalling and breach of the tunnel liner which, depending on the surrounding media, can result in local damage and progressive collapse of the tunnel. Current approaches for evaluating blast-induced damage to a tunnel liner either require significant computational effort or oversimplification such that accurate spatial distributions of damage cannot be obtained. This study presents an effective approach to predict and map the damage to a reinforced concrete liner of a roadway tunnel from various explosive threat sizes and tunnel geometries. A literature review of existing studies is conducted, and potential scenarios of blast events are examined with varying charge position and size. Rectangular, horseshoe, and circular tunnel geometries, each with the same traffic throughput, are evaluated. An efficient analytical approach to determine the spatial distribution of blast-induced spall and breach damage is presented and shows good agreement with numerical models analyzed in LS-DYNA. The proposed approach is then used to examine the relationship between increasing blast hazard intensity and the extent of spall and breach damage. Inflection points in this relationship can be used to identify hazard levels at which a progressive collapse evaluation would be warranted.

Extreme threats such as blast, fire, and vehicle accidents pose challenges to the safety of transportation infrastructure. The threats themselves can result in direct loss of life, and can lead to localized damage and catastrophic outcomes such as progressive collapse. Bridges and tunnels are typically considered critical to our national infrastructure as the incapacity or damage of such structures would jeopardize regional security as well as economic functionality. A Presidential Policy Directive (*1*) has specifically stipulated the need to strengthen the security and resilience of our nation's critical infrastructure. Furthermore, a National Cooperative Highway Research Program (NCHRP) study of transportation tunnel security recently highlighted the threats posed by intentional and accidental blast (*2*).

Roadway tunnels are particularly vulnerable to terrorist attack because most have uncontrolled public access to vehicles ranging from motorbikes and passenger cars to large cargo-hauling trucks. Though the risk of terrorist attack is low, consideration of these threats is amplified as (1) they typically carry high consequence; (2) they are attracted to high-impact, high-visibility targets, particularly those with lower levels of perceived security; and (3) public fear or dread of these events contributes to "probability neglect," which is the goal of most terrorist activities (*3*). Between 1970 and 2015, the United States alone has seen 89 terrorist attacks targeting the transportation sector, 64% of which were carried out through bombings (*4*). Tunnels specifically may pose an even greater concern when considering the additional confinement during a blast as well as the enclosed space in which to operate rescue and repair efforts. Though many recent incidents around the world have involved hand-carried threats in subway tunnels (*5*), U.S. government agencies have been focusing their mitigation efforts more on the threat posed by vehicle-borne improvised explosive devices (VBIEDs) (*2*).

The aforementioned NCHRP effort has highlighted a need for improved assessment of transportation tunnel vulnerability to blast (*2*). In particular, few available approaches can provide reasonably conservative, computationally efficient evaluations of the spatial distribution and structural consequences of blast-induced damage.

[1]ATLSS Research Center, Lehigh University, Bethlehem, PA

**Corresponding Author:**
Address correspondence to Spencer Quiel: squiel@lehigh.edu

Numerous research studies have examined the propagation of blast-induced pressure loads in tunnels and other confined structures via experiments (*6*, *7*) and high-fidelity computational fluid dynamic analysis (*8–10*). The results of these studies have made strides in establishing the temporal and spatial distribution of blast pressure on tunnel liners. Some studies have used finite element analyses to model the spalling and/or breach effects of these blast loads on the tunnel liners, which are typically constructed of reinforced concrete (*7*, *11*, *12*). Previous studies have collectively demonstrated that shock-induced cracking, spalling, and breach is the dominant failure mode of tunnel liners in response to blast pressure, as opposed to the flexural response that is more typical of many building or bridge structures.

The most significant contributions made by previous research in this field have typically pertained to the use of high-fidelity modeling tools, which are computationally demanding and are not conducive to parametric assessment of tunnels subjected to blast loads. This study addresses the need for assessment tools of "intermediate" complexity, which can deliver accurate, conservative assessments of blast-induced damage at a fraction of the computational effort of high-fidelity models. A similar approach to intermediate models has been developed previously by one of the authors for the vulnerability of bridges to unconfined hydrocarbon fires (*13*). This study will focus on the levels of spall and breach damage induced by VBIED hazards that are traveling in typical roadway tunnels. Three tunnel prototypes with varying geometry but identical vehicle throughput are evaluated for a range of blast hazards. Spatial contours of damage are developed using the proposed approach and are compared with results obtained from high-fidelity computational analysis.

The proposed approach can be used to efficiently establish the relationship between blast hazard intensity and the extent of damage for a range of tunnel design parameters or conditions. The extent of tunnel damage can then be used as the initial conditions for progressive collapse assessment of the tunnel liner. A similar approach for mapping blast-induced damage as a prelude to progressive collapse evaluation was used previously by the authors for building frame structures (*14*). The proposed approach is used to develop threat versus damage curves which indicate appropriate hazard levels for which progressive collapse-resistant design may be feasible and/or warranted. Increases in tunnel wall thickness as a function of tunnel geometry are also developed to illustrate the relative amount of concrete needed to prevent blast-induced breach of the tunnel liner.

## Blast Evaluation of Tunnels

This study will focus on VBIED blast hazards and the associated spall and breach damage to the liner of common reinforced concrete tunnels. Detonation of explosive materials generates a rapidly expanding pressure wave that radiates spherically from the charge. These shock waves can generate a significant pressure impulse when they strike structural components in their path. This short-duration load is termed reflected pressure and varies in magnitude and duration depending on the size of explosive, standoff distance, and the spatial orientation of the loaded component to the shock wave front. Blast demands are generally divided into contact, close-in, and far-field events. Far-field events are generally assumed to occur at a scaled distance of greater than $5.0\,\text{ft/lb}^{1/3}$—scaled distance is the standoff distance (in feet) divided by the cube root of the charge size (in lbs of TNT). Contact and close-in events occur at smaller scaled distances. The response of a conventional reinforced concrete structural component under a far-field blast demand is typically governed by flexural and shear response. Tunnel liners are backed by the surrounding media and are thus resistant to flexural modes of damage. The response to a close-in event is controlled by stress wave propagation through the thickness of the component, thereby resulting in spalling, cracking, and breach. In tunnel liners under blast loading, the pressure demand generates a compression wave through the thickness of the concrete liner that can cause spalling as the compression wave reflects off the back and/or front surfaces, that is, causing the stress to exceed the dynamic strength capacity. If the depth of spall approaches the liner thickness, a breached opening can be generated through the component. Tunnel liners may experience an even more pronounced spall and breach damage response if the surrounding media has a higher impedance to the shockwave propagation. Tunnel enclosures limit the available standoff and can result in reflection of the blast front such that each surface can experience multiple pressure impulses. Therefore, the primary concern for tunnel liners subject to blast loads is breach and spall damage.

Evaluations of the damage resulting from blast events have been conducted via either experimental studies or complex numerical modeling. Full-scale blast events are very expensive to conduct, and alternative means (such as scaled tests via small charges, shock tubes, or impact hammers) are often used for damage analysis. Experimental studies such as Colombo et al. (*15*) focused mainly on the development of new tunnel liner technologies rather than the evaluation of existing inventories.

## Proposed Framework for Blast Damage Assessment

UFC 3-340-02, a U.S. Government manual for blast resistant structural design (*16*), utilizes an empirical approach for calculating spall and breach in concrete

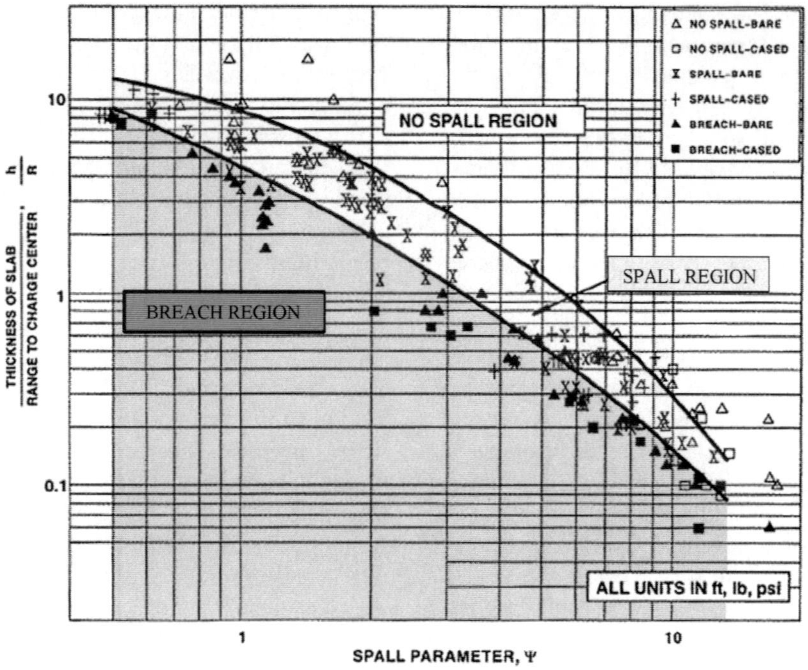


**Figure 1.** Spall and breach curves and supporting test data from Figure 4-65a, UFC 3-340-02 (*16*).

slabs caused by blast loading as a function of concrete compressive strength, as shown in Figure 1. The empirical equations below are valid for a wide range of data as shown in (*16*):

$$\textit{Spall Threshold}: \frac{h}{R} = \frac{1}{a + b\psi^{2.5} + c\psi^{0.5}} \tag{1}$$

$$\textit{Breach Threshold}: \frac{h}{R} = \frac{1}{a + b\psi + c\psi^{2}} \tag{2}$$

where $h$ = concrete thickness (ft); $R$ = standoff from slab face to charge center of gravity (ft); and $\Psi$ = spall parameter. In Equation 1, $a$ = −0.02511; $b$ = 0.01004; $c$ = 0.13613; in Equation 2, $a$ = 0.028205; $b$ = 0.144308; $c$ = 0.049265. For the non-contact and contact charges, the spall parameter $\Psi$ is calculated via the following expressions:

$$\textit{Non-Contact Charges}: \quad \psi = R^{0.926} f_c'^{0.266} W_{adj}^{-0.353} (W_{adj}/(W_{adj} + W_c))^{0.333} \tag{3}$$

$$\textit{Contact Charges}: \psi = 0.527 R^{0.972} f_c'^{0.308} W_{adj}{}^{-0.341} \tag{4}$$

where $f_c'$ = concrete compressive strength (psi); $W_c$ = steel casing weight (lb); $W_{adj}$ = adjusted charge weight (lb) = $B_f\, C_f\, W$; $W$ = equivalent TNT charge weight (lb); $B_f$ = burst configuration factor (1.0 for surface bursts, 0.5 for free air bursts); and $C_f$ = cylindrical charge factor, which is calculated as follows:

**Table 1.** Parametric Ranges for Spall Prediction from Table 4-15a, UFC 3-340-02 (*16*)

| Parameter | Max. | Min. | Avg. |
|---|---|---|---|
| $R$, in | 360 | 0.1 | 21.0 |
| Charge weight, $W$, lb | 2299 | 0.03 | 24.4 |
| Case length, in | 60.0 | 0.80 | 8.8 |
| Case diameter, in | 18.0 | 0.80 | 4.0 |
| Case thickness, in | 0.62 | 0.00 | 0.05 |
| $R/W^{1/3}$, in/lb$^{1/3}$ | 12.1 | 0.008 | 0.70 |
| Concrete thickness, $T$, in | 84.0 | 2.00 | 9.23 |
| $f_c'$, psi | 13815 | 1535 | 5067 |

$$\text{For } L > D \text{ and } R/W^{0.333} < 2.0,$$

$$\text{Else: } C_f = 1 + 2\left(LD/\left(\pi\left(3LD^2/16\right)^{0.667}\right) - 1\right)\left(1 - R/\left(2W^{0.333}\right)\right) \tag{5}$$

$$C_f = 1.0 \tag{6}$$

where $L$ = charge length; $D$ = charge diameter. Dynamic strength increases caused by strain rate effects are included in the expression of $\Psi$ (*16*). As summarized in Table 1, the applicability of the model is limited to the range of test data on which the equations were founded.

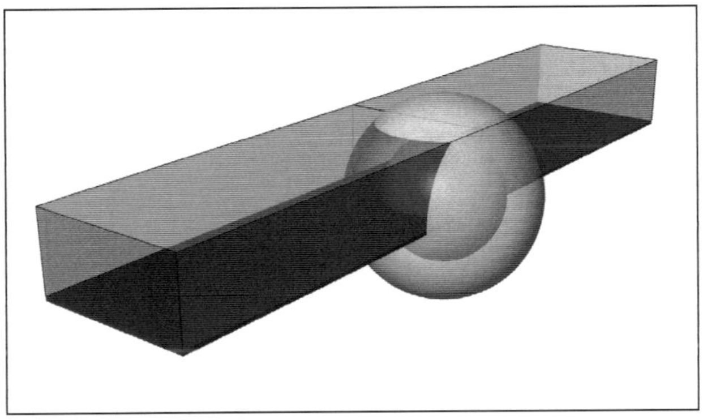

**Figure 2.** Mapping of spherical damage caused by blast in a rectangular tunnel.

In these equations, $W_{adj}$ is the weight of a hemispherical surface charge that applies an explosive impulse at the target equal to that of the actual charge.

In this study, breach and spall thresholds are applied to a tunnel geometry as spheres with radii explicitly solved from Equation 1. The intersection of these spheres with the tunnel liner denotes the extent of breach and spall damage. For this study, a parametric computer-aided drafting environment (*17*, *18*) is used to generate an extruded cross-section of the representative tunnel geometry and analyze its intersection with the blast-induced damage spheres (yellow for breach and red for spall), as demonstrated in Figure 2. This approach has also been implemented in Matlab (*19*), which is more conducive to a parametric study for a range of threat sizes or tunnel geometries.

## Tunnel Prototypes

To demonstrate the proposed approach, three tunnel geometries are considered as representative examples in the current U.S. tunnel inventory: rectangular, horseshoe, and circular. Each tunnel supports two lanes of traffic moving in the same direction. Lane width and shoulder size vary in all three tunnels. Cross-sectional details and typical threat locations are shown in Figure 3. The cross-sectional openings and wall thicknesses are based on tunnels currently in service in the United States. The horseshoe and circular liner thicknesses are scaled to approximately 22 times the effective diameter of the tunnel opening (based on the cross-sectional area). The rectangular tunnel liner thickness is also based on the effective diameter-opening approach but is further increased to account for the additional bending action. These systems are examined for detonation of an explosive charge within the tunnel at the center of the right lane, as shown in Figure 3. Reduction in pressure demands at the entry/exit and damage to the end portals are not considered.

FEMA 426 (*20*) specifies a range of design values for vehicle threats (ranging from sedans to vans and light cargo trucks) from 500 to 4000 lbs of TNT. Three charge weights of 500, 1000, and 2000 lbs of TNT (labeled A, B, and C, respectively) were chosen to conform both to the FEMA hazard scale as well as the charge sizes within the parametric range of damage predictions shown in Table 1 from UFC 3-340-02. These threat sizes are also similar to those selected in previous studies of blast effects in tunnels (*11*).

The evaluation performed in this study conservatively assumes that the charges will be located on the road surface ($B_f$ = 1.0) owing to the uncertainty of the vehicle size and to avoid the complexities associated with air burst (such as complex reflection). The yellow hemispheres in Figure 4 indicate the charge location in the center of the right lane, consistent with Figure 3. Different vehicle heights and delivery methods can be accounted for by varying the elevation of the explosive.

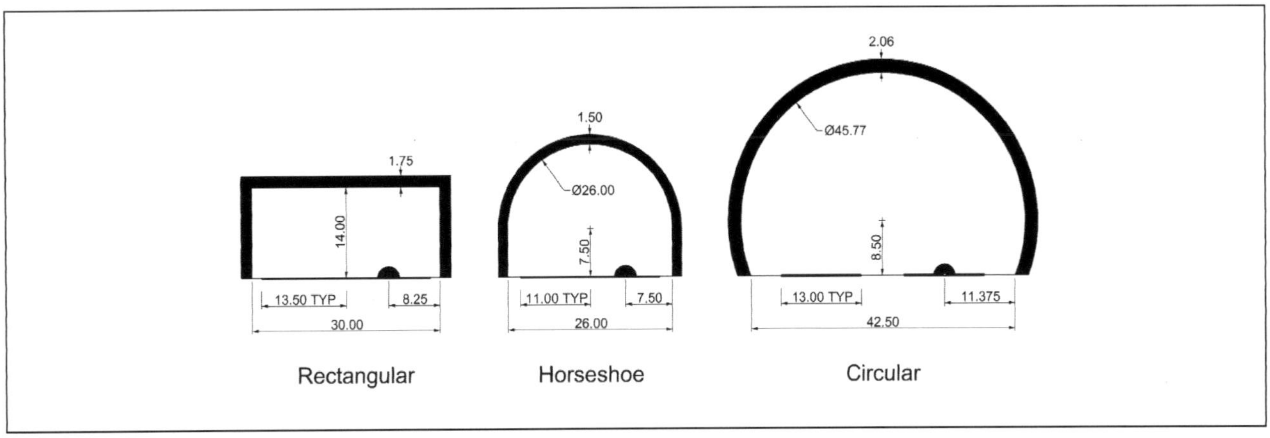


**Figure 3.** Tunnel geometries and details (all dimensions in feet = 0.3048 m).

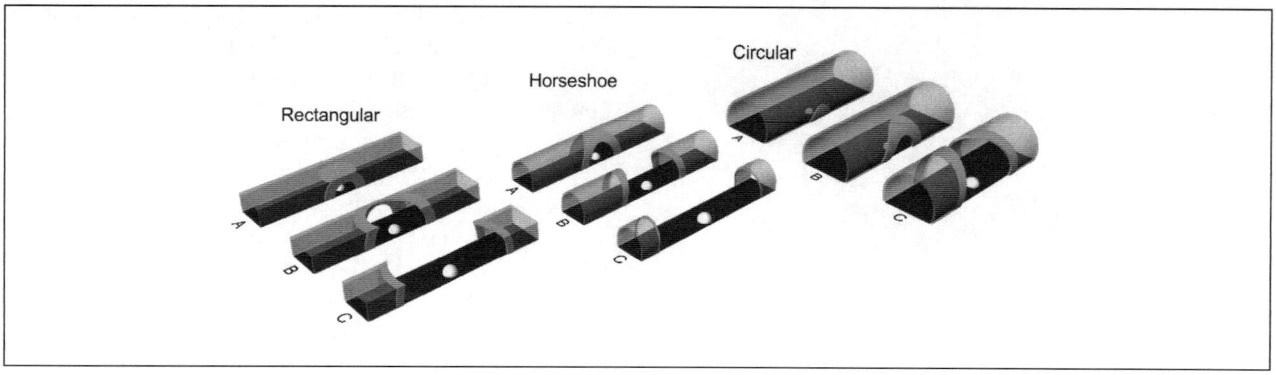


**Figure 4.** Tunnel spall and breach damage for threats A, B, and C.

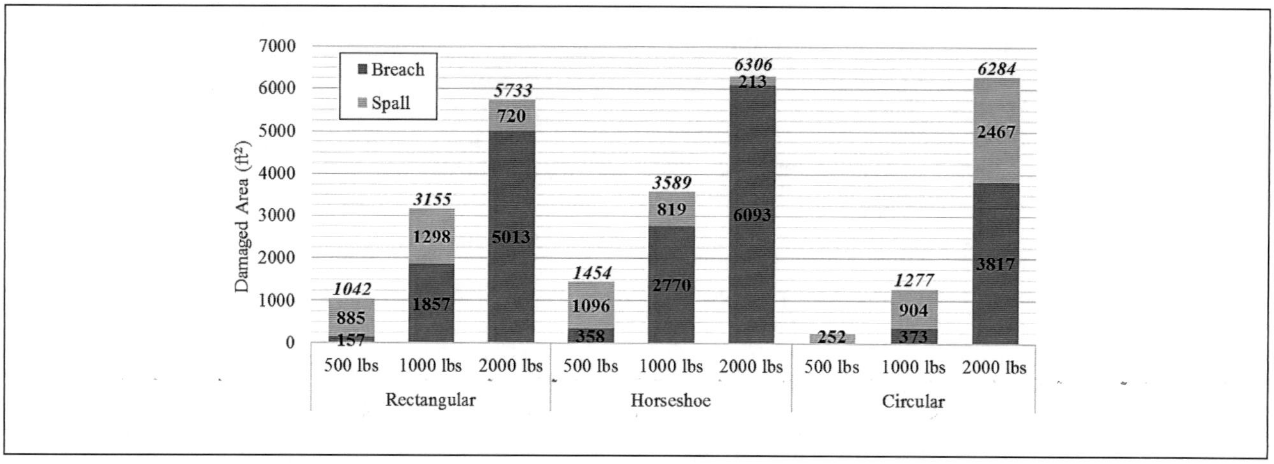


**Figure 5.** Summary of spall and breach damage experienced by each tunnel type for each blast threat.

Breached areas are removed from the tunnel geometry, whereas areas outside the breach radius yet still affected by spalling are represented in red. Road surfaces are shown strictly for clarity—this visualization only considers damage to the tunnel walls. Crater formation at the road surface and blast effects on tunnel lining below the road surface are outside the scope of this study but will be considered in future work.

As shown in Figure 4, the circular tunnel exhibits the least damage when exposed to threats A and B, whereas the rectangular and horseshoe tunnel experienced significant breach when exposed to the same threat levels. A summary of affected areas and threat size for each tunnel is tabulated in Figure 5. Both the rectangular and horseshoe tunnels experience breach at all threat levels as well as a total loss of cross-section for the larger threats—in these cases, a cave-in of the surrounding media as a result of loss of support would be very likely. Damage cases in which up to half of the cross-section is lost would be candidates for progressive collapse evaluation to prevent further loss of the cross-section. Though all three tunnels have the same throughput, the circular tunnel suffered significantly less liner damage because of a larger standoff distance from the charge via the increased shoulders, as shown in Figure 3.

## Comparison with Computational Models

To validate the proposed approach, high-fidelity analyses in LS-DYNA (*21*) were performed with the rectangular tunnel prototype in Figure 3. The explosive threats are assumed to be located on the ground (i.e., as a surface charge) to eliminate the influence of partial reflection if the charge were elevated in a vehicle. The blast effects resulting from the surface charge assumption are conservative and produce an upper bound of applied pressure. The Multi-Material Arbitrary Lagrangian Eulerian method is chosen in LS-DYNA to simulate the blast event. Concrete is modeled with Lagrangian solid elements with an approximate discretization dimension of 4 inches. The tunnel liner is modeled with five solid elements through its thickness, and about one million total solid elements are used to model a 50-foot length of the tunnel. The outside of the tunnel is approximated as

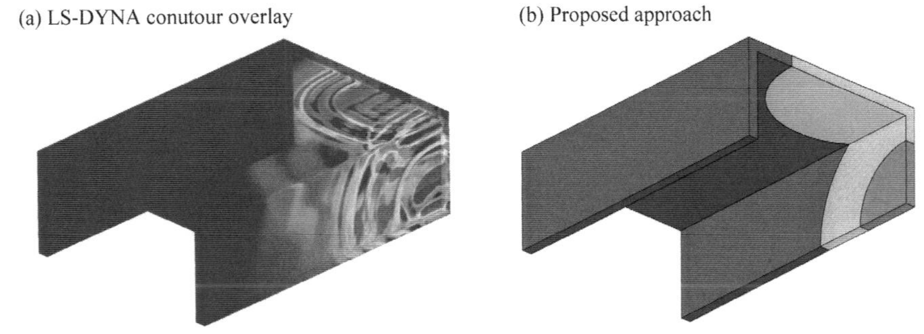


**Figure 6.** Comparison of spall and breach damage contours from (*a*) computational (LS-DYNA) analysis; and (*b*) the proposed approach for the rectangular tunnel subjected to the 500-lb threat.

a rigid boundary. This assumption results in a conservative evaluation of damage by ensuring maximum energy will be reflected back into the liner, providing an upper bound for blast damage. In reality, the surrounding media will absorb some of the pressure shock depending on its impedance—though this energy dissipation is neglected in this analysis, it can be considered via a more complex numerical analysis. A real tunnel system will interact with its backfill substrates, potential groundwater, and unstable sediment and/or rock formations. Variation in the surrounding media can affect the response of the liner to blast depending on the longitudinal location of the threat—critical locations could then be identified and prioritized for strengthening.

The concrete material behavior is approximated as linear elastic with an elastic modulus of 3600 ksi (approximately corresponding to a compressive strength of 4000 psi). As the damage response is governed by shock wave propagation and associated tensile concrete failure through the liner thickness, the steel reinforcement is neglected in the current analyses. Symmetry about the charge location is used to decrease the computational expense. The explosive charge is modeled with LS-DYNA Material Type 8, *MAT_HIGH_EXPLOSIVE_BURN. The JWL equation of state is used to link the explosive reaction pressure to the air volume elements. Both the charge and air domains are constructed with solid elements.

The 500-lb charge is used as a demonstration to compare the proposed approach with the computational analysis. The analysis in LS-DYNA takes approximately 4 hours to analyze 3 milliseconds of the close-in blast (which captures the peak exposure) when performed on a modern desktop computer with i7 quad core processor, 16 GB of RAM, and solid state hard drive. Using the same hardware, the proposed approach takes approximately 10 seconds to analyze the same tunnel segment. Figure 6 compares the results of the computational model and the proposed approach. The contour in Figure 6*a* represents an overlay of maximum principle tensile stresses on the interior face of the tunnel liner at various time steps in the analysis. The red shaded areas indicate that the tensile stress has exceeded the concrete tensile capacity (taken as 316 psi). Breach is indicated at regions where the tensile capacity has been reached through the liner thickness during the analysis. Spalled regions have reached the tensile capacity only on the interior face owing to thru-thickness shock propagation. Note that more advanced concrete damage models are available, but they were not chosen in this particular study because of their complexity, increased computation time, and limited accessibility to practicing engineers. Rather, a simplified failure criterion that provides an upper bound of concrete damage is used.

Figure 6 shows good agreement of the spatial distribution of damage between the LS-DYNA analysis and the proposed approach. The overall areas of spall and breach damage agree within 10%, with the computational model indicating slightly more breach and the proposed approach slightly more spall. This variation is most likely caused by the pressure variability in the computational model resulting from confinement and reflection of the shock wave. These effects will be considered in future study.

## Parametric Study

As a result of its computational efficiency, the proposed approach readily allows for parametric variation of the tunnel geometry and rapid determination of damaged regions for a given threat. In this way, tunnels with non-uniform cross-sections and/or cross-passageways could be quickly assessed with the same framework. This approach provides two important tools to the designer. First, the area contour of damage can be quantified for a given threat. The damaged zone can then be used as the input for a progressive collapse assessment of the tunnel liner. Second, the approach can be used to examine a

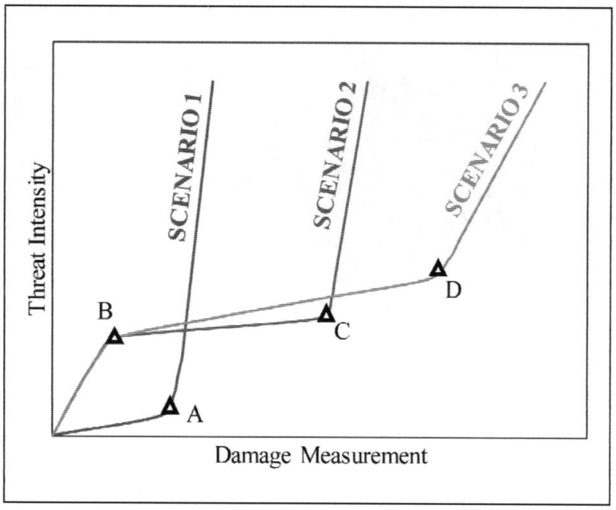


**Figure 7.** Typical relationships of spall+breach damage versus charge size.

range of threats and assess the correlation of increasing damage with increases in blast intensity.

Several typical relationships between damage and demand (i.e., breach/spall area versus charge size) are illustrated in Figure 7. In scenario 1, damage increases very quickly at low threat levels but slows down considerably at point A. For this scenario, the system could conceptually be designed for threat intensity A, as larger threats would not significantly increase the damage. Scenarios 2 and 3 have an initial region where damage accumulates slowly with increasing threat size. This could represent a small charge size which would have minimal damage until the spall threshold is reached. Both scenarios 2 and 3 reach a critical point B, at which the gradient decreases and damage starts to increase with small changes in threat size. Then at points C and D, the rate of damage decreases as larger threats become less effective at inflicting additional damage. Point B or points C/D could be used for design depending on the allowable level of damage for the system.

This approach is implemented in Figure 8 for the three prototype tunnels. The solid curves in Figure 8 show the relationship between charge weight and damaged area, and the dotted lines represent the gradient or slope of the threat-to-damage relationship. For all the three tunnels, the TNT weight was progressively increased from 200 to 2300 lbs of TNT, which is the maximum value in the valid range of the empirical model in UFC 3-340-02 (see Table 1).

The dotted line for the rectangular tunnel in Figure 8*a* shows that the damage gradient initially reduces with increasing blast load, reaching the minimum at approximately 500 lbs and then increasing again at about 900 lbs. After this point, the damage versus charge relationship

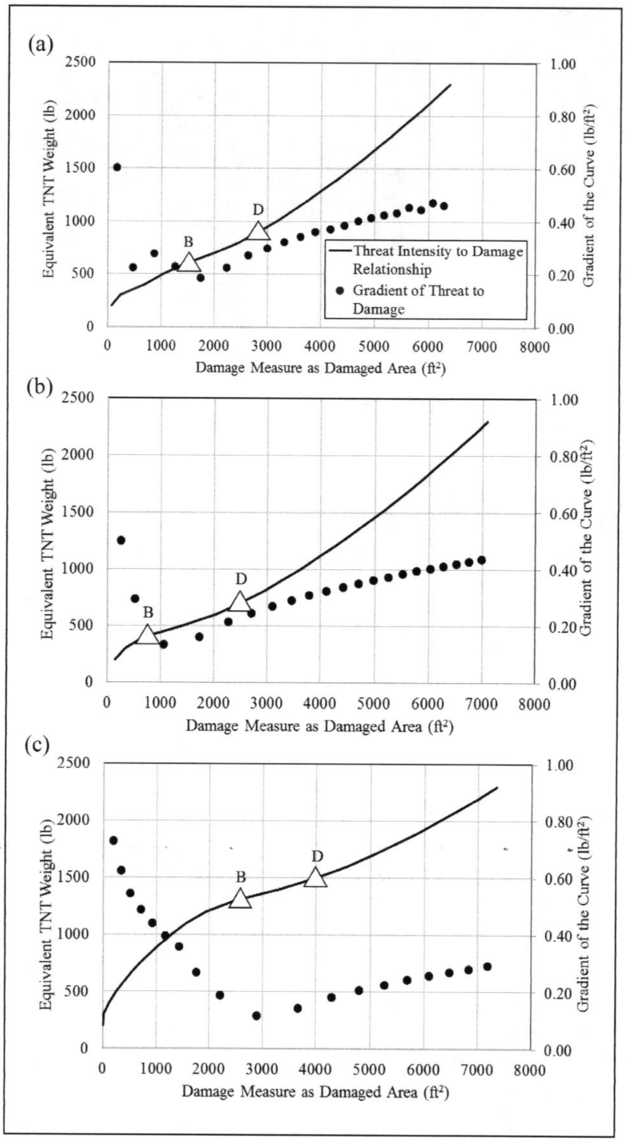


**Figure 8.** Damage area (spall + breach) versus charge weight for each tunnel type: (*a*) rectangular; (*b*) horseshoe; and (*c*) circular.

most resembles scenario 3 in Figure 7. Point B corresponds to the gradient value just before its minimum. Point D corresponds to the subsequent gradient value after which the tangential slope of the gradient line demonstrably decreases. For the rectangular tunnel, Point D at 900 lbs could be an optimal value for design because the damage increases more significantly for a modest increase in threat size up to this point. This is consistent with the results shown in Figures 4 and 5, which showed significant increases in breach area as the threat is increased from 500 to 1000 lbs. The horseshoe and circular tunnels show similar behavior to that of the rectangular tunnel, with a more pronounced change of gradient beyond point D. The critical threat size range in

**Table 2.** Tunnel Performance and Retrofit Requirements for the 1000-lb Threat

| Tunnel type | Design/retrofit | Wall thickness (ft)[a] | Total damaged area ($ft^2$)[b] | Breaching area ($ft^2$)[b] |
|---|---|---|---|---|
| Rectangular tunnel | Design | 1.75 | 3155 | 1857 |
| | Retrofit | 3.47 | 837 | 0 |
| Horseshoe tunnel | Design | 1.50 | 3589 | 2770 |
| | Retrofit | 3.64 | 467 | 0 |
| Circular tunnel | Design | 2.06 | 1277 | 373 |
| | Retrofit | 2.90 | 534 | 0 |

[a] 1 ft = 0.3048 m.
[b] 1 $ft^2$ = 0.093 $m^2$.

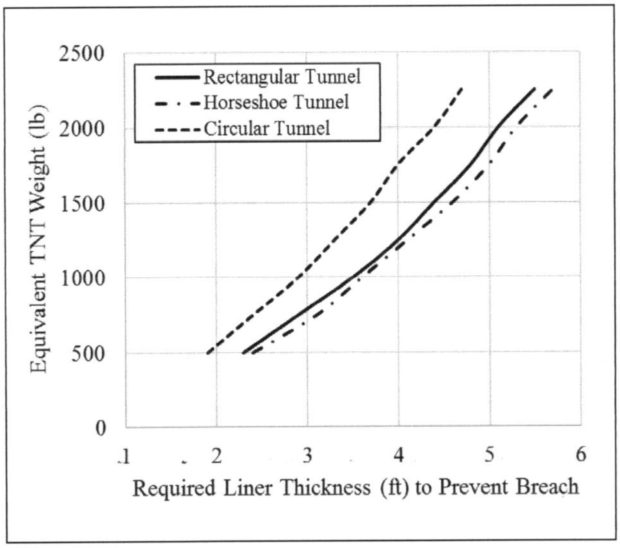


**Figure 9.** Tunnel liner thicknesses needed to prevent breach.

which breaching rapidly increases is approximately 400 to 700 lbs of TNT for the horseshoe tunnel and 1300 to 1500 lbs of TNT for the circular tunnel, which are marked in Figures 8*b* and 8*c* between points B and D. The hazard level at point D could be used to establish the initial conditions for progressive collapse evaluation of the tunnel. Up to this point, enough of the cross-section remains such that an evaluation of the damaged liner to resist subsequent progressive collapse could be warranted and effective. Threat sizes beyond this point will increase the damage such that the cross-section circumference becomes completely breached and a subsequent cave-in becomes likely.

To avoid breach, the wall thickness could be increased via shotcrete, additional precast concrete panels, or other means. As a simple illustration in Table 2, the horseshoe and rectangular systems wall thickness can be increased to 3.63 ft and 3.47 ft, respectively, to eliminate the occurrence of breach under the 1000 lbs VBIED, though spalling will still occur. This approach can be extended to any charge size and tunnel geometry to provide the owner with a means to estimate a target level of design performance. As an example, the liner thickness needed to eliminate breaching is plotted for each tunnel geometry for increasing charge size in Figure 9. The results show that for standard VBIED sizes, it is possible to mitigate breach damage via increases in wall thickness.

## Conclusions

An examination of concrete tunnel liner vulnerability to blast was conducted for three prototype systems subjected to a range of blast load intensities. The tunnels all provide access to two lanes of traffic but have varying geometries that include a rectangular, horseshoe, and circular shape corresponding to three tunnel types currently in service in the U.S. A simplified analysis approach is presented, and numerical modeling is performed as validation. The numerical model was developed to conservatively evaluate the maximum blast damage. The results of the study are as follows:

1. Empirical equations in UFC 3-340-02 can reasonably predict the upper bound of concrete liner breach and spall damage caused by blast loads. This approach is highly efficient, and the results are comparable to numerical models; however, the approach relies on several conservative assumptions and should be considered as an upper bound for damage prediction.
2. Three existing tunnels and three charge weights are analyzed. As expected, the results show that the spall and breach damage is correlated to standoff distance for a specific charge weight.
3. The results also indicate that a moderately sized VBIED can produce a significant amount of breaching in a tunnel system.
4. The approach can be readily implemented to estimate the amount of liner thickness retrofit needed to mitigate damage under a specific threat.
5. The approach can also be used to determine the minimum recommended design threats based on the gradient of the threat-to-damage relationship.

Inflection points in this relationship can be used to identify hazard levels at which a progressive collapse evaluation would be warranted.

Tunnel damage under blast is a developing topic. A reliable and efficient methodology is needed for engineering design and analysis. Future efforts include improved analytical approaches for increasingly accurate and efficient predictions, as well as coupling the effects of blast with associated hazards such as fire.

## Acknowledgments

Financial support provided by the U.S. Department of Transportation through the University Transportation Center for Underground Transportation Infrastructure (UTC-UTI) is gratefully acknowledged. Kyle Root also received a tuition scholarship from Lehigh University to support his participation in this project.

## Author Contributions

The authors confirm contribution to the paper as follows: study conception and design: all, data collection: all, analysis and interpretation of results: all, and draft manuscript preparation: all. All authors reviewed the results and approved the final version of the manuscript.

## References

1. PPD-21. *Presidential Policy Directive – Critical Infrastructure Security and Resilience*, The White House, Office of the Press Secretary, Washington, D.C., 2013.
2. Transportation Research Board. *TCRP Report 86/NCHRP Report 525: Public Transportation Security, Making Transportation Tunnels Safe and Secure*. Transportation Research Board of the National Academies, Washington, D.C., 2006.
3. Stewart, M. G., M. D. Netherton, and D. V. Rosowsky. Terrorism Risks and Blast Damage to Built Infrastructure. *Natural Hazards Review*, Vol. 7, No. 3, 2006, pp. 114–122.
4. Erin, M. *Terrorist Attacks Targeting Critical Infrastructure in the United States, 1970-2015*. Report. National Consortium for the Study of Terrorism and Responses to Terrorism, University of Maryland, College Park, MD, 2016.
5. Tim, L., E. Burrows, and A. Dewan. St. Petersburg Metro Explosion: At Least 11 Dead in Russia Blast. *CNN*, http://www.cnn.com/2017/04/03/europe/st-petersburg-russia-explosion/index.html. Accessed August 2, 2017.
6. Edri, I., Z. Savir, V. R. Feldgun, Y. S. Karinski, and D. Z. Yankelevsky. On Blast Pressure Analysis due to a Partially Confined Explosion: I. Experimental Studies. *International Journal of Protective Structures*, Vol. 2, No. 1, 2011, pp. 1–20.
7. Stolz, A., and M. L. Ruiz-Ripoll. Experimental and Computational Characterization of Dynamic Loading and Structural Resistance of Tunnels in Blast Scenarios. *Fire Technology*, Vol. 52, No. 5, 2016, pp. 1595–1618.
8. Feldgun, V. R., A. V. Kochetkov, Y. S. Karinski, and D. Z. Yankelevsky. Internal Blast Loading in a Buried Lined Tunnel. *International Journal of Impact Engineering*, Vol. 35, No. 3, 2008, pp. 172–183.
9. Pennetier, O., M. William-Louis, and A. Langlet. Numerical and Reduced-Scale Experimental Investigation of Blast Wave Shape in Underground Transportation Infrastructure. *Process Safety and Environmental Protection*, Vol. 94, No. C, 2015, pp. 96–104.
10. Uystepruyst, D., and F. Monnoyer. A Numerical Study of the Evolution of the Blast Wave Shape in Rectangular Tunnels. *Journal of Loss Prevention in the Process Industries*, Vol. 34, 2014, pp. 225–231.
11. Choi, S., J. Wang, G. Munfakh, and E. Dwyre. 3D Nonlinear Blast Model Analysis for Underground Structures. *GeoCongress* 2006, 2006, pp. 1–6.
12. Buonsanti, M., and G. Leonardi. 3-D Simulation of Tunnel Structures Under Blast Loading. *Archives of Civil and Mechanical Engineering*, Vol. 13, No. 1, 2013, pp. 128–134.
13. Quiel, S. E., T. Yokoyama, L. S. Bregman, K. A. Mueller, and S. M. Marjanishvili. A Streamlined Framework for Calculating the Response of Steel-Supported Bridges to Open-Air Tanker Truck Fires. *Fire Safety Journal*, Vol. 73, 2015, pp. 63–75.
14. Gombeda, M. J., C. J. Naito, S. E. Quiel, and C. T. Fallon. Blast-Induced Damage Mapping Framework for use in Threat-Dependent Progressive Collapse Assessment of Building Frames. *Journal of Performance of Constructed Facilities*, Vol. 31, No. 2, 2017, pp. 0401-6089.
15. Colombo, M., P. Martinelli, and M. di Prisco. On the Blast Resistance of High Performance Tunnel Segments. *Materials and Structures*, Vol. 49, No. 1, 2016, pp. 117–131.
16. UFC 3-340-02. *Unified Facilities Criteria. Structures to Resist the Effects of Accidental Explosions*. Department of the Army, the NAVY and the Air Force, Washington, D.C., 2008.
17. Robert McNeel & Associates. *Rhinoceros*, Vol. 5.14.522.8390, 2017.
18. Robert McNeel & Associates. *Grasshopper*, Vol. 0.9.0076, 2014.
19. Mathworks. *Matlab, version R2016b*. Mathworks, Inc., Natick, MA, 2016.
20. Federal Emergency Management Agency. *FEMA 426 - Reference Manual to Mitigate Potential Terrorist Attacks Against Buildings*, Federal Emergency Management Agency, Washington, D.C., 2003.
21. Livermore Software Technology Corporation. *LS-DYNA Manual*. Livermore Software Technology Corporation, Livermore, CA, Vol. 199, 1998.

*The Standing Committee on Tunnels and Underground Structures (AFF60) peer-reviewed this paper (18-05880).*

*The opinions expressed in this paper are those of the authors and not the U.S. Department of Transportation.*

TRR
JOURNAL OF THE TRANSPORTATION RESEARCH BOARD

Article

Transportation Research Record
2018, Vol. 2672(41) 165–176

DOI: 10.1177/0361198118780878
journals.sagepub.com/home/trr
SAGE

# Effects of Imperfections in Heat Straightening Repair of Steel Beam Bridges

**Young Moo Sohn**[1], **Amit H. Varma**[2], **and Robert J. Connor**[3]

**Abstract**

A 40 ft. long two-span continuous steel bridge with two composite beams was constructed in the laboratory and subjected to damage followed by heat straightening repair. A36 steel section (W30 × 90) was used for the main girders (beams). Four spans (specimens) of the test bridge were statically damaged at each midspan using a hydraulic actuator, and subsequently repaired by applying Vee heats and restraining forces in the damaged region. Restraining force magnitude (corresponding to 0.4 $M_p$: 6.2 kips and 0.6 $M_p$: 9.5 kips), maximum heating temperature (800°F, 1200°F, and 1400°F), and the number of multiple damage-repair cycles (one and three cycles) were considered as the test parameters. The steel material properties were measured by taking samples from the repaired areas, and compared with undamaged steel material properties. Samples taken from specimens subjected to overheating (up to 1400°F) had similar structural properties and fracture toughness values as those taken from specimens subjected to normal heating (up to 1200°F). Specimens repaired with overstraining (0.6 $M_p$) combined with underheating (up to 800°F) required the largest number of heating cycles to fully repair the same damage. The fracture toughness of samples taken from specimens subjected to multiple (three times) damage-repair cycles was lower (decreased to about 84%) than the fracture toughness of samples taken from specimens subjected to only one damage-repair cycle. Therefore, multiple heat straightening repairs of a damaged beam should be performed with caution. With reference to serviceability performance for AASHTO HL-93 live load, the midspan deflections of beam specimens subjected to damage and heat straightening repair were comparable to those of undamaged beam specimens.

Members of steel bridges are occasionally damaged by impacts from over-height vehicles. Heat straightening is one of the most effective and efficient methods to repair the damage (permanent deformations) caused by over-height vehicle impacts. An evaluation of the Indiana heat straightening repair database indicated that the most frequently damaged-repaired bridges in Indiana were made of A36 steel (*1*, *2*).

Heat straightening repair consists of two stages. In the first stage, restraining forces opposite to the direction of damage are applied and maintained through the second stage. In the second stage, heating is applied in appropriate patterns like Vee, strip, line, or spot heat on the damaged portions of the steel bridge girder using oxygen-fuel torches. The combination of restraining force and heating produces thermoplastic strains and deformations that are opposite to the permanent damage deformations. Several repetitions of the combined restraining force and heating are required to reverse (repair) the permanent deformations to acceptable levels within tolerance (*3–5*).

In previous research (*2*), the most detrimental deviations commonly found in field applications of heat straightening were: (a) underheating below 1200°F (about 800°F), (b) overheating above 1200°F (about 1400°F), (c) overstraining above the recommended restraining force limit, and (d) multiple heat straightening of the same beam more than two times.

There is a lack of data regarding the effects of these deviations/imperfections in the heat straightening repair process on the conditions and serviceability of the damaged-repaired beam bridges. This information is needed to develop tolerances for imperfections in the heat straightening repair process. Bridge rehabilitation engineers need these imperfection tolerances to approve heat straightening implemented in the field and to develop confidence in the condition of the bridge girder after repair.

[1]Engineering Department, Central Connecticut State University, New Britain, CT
[2]Bowen Laboratory, Lyles School of Civil Engineering, Purdue University, West Lafayette, IN
[3]Lyles School of Civil Engineering, Purdue University, West Lafayette, IN

**Corresponding Author:**
Address correspondence to Young Moo Sohn: ymsohn@ccsu.edu

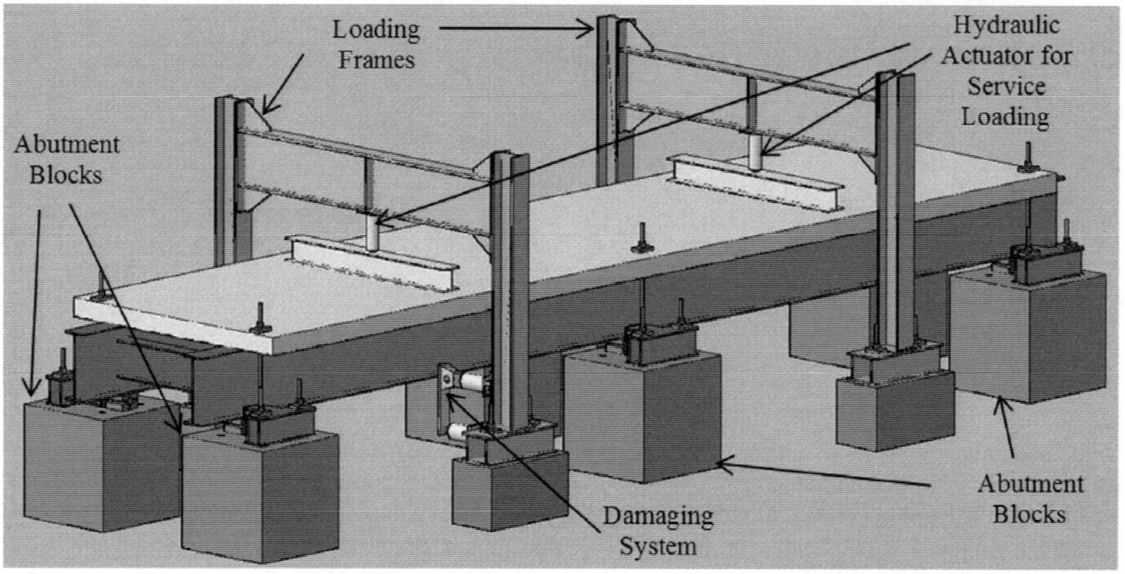


**Figure 1.** Test setup.

## Background

The primary focus of prior research on heat straightening repair has been on: (i) developing effective heating patterns and techniques, (ii) determining thermo-plastic deformations, (iii) determining the structural properties of steels after heat straightening, and (iv) developing guidelines implementing the heat straightening repair process. It is important to note that most prior research has not included the effects of imperfections in the heat straightening repair process other than Sohn and Varma (*1*, *2*).

Kowalkowski and Varma (*6–8*) have conducted detailed experimental investigations to evaluate the effects of multiple damage–heat straightening repair cycles on the structural properties, fracture toughness, and microstructure of A36, A588, and A7 bridge steels.

Sohn and Varma (*1*, *2*) have determined the effects of realistic imperfections including multiple damage–heat straightening repair cycles, overheating, underheating, large damage, and overstraining and their combinations on the structural properties, fracture toughness, and microstructure of A36 bridge steel beams. In summary, the major findings were that multiple damage-repair cycles (up to three times) and overheating (temperatures up to 1400°F) could be acceptable imperfections in heat straightening repair. These findings have not yet been validated for large-scale steel bridges, which is the focus of this paper.

## Experimental Investigations

Experimental investigations were conducted to evaluate the effects of damage followed by realistic heat straightening repair (with imperfections) on the material properties and serviceability of large-scale steel beam bridges. The test bridge was constructed following an analysis of an Indiana Department of Transport database developed during previous research (*1*, *2*). The analysis indicated that: (i) A36 was the most frequently damaged-repaired steel type, (ii) composite continuous bridges were the most frequently damaged-repaired structure type, and (iii) considering elastic rebound of the damaged steel, 30 times the yield strain ($\varepsilon_y$) was typically the maximum average damage amount. Therefore, the experimental investigations were conducted using: (a) A36 W30 × 90 steel beams for the main bridge girder, (b) full composite continuous bridge, (c) 20 ft. span length (equivalent to the cross beam spacing), and (d) 8 in. concrete slab thickness. The applied maximum damage strain ($\varepsilon_d$) was approximately equal to 30 times $\varepsilon_y$.

Figure 1 shows the test setup for conducting the experimental investigations. As shown, the test setup represents a real scale model of steel beam bridges damaged-repaired most frequently in Indiana. Bridges with other boundary conditions, or support/restraint conditions, and different designs (noncomposite, plate girders, etc.) were not included in this study, but are recommended for future research. The test bridge is composed of two steel rolled beams with a concrete deck supported on six concrete support blocks as shown in Figure 1. A total of four specimens were tested as described in Table 1. Pin bearings were used at the middle supports and roller bearings were used at both ends.

Each of the four beam spans were considered as specimens, and subjected to damage and heat straightening repairs simulating field activities and conditions. The specimens were damaged at midspan. After the damage, Vee heats were

**Table 1.** Test Matrix

| Specimen name | Span name | Cycles (number) | Damage strain ($\varepsilon_d$) | Restraining moment ($M_r$) | Maximum temperature ($T_{max}$) | Comment |
|---|---|---|---|---|---|---|
| 3-1400-0.4-30 | S1 | 3 | 30 $\varepsilon_y$ | 0.40 $M_p$ (6.32 kips) | 1400°F | Effects of multiple damage-repair cycles & overheating |
| 1-1400-0.4-30 | S2 | 1 | 30 $\varepsilon_y$ | 0.40 $M_p$ (6.32 kips) | 1400°F | Control specimen (over heating) |
| 3-1200-0.4-30 | N1 | 3 | 30 $\varepsilon_y$ | 0.40 $M_p$ (6.32 kips) | 1200°F | Effects of multiple damage-repair cycles |
| 1-800-0.6-30 | N2 | 1 | 30 $\varepsilon_y$ | 0.60 $M_p$ (9.48 kips) | 800°F | Effects of overstraining & underheating |

(a) (b) (c) (d)

**Figure 2.** Test bridge construction: (*a*) main girder fabrication, (*b*) slab formwork and rebar placement, (*c*) slab curing, and (*d*) test setup.

applied at the bottom flange of the girders to repair the damage. Hydraulic actuators attached to the midspan loading frames were used to apply the damage and restraining forces to the specimens.

After the specimens were subjected to damage and heat straightening repair, the serviceability performance of the repaired bridge was evaluated by subjecting it to static loads simulating the AASHTO 2007 HL-93 live loading. As shown in Figure 1, two hydraulic actuators were attached to the midspan loading frames over the bridge specimen to apply concentrated loads to the bridge.

Finally, the material properties including the structural properties and fracture toughness of the damaged-repaired beam specimens were determined experimentally by taking and testing samples. These material properties were measured by conducting ASTM tests (*9–11*) on standard coupons fabricated from material samples taken from each damaged-repaired beam specimen.

Photographs of the test bridge in the construction stage are shown in Figure 2. The test bridge was constructed as follows: (i) steel members (main girders and cross beams) were assembled on the laboratory floor and lifted up on the

**Figure 3.** Damaging procedure: (*a*) damaging system and (*b*) actual damage of specimen.

abutment blocks (Figure 2*a*), (ii) slab formworks were attached to steel members and steel reinforcements (including shear studs) were placed at the designed positions (Figure 2*b*), (iii) 4000 psi concrete was cast and cured (Figure 2*c*), and (iv) other test setups including loading frames and counter blocks were placed and post-tensioned to the laboratory floor (Figure 2*d*).

## Test Matrix

Following the findings from previous research (*1*, *2*), the damage and repair parameters considered in the test were: (a) damage strain magnitude equal to 30 $\varepsilon_y$, (b) the restraining force (40% and 60% of plastic moment ($M_p$) force), (c) the maximum heating temperature (800°F, 1200°F, and 1400°F), and (d) the number of multiple damage-repair cycles.

## Experimental Procedure

*Damage Procedure.* A specially designed loading beam, connected to two 100-ton capacity hydraulic rams, was used to damage the bottom flange of the test specimen. The two rams were acting together to cause the lateral force, damaging the test specimen horizontally. A photograph of this damaging system is shown in Figure 3*a*. The actual operation of this damage system is shown in Figure 3*b*.

The damage of the bridge specimen during the first damage cycle was monitored carefully using strain gages and displacement transducers attached to the midspan of each specimen. The damaging force applied at the midspan by the hydraulic actuators was slowly increased until the targeted damage strain was reached at the midspan. For multiple damage and repair specimens (S1 and N1), the subsequent damage cycles were applied while monitoring the lateral displacement at the midspan. These damage cycles were completed when the lateral displacement of the midspan reached the same maximum lateral displacement as the first damage cycle.

*Repair Procedure.* An oxygen-propylene torch was used for heating the specimens. The heating technician used a handheld infrared temperature gun to monitor the heating temperature of the steel. A spot heat was applied at the apex of the Vee to reach the targeted heating temperature. After reaching the targeted temperature, the heating torch was moved toward the flange edge following a serpentine path. Vee heat was only applied to the top surface of the bottom flange during the first couple of cycles. However, after noticing excessive upward distortions, Vee heats were also applied to the bottom surface of the bottom flange to reduce distortions.

Vee heat regions of the test specimens are shown in Figure 4*a*. These Vee zones were located within the central plastic strain region of the bridge specimen. The nomenclature of each Vee region is also shown in Figure 4*a*. As seen in Figure 4*a*, the Vee angle used for the test was about 53°. This angle was approximately equal to the Vee angle used to repair the damaged bridge in the heat straightening field visit. Figure 4*b* shows the actual restraining system used to apply restraining force to the damaged specimen. Figure 4*c* shows the Vee heat application using the oxygen-propylene heating torch. The non-contact infrared temperature gun used to check the applied temperature is also shown in Figure 4*c*.

## Evaluation of Serviceability Performance

After the specimens were damaged and heat straightening repaired, the serviceability performance of the repaired beams was determined by subjecting them to static loads simulating the AASHTO 2007 HL-93 live load which can cause the maximum moment at each midspan (damaged-repaired location).

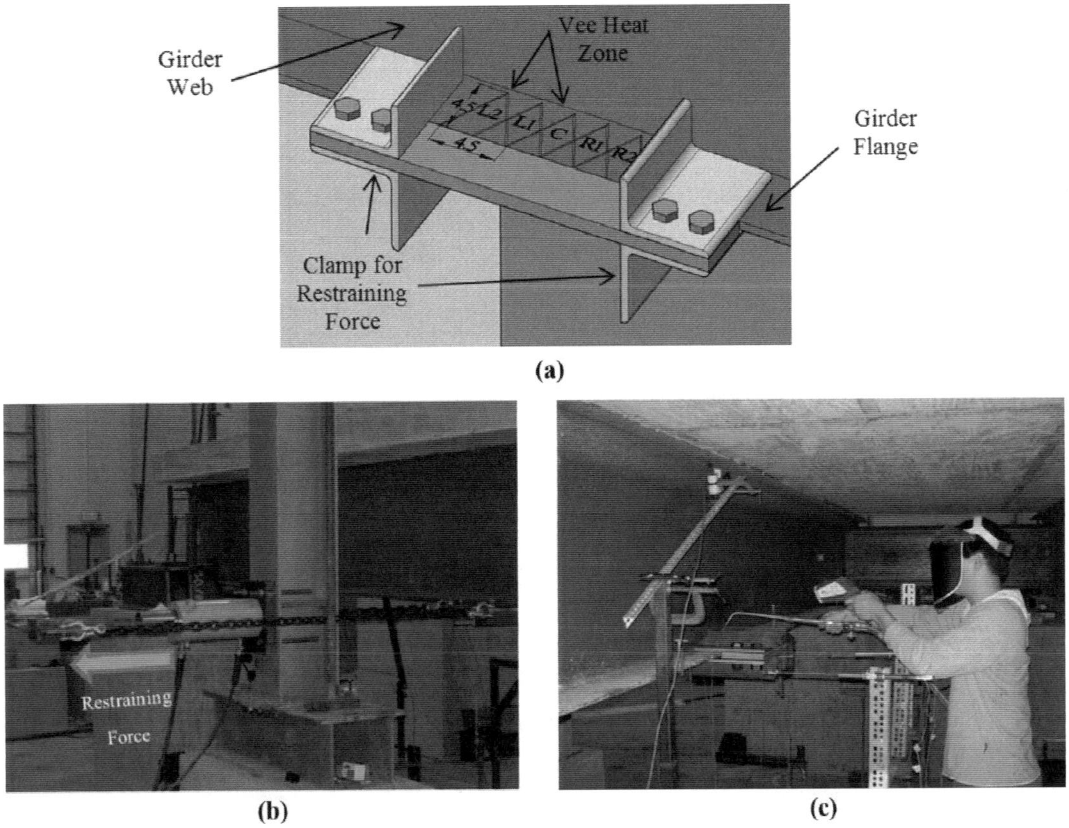


**Figure 4.** Heat straightening repair procedure: (*a*) Vee heat locations and nomenclature, (*b*) restraining system, and (*c*) Vee heat application.

### *Material Testing Procedures*

Charpy V-notch (CVN) fracture toughness coupons and uniaxial tension coupons were taken from the damaged-repaired flange area of each specimen. In addition to the four damaged-repaired specimens, one steel plate from the undamaged bottom flange was prepared to take additional material test coupons.

As shown in Figure 5, two uniaxial tension coupons were taken from the midspan of each test specimen. All tension coupons were taken from the Vee heated regions at the midspan (L1, C, and R1) between 1.6 in. and 3.4 in. from the outer surface of the bottom flange. The actual location and nomenclature for each tensile coupon is also shown in Figure 5.

Fracture toughness tests and Rockwell hardness tests were conducted on the CVN coupons of each test specimen. The CVN coupons of the test specimens are shown in Figure 5. The CVN coupon specimens were removed within a distance of 1.6 in. of the flange edge. As seen in Figure 5, the v-notches of the CVN coupons were fabricated facing the edge of the bottom flange plate. The location and nomenclature of each CVN coupon are also shown in Figure 5.

## Experimental Behavior

### *Damage Process*

Several sensors were used to monitor the specimen behavior during the first damage cycle of each specimen. After the first damage cycle, all the sensors, except strain gages, were attached to the test bridge specimen to monitor the behavior during the following cycles. Damage cycles 2 and 3 were conducted under displacement control, where the displacement at the midspan was monitored and controlled to reach the displacement from the first damage cycle.

Figure 6 shows the lateral force–midspan displacement (P-Δ) behavior of specimen N1 caused by the lateral damaging force. All three graphs in Figure 6 show similar behavior in the damage cycles. The stiffness of specimen N1 in each damage-repair cycle was not severely affected by the damage-repair process.

### *Repair Process*

Figure 7 shows the behavior of specimen N1 in the initial steps of the first repair cycle. Figure 7*a* includes the restraining force applied by the hydraulic ram in kips, which is

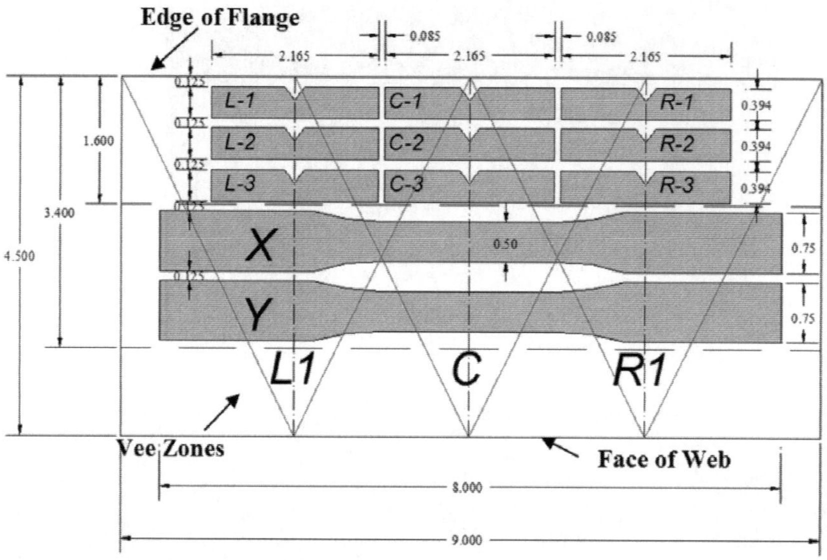


**Figure 5.** Nomenclature and location of material coupons.

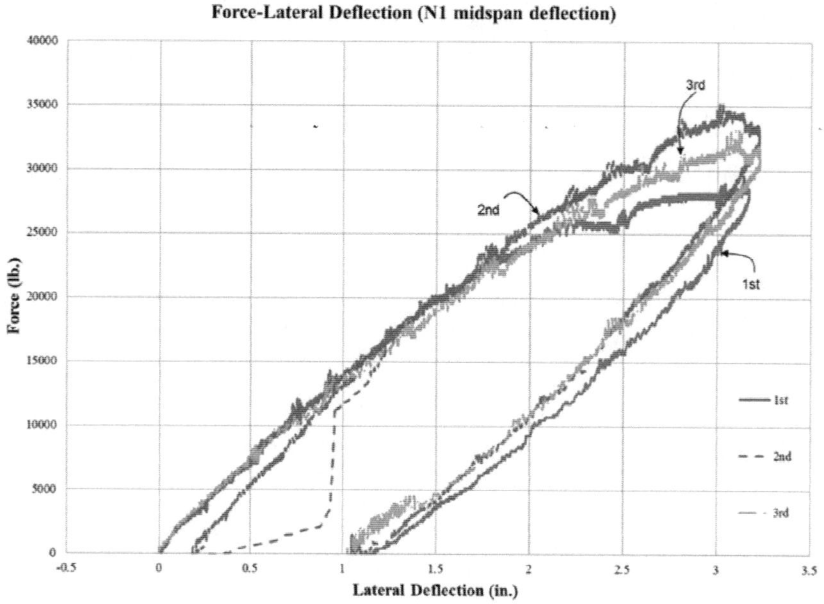


**Figure 6.** Load-displacement diagram during all damage cycles of specimen N1.

divided by 3 (i.e., 1.0 on the graph shows 3 kips of force) and the temperature of the steel in °F, which is divided by 1000°F (i.e., 1.0 on the graph corresponds to 1000°F). Figure 7*b* includes the displacements from three interior displacement transducers in inches.

Repair cycles of specimen N1 were conducted by applying an external force corresponding to the restraining moment of 0.4 $M_p$. Vee heats were initially applied at the center (C) Vee region of the bottom flange shown in Figure 4*a*. After that, other Vee regions (L1, R1-L2, R2) were heated to minimize the out-of-plane distortion of the damaged flange. Each Vee region was heated to a maximum temperature of 1200°F.

After applying Vee heats to straighten the specimen, localized damage of the flange bulges were still left in the midspan at the specimen. These localized damages were repaired by the line heat method with additional jacking force.

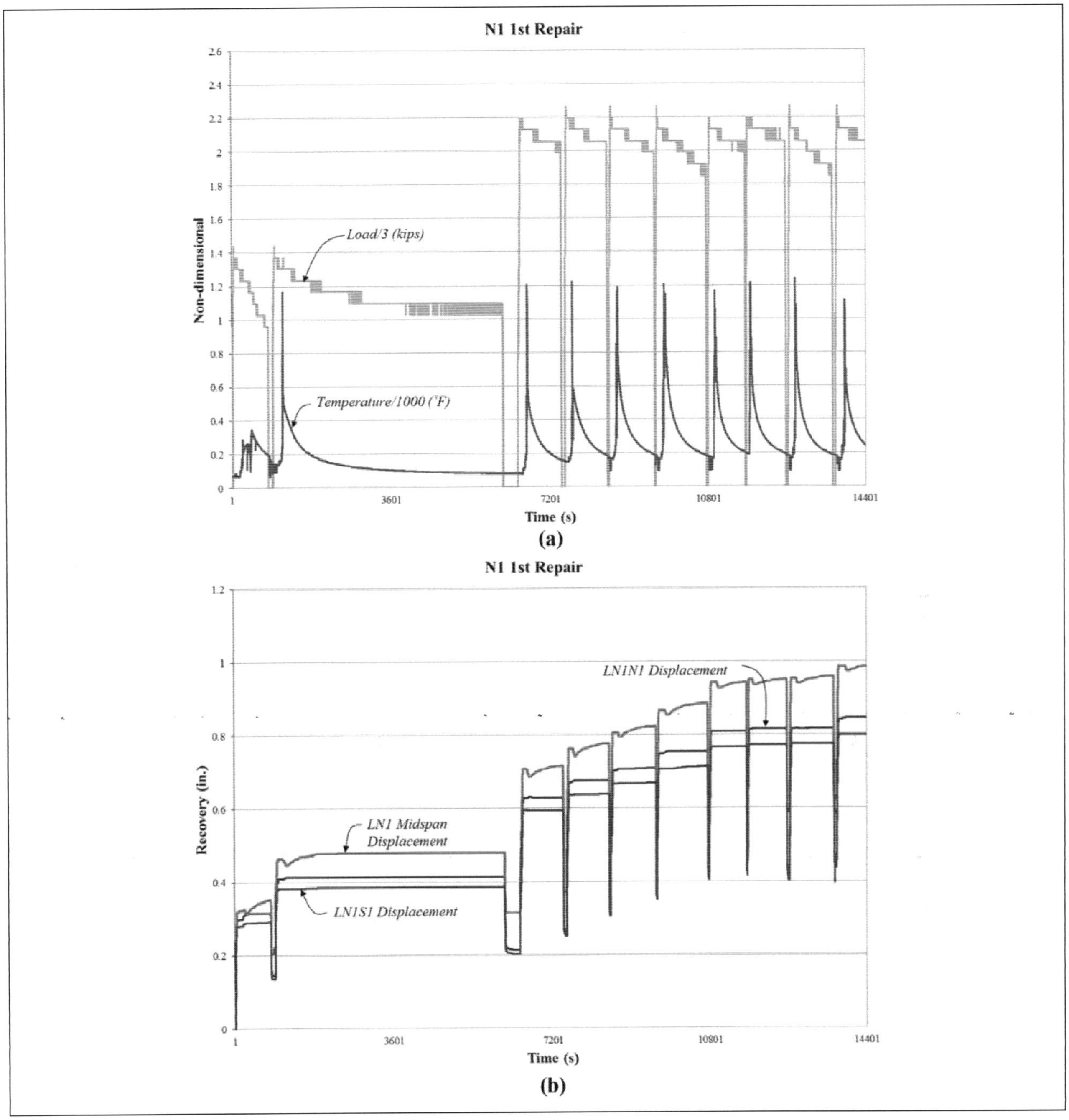


**Figure 7.** Experimental behavior during the repair cycle of specimen N1: (*a*) restraining force and heating temperature, and (*b*) repair progress.

### *Serviceability Behavior*

The serviceability performance of the test bridge during the damage-repair process was determined by simultaneously subjecting each midspan of the test bridge to 55.8 kips of static loading. All vertical midspan displacements of the test bridge were measured and compared to evaluate the influence of damage-repair cycles on serviceability performance.

## Experimental Results

### *Repair Speed*

Table 2 presents the number of heating cycles needed to fully straighten each damaged specimen. To evaluate the effects of overheating on the number of required heating cycles, specimen S1 (heated up to 1400°F) is compared with specimen

**Table 2.** Required Heating Cycles

| | Specimens | | Heating cycles | | | Normalized values | | |
|---|---|---|---|---|---|---|---|---|
| | ID | Designation | 1st | 2nd | 3rd | 1st | 2nd | 3rd |
| Test results | S1 | 3-1400-0.4-30 | 32 | 39 | 58 | – | – | – |
| | S2 | 1-1400-0.4-30 | 24 | – | – | – | – | – |
| | N1 | 3-1200-0.4-30 | 42 | 50 | 66 | – | – | – |
| | N2 | 1-800-0.6-30 | 410 | – | – | – | – | – |
| Effects of overheating | N1 | 3-1200-0.4-30 | 42 | 50 | 66 | 1.00 | 1.00 | 1.00 |
| | S1 | 3-1400-0.4-30 | 32 | 39 | 58 | 0.76 | 0.78 | 0.88 |
| Effects of underheating & overstraining | N1 | 3-1200-0.4-30 | 42 | – | – | 1.00 | – | – |
| | N2 | 1-800-0.6-30 | 410 | – | – | 9.76 | – | – |
| Effects of multiple damage-repair cycles | N1 | 3-1200-0.4-30 | 42 | 50 | 66 | 1.00 | 1.19 | 1.57 |
| | S1 | 3-1400-0.4-30 | 32 | 39 | 58 | 1.00 | 1.22 | 1.81 |

N1 (heated up to 1200°F). The number of heating cycles is normalized with respect to the number of heating cycles required for the normal-heating specimens (heated up to 1200°F) and included in Table 2.

The overheated specimens required a much lower number of heating cycles to fully straighten the damaged steel beams. The required number of heating cycles reduced to 78% (average value) of the heating cycles of the normal-heating specimens.

Table 2 includes the effects of the underheating and overstraining on the required number of heating cycles to repair the damaged steel specimen. Specimens N2 and N1 are compared in Table 2 to evaluate the difference in the required heating cycles. Specimen N1 was repaired using 1200°F heating temperature and 0.4 $M_p$ restraining force, while specimen N2 was repaired using 800°F heating temperature and 0.6 $M_p$ restraining force. Table 2 indicates that specimen N2 required many more heating cycles to fully straighten the damaged steel beams. Specimen N2 needed almost 10 times the number of heating cycles required to repair the damage to specimen N1.

Table 2 shows the effects of the multiple damage-repair cycles on the required heating cycles. As seen in the table, the required number of heating cycles increases with the number of multiple damage-repair cycles. Compared with the first damage-repair cycle, 19% and 22% more heating cycles were required in the second damage-repair cycle. In the third damage-repair cycle, 57% and 81% more heating cycles were required than in the first damage-repair cycle to complete the repair processes.

## *Fracture Toughness Tests*

CVN specimens were removed from the midspan of the bottom flange of all test specimens as shown in Figure 5. One additional flange plate from the undamaged part of the test bridge was prepared to fabricate additional Charpy test coupons to determine the undamaged fracture toughness of the test bridge. The CVN impact fracture tests were conducted at 40°F as recommended for zone 2 by the AASHTO (2007) specification.

The results of the fracture toughness tests on CVN specimens taken from the bridge specimen are listed in Table 3. Averages and standard deviations were taken both across the width of the flanges (i.e., averages of 1, 2, and 3) and along the flange length (i.e., averages of L1, C, and R1). The overall average values and standard deviations of all nine Charpy specimens are also provided in Table 3. These fracture toughness values were normalized with respect to the average (mean) fracture toughness of the undamaged plate, which is also presented in Table 3.

The overall fracture toughness of all damaged-repaired specimens was much lower than the undamaged steel toughness (27% of undamaged fracture toughness value). Compared with the normal-heating specimen (1200°F), the fracture toughness of the overheated specimen (1400°F) increased (comparison of S1 with N1). The 1-800-0.6-30 specimen N2 (with underheating and overstraining) had almost the same fracture toughness as the 1-1400-0.3-30 specimen S2 (with overheating). The fracture toughness of the multiple damage-repair specimen (S1) decreased to 84% of the single damage-repair specimen (S2) toughness (comparison of S1 with S2). The fracture toughness values decreased more in Charpy specimens closer to the flange-web junction. Five individual Charpy coupons showed their fracture toughness values smaller than the fracture critical member limitation (Zone 2 of 25 ft•lb) of the AASHTO 2007 bridge specification. Three of those were slightly lower than the non-fracture critical member limitation (15 ft•lb).

## *Uniaxial Tension Tests*

Uniaxial tension tests were conducted according to ASTM E8 on tension coupons from each specimen. The test results are shown in Table 4, which includes the yield stress ($\sigma_y$), the ultimate stress ($\sigma_t$), the percent elongation, and the percent reduction (area). These test results were normalized with respect to the corresponding values from undamaged steel tests, and these normalized results are included in Table 4.

The yield stress of all the damaged-repaired specimens increased (from 13% to 41%). For overheated specimens (S1 and S2), the yield stress of the multiple damage-repair specimen

**Table 3.** Fracture Toughness Test Results

| | Specimen S1 (ft.·lbs.) | | | | | Normalized S1 (ft.·lbs.) | | | | |
|---|---|---|---|---|---|---|---|---|---|---|
| | L1 | C | R1 | Avg. | Std. dev. | L1 | C | R1 | Avg. | Std. dev. |
| 1 | 98 | 56 | 64 | 72.7 | 18.21 | 0.40 | 0.23 | 0.26 | 0.30 | 0.07 |
| 2 | 220 | 26 | 63 | 103.0 | 84.10 | 0.90 | 0.11 | 0.26 | 0.42 | 0.34 |
| 3 | 41 | 15 | 12 | 22.7 | 13.02 | 0.17 | 0.06 | 0.05 | 0.09 | 0.05 |
| Avg. | 119.7 | 32.3 | 46.3 | 66.1 | | 0.49 | 0.13 | 0.19 | 0.27 | |
| Std. dev. | 74.67 | 17.33 | 24.28 | | 60.18 | 0.31 | 0.07 | 0.10 | | 0.25 |
| | Specimen S2 (ft.·lbs.) | | | | | Normalized S2 (ft.·lbs.) | | | | |
| 1 | 95 | 92 | 100 | 95.7 | 3.30 | 0.39 | 0.38 | 0.41 | 0.39 | 0.01 |
| 2 | 59 | 104 | 94 | 85.7 | 19.29 | 0.24 | 0.43 | 0.39 | 0.35 | 0.08 |
| 3 | 39 | 74 | 49 | 54.0 | 14.72 | 0.16 | 0.30 | 0.20 | 0.22 | 0.06 |
| Avg. | 64.3 | 90.0 | 81.0 | 78.4 | | 0.26 | 0.37 | 0.33 | 0.32 | |
| Std. dev. | 23.17 | 12.33 | 22.76 | | 22.70 | 0.10 | 0.05 | 0.09 | | 0.09 |
| | Specimen N1 (ft.·lbs.) | | | | | Normalized N1 (ft.·lbs.) | | | | |
| 1 | 73 | 71 | 17 | 53.7 | 25.94 | 0.30 | 0.29 | 0.07 | 0.22 | 0.11 |
| 2 | 84 | 83 | 40 | 69.0 | 20.51 | 0.34 | 0.34 | 0.16 | 0.28 | 0.08 |
| 3 | 33 | 34 | 12 | 26.3 | 10.14 | 0.14 | 0.14 | 0.05 | 0.11 | 0.04 |
| Avg. | 63.3 | 62.7 | 23.0 | 49.7 | | 0.26 | 0.26 | 0.09 | 0.20 | |
| Std. dev. | 21.91 | 20.85 | 12.19 | | 26.65 | 0.09 | 0.09 | 0.05 | | 0.11 |
| | Specimen N2 (ft.·lbs.) | | | | | Normalized N2 (ft.·lbs.) | | | | |
| 1 | 103 | 84 | 153 | 113.3 | 29.10 | 0.42 | 0.34 | 0.63 | 0.46 | 0.12 |
| 2 | 25 | 91 | 99 | 71.7 | 33.16 | 0.10 | 0.37 | 0.41 | 0.29 | 0.14 |
| 3 | 9 | 55 | 82 | 48.7 | 30.14 | 0.04 | 0.23 | 0.34 | 0.20 | 0.12 |
| Avg. | 45.7 | 76.7 | 111.3 | 77.9 | | 0.19 | 0.31 | 0.46 | 0.32 | |
| Std. dev. | 41.06 | 15.58 | 30.27 | | 40.84 | 0.17 | 0.06 | 0.12 | | 0.17 |
| | Undamaged specimen (ft.·lbs.) | | | | | Normalized undamaged (ft.·lbs.) | | | | |
| 1 | 246 | 238 | 248 | 244.0 | 4.32 | 1.01 | 0.98 | 1.02 | 1.00 | 0.02 |
| 2 | 235 | 249 | 249 | 244.3 | 6.60 | 0.96 | 1.02 | 1.02 | 1.00 | 0.03 |
| 3 | 234 | 247 | 249 | 243.3 | 6.65 | 0.96 | 1.01 | 1.02 | 1.00 | 0.03 |
| Avg. | 238.3 | 244.7 | 248.7 | 243.9 | | 0.98 | 1.00 | 1.02 | 1.00 | |
| Std. dev. | 5.44 | 4.78 | 0.47 | | 5.97 | 0.02 | 0.02 | 0.00 | | 0.02 |

increased less than the single damage-repair specimen. The yield stress and ultimate stress of overheated specimens (S1 and S2) increased less than the normal and underheated specimens (N1 and N2). Damage-heat straightening repairs reduced the percent elongation (ductility) of A36 steel. The reduced ductility ranged approximately from 69% to 84% of the undamaged steel. The decrease in the percent elongation of normal and underheated specimens (N1 and N2) was larger than that of overheated specimens (S1 and S2). The reduction in area decreased by 3–7% because of the damage–heat straightening repairs.

## *Rockwell Hardness Tests*

Rockwell hardness tests were conducted on the surfaces of the CVN coupons according to ASTM E18. Four points were taken on every specimen. Table 5 shows the Rockwell hardness test results, and the normalized values (normalized with respect to undamaged steel values). The Rockwell hardness increased in all damaged-repaired specimens by about 8.4–13.1%. Multiple damage-repair specimens (S1 and N1) showed more increase in the Rockwell hardness than single damage-repair specimens (S2 and N2).

## *Serviceability Performance*

The serviceability performance of the test bridge was determined by subjecting it to static loads simulating the AASHTO 2007 HL-93 live load which causes maximum moment at each midspan (damaged-repaired point).

The AASHTO 2007 bridge specification limited live-load deflections to L/800 for ordinary bridges and L/1000 for bridges in urban areas that are subject to pedestrian use (*12*). Using these limitations with a 20 ft. (240 in.) span length, allowable maximum deflection values of the test bridge were

**Table 4.** Normalized Uniaxial Tension Test Results

| Specimen | | Coupon | Yield stress (ksi) | | Ultimate stress (ksi) | | Elongation (%) | | Reduction in area (%) | |
|---|---|---|---|---|---|---|---|---|---|---|
| ID | Designation | ID | Actual | Normalized | Actual | Normalized | Actual | Normalized | Actual | Normalized |
| S1 | 3-1400-0.4-30 | 1-XL | 46.20 | 1.13 | 65.00 | 1.09 | 34.00 | 0.80 | 63.00 | 0.93 |
| | | 1-XR | 45.90 | 1.12 | 68.50 | 1.15 | 33.00 | 0.78 | 66.00 | 0.98 |
| | | avg. | 46.05 | 1.13 | 66.75 | 1.12 | 33.50 | 0.79 | 64.50 | 0.96 |
| S2 | 1-1400-0.4-30 | 2-XL | 47.60 | 1.16 | 64.50 | 1.08 | 36.00 | 0.85 | 67.00 | 0.99 |
| | | 2-XR | 50.00 | 1.22 | 66.00 | 1.10 | 35.00 | 0.82 | 64.00 | 0.95 |
| | | avg. | 48.80 | 1.19 | 65.25 | 1.09 | 35.50 | 0.84 | 65.50 | 0.97 |
| N1 | 3-1200-0.4-30 | 3-XL | 56.00 | 1.37 | 69.50 | 1.16 | 31.00 | 0.73 | 64.00 | 0.95 |
| | | 3-XR | 59.50 | 1.45 | 73.00 | 1.22 | 28.00 | 0.66 | 64.00 | 0.95 |
| | | avg. | 57.75 | 1.41 | 71.25 | 1.19 | 29.50 | 0.69 | 64.00 | 0.95 |
| N2 | 1-800-0.6-30 | 4-XL | 52.00 | 1.27 | 66.50 | 1.11 | 34.00 | 0.80 | 63.00 | 0.93 |
| | | 4-XR | 56.00 | 1.37 | 69.50 | 1.16 | 32.00 | 0.75 | 63.00 | 0.93 |
| | | avg. | 54.00 | 1.32 | 68.00 | 1.14 | 33.00 | 0.78 | 63.00 | 0.93 |
| UN | Undamaged | 5-XL | 42.00 | 1.03 | 61.00 | 1.02 | 43.00 | 1.01 | 66.00 | 0.98 |
| | | 5-XR | 39.80 | 0.97 | 58.50 | 0.98 | 42.00 | 0.99 | 69.00 | 1.02 |
| | | avg. | 40.90 | 1.00 | 59.75 | 1.00 | 42.50 | 1.00 | 67.50 | 1.00 |

**Table 5.** Rockwell Hardness Test Results

| | | | Rockwell hardness point | | | | | |
|---|---|---|---|---|---|---|---|---|
| Specimen ID | Specimen designation | Coupon ID | 1 | 2 | 3 | 4 | Avg. | Normalized |
| S1 | 3-1400-0.3-30 | 1-C-2 | 79.5 | 79.0 | 78.5 | 79.0 | 79.0 | 1.13 |
| S2 | 1-1400-0.3-30 | 2-C-2 | 75.0 | 76.0 | 76.0 | 76.0 | 75.8 | 1.08 |
| N1 | 3-1200-0.3-30 | 3-C-2 | 77.5 | 79.0 | 78.0 | 78.0 | 78.1 | 1.12 |
| N2 | 1-800-0.6-30 | 4-C-2 | 76.5 | 77.0 | 77.0 | 76.5 | 76.8 | 1.10 |
| UN | Undamaged | 5-C-2 | 70.5 | 71.0 | 69.0 | 69.0 | 69.9 | 1.00 |

0.3 in. (L/800) and 0.24 in. (L/1000). The maximum midspan deflections in all cases were not larger than 0.045 in. which is much smaller than the AASHTO deflection limit of 0.3 in. or 0.24 in. Therefore, the maximum midspan deflection after the damage-repair process was within the AASHTO specification deflection limit.

Detailed graphs of the midspan deflection during all the stages of testing are shown in Figure 8. One notable feature in the graph was the increase of the midspan deflection right after each damage process. These increases vary from about 25–76%. However, all the deflections are very small, and the changes in deflection are effectively negligible when compared with the deflection limits. This indicates that there is no impact of the damage and repair process on the overall stiffness of the bridge. Since the deflections under live load would not be changed, concerns about the long-term performance of the bridge due to the damage and repair cycle as a result of the impact are not expected. This is also corroborated by the excellent in-service performance of many bridges for which heat straightening repairs have been made. It is also worth noting that while repairs should be made, short-term serviceability issues would not be expected if the bridge is permitted to remain in service in the damaged state for some finite period.

## Conclusion

The maximum heating temperature of 1400°F may be an acceptable imperfection. In this research, overheating specimens (heated up to 1400°F) and normal-heating specimens (heated up to 1200°F) showed similar test results in terms of structural properties and fracture toughness.

The yield stress of all the damaged-repaired specimens increased from 13% to 41% of the undamaged steel. The ultimate stress increased from 9% to 19% of the undamaged steel. The percent elongation decreased to 69–84% of the undamaged steel ductility. The increase in the yield and ultimate stress and the decrease in the ductility was lower for the specimens subjected to overheating (1400°F) than for the specimens subjected to normal heating (1200°F) or underheating (800°F).

The fracture toughness of the multiple damaged-repaired beam (S1) decreased to 84% of the single damage-repaired specimen (S2), and the single damaged-repaired beams' values decreased to 32% of the undamaged specimen. Therefore, multiple heat straightening repairs of a damaged beam should be done with caution. Note, these conclusions are specific to the steel types considered in the study.

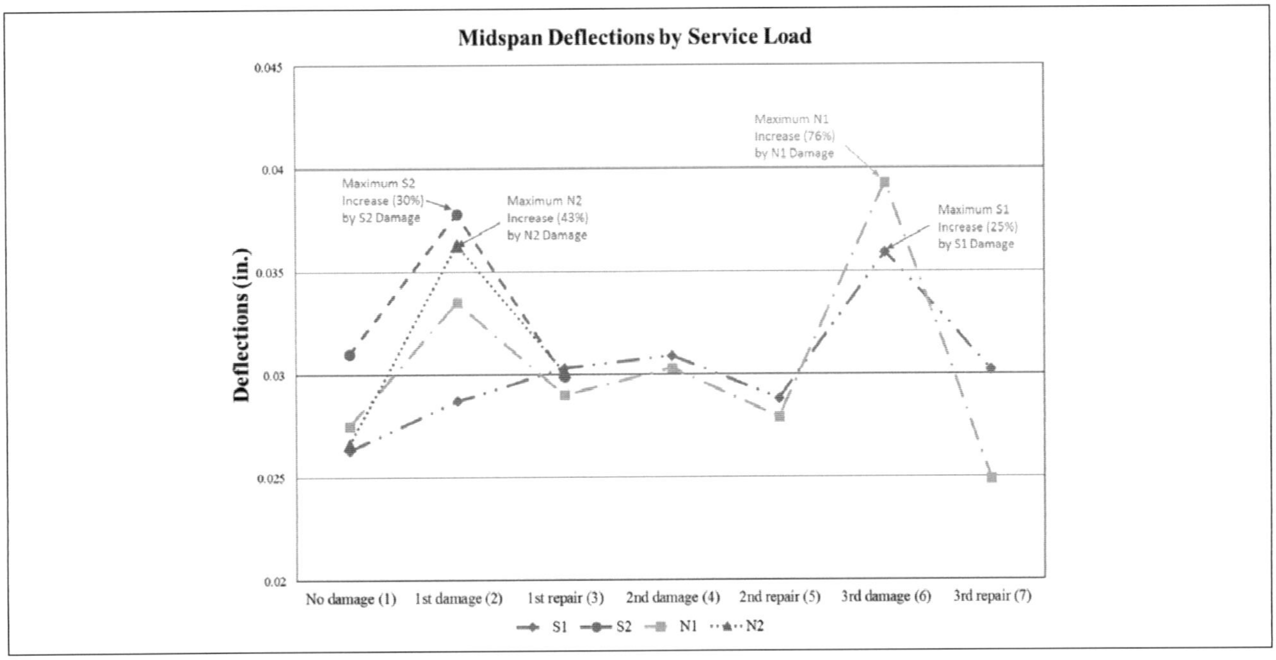


**Figure 8.** Midspan deflections by service load.

The combinations of underheating and overstraining (specimen N2) required 410 heating cycles to fully repair the same damage. This implies that the repair speed of 1-800-0.6-30 specimen (N2) was the slowest. Underheating combined with overstraining is extremely inefficient for performing heat straightening repair.

The serviceability performance test showed that the bridge deflections under the AASHTO HL-93 live loading may increase slightly right after each damage process. However, since the deflection magnitudes and the associated differences observed were very small with respect to the deflection limits, there are no concerns about the effects of damage–heat straightening repair on the long-term serviceability of the bridge. It is worth noting that the effects on stiffness observed in this study are not limited to A36 steel as stiffness is independent of steel type and other steels can be expected to demonstrate similar performance (in terms of stiffness) before or after the repair.

## Limitations

This study focusses on A36 steel because it was identified as the most frequently damaged and heat straightened steel type in Indiana. While the findings, conclusions, and recommendations of this study are limited to A36 steel, the authors believe they could be cautiously applied to other steel grades such as A572, A588, and HPS 70W without additional experimental correlations. In particular, observations regarding damage and repair cycles on stiffness are not limited to A36 steel as stiffness is independent of steel type.

Field performance of a bridge that has been heat straightened in the past seems to further support this conclusion.

Welded details that may be susceptible to fatigue issues were not included in this study. Therefore, these findings, conclusions, and recommendations should not be extended to regions near fatigue critical welded details without additional experimental correlations.

The experimental investigations in this study were conducted inside the laboratory with a limited number of specimens using static damage loading. Additional research is recommended to further validate conclusions in this paper as there could be local damage induced from actual dynamic impacts at the flange-to-deck interface, especially after repeated impacts. The specimens were exposed to conventional air cooling after heating. Other cooling methods like water mist and forced air cooling were not included in this study.

## Acknowledgments

This research was sponsored by the Indiana Department of Transportation through the Joint Transportation Research Program and conducted at Bowen laboratory of Purdue University.

## References

1. Sohn, Y. M. *Effects of Realistic Heat Straightening Repair on the Properties and Serviceability of Damaged Steel Beam Bridges*. PhD dissertation. Purdue University, School of Civil Engineering, West Lafayette, Ind., 2012.
2. Sohn, Y. M., and A. H. Varma. Effects of Realistic Heat Straightening Repair on Damaged Steel Beams. *Transportation Research Record: Journal of the Transportation Research Board*, 2014. 2406: 68–78.

3. Avent, R. R., and D. J. Mukai. What You Should Know about Heat Straightening Repair of Damaged Steel. *Engineering Journal*, Vol. 38, No. 1, 2001, pp. 27–49.
4. Avent, R. R., D. J. Mukai, P. F. Robinson, and R. J. Boudreaux. Heat Straightening Damaged Steel Plate Elements. *Journal of Structural Engineering*, Vol. 126, No. 7, 2000, pp. 747–754.
5. Avent, R. R., and D. J. Mukai. *Heat-Straightening Repairs of Damaged Steel Bridges - A Technical Guide and Manual of Practice*. Report No. FHWA-IF-99-004. Federal Highway Administration, U.S. Department of Transportation, 1998.
6. Kowalkowski, K. J. *Effects of Multiple Damage-Heat Straightening Repair Cycles on the Structural Properties and Serviceability of Steel Beam Bridges*. PhD dissertation. Purdue University, School of Civil Engineering, West Lafayette, Ind., 2005.
7. Kowalkowski, K. J., and A. H. Varma. Structural Properties of Steels Subjected to Multiple Cycles of Damage Followed by Heating Repair. *Journal of Structural Engineering*, Vol. 133, No. 2, 2007, pp. 283–296.
8. Kowalkowski, K. J., and A. H. Varma. Effects of Multiple Damage-heat Straightening Repairs on Steel Beams. *Transportation Research Record: Journal of the Transportation Research Board*, 2007. 2028: 67–77.
9. American Society for Testing and Materials. *Standard Test Methods for Tension Testing of Metallic Materials*. American Society for Testing and Materials, West Conshohocken, Pa., 2004.
10. American Society for Testing and Materials. *Standard Test Methods for Rockwell Hardness and Rockwell Superficial Hardness of Metallic Materials*. American Society for Testing and Materials, West Conshohocken, Pa., 2005.
11. American Society for Testing and Materials. *Standard Test Methods for Notched Bar Impact Testing of Metallic Materials*. American Society for Testing and Materials, West Conshohocken, Pa., 2006.
12. Association of State Highway Transportation Officials. *LRFD Bridge Design Specifications – 4th edition*. Association of State Highway Transportation Officials, Washington, D.C., 2007.


*The Standing Committee on Steel Bridges (AFF20) peer-reviewed this paper (18-01985).*

Article

TRR
JOURNAL OF THE TRANSPORTATION RESEARCH BOARD

Transportation Research Record
2018, Vol. 2672(41) 177–185

DOI: 10.1177/0361198118781685
journals.sagepub.com/home/trr
SAGE

# Assessment of Shear Connection through Composite Beam Modeling

**Matthew Sjaarda[1], Scott Walbridge[1], and Jeffrey S. West[1]**

## Abstract

Steel-concrete composite construction is used extensively in bridges across North America. The welded shear stud is the standard connector used today, but other connectors, such as the through-bolt connector, may have advantages for precast construction or applications where better fatigue resistance or deconstructability is desired. The standard method of assessing the performance of a shear connector is through the use of push tests. However, the load-slip curves that result from these tests do not accurately predict load-slip behavior at the shear interface of the beams and girders they are meant to simulate. In this paper, a model is presented that predicts composite beam behavior using elastic material properties and nonlinear shear connector load-slip curves. The finite element (FE) model features link connector elements between a steel beam and concrete slab that can be programmed to simulate different connector types. Although the model can be used with push test load-slip curves as inputs, it is shown that a much better prediction can be made using force-deformation data from experimental beam tests or FE analysis. Results are discussed for stud connectors and through-bolt connectors, and it is shown that while through-bolts allow more interfacial slip and overall deflection, material stresses and composite interaction are not affected as much as might be expected. The outcome of this work is a comparison tool which can be used to assess the viability of current and future shear connection alternatives with the goal of achieving an economical and structurally sound shear connector.

Steel-concrete composite construction is common because it is a structurally efficient use of material, and the resulting geometries are well suited for many structural applications. The role of the shear connector is to tie the materials together at the interface, allowing the transfer of longitudinal shear forces. Mechanical shear connectors are used extensively in composite construction for building floor systems and bridge deck systems. The welded shear stud is by far the most common connector type currently used in new construction; the welding process is completed using efficient stud-gun welding technology, and the connection between steel beams or girders and cast-in-place concrete is stiff and ductile. However, the welded stud connection has drawbacks when considering fatigue performance, disassembly (for bridge re-decking or replacement), and precast construction. The through-bolt connection may have certain advantages over the welded shear stud, including fatigue performance and rapid non-destructive deconstruction, but proper validation of performance is required. This connection features high strength through-bolts that clamp precast deck panels to beams or girders, relying on friction and dowel action to transmit interfacial shear.

A common test to assess the performance of a shear connector is called a push test. Push tests feature a vertical steel beam connected to two concrete slabs with shear connectors that are tested in direct shear. Although widely used, the push test fails to take into account bending, interfacial friction, and force redistribution between connectors. The output from a push test is a load-slip curve that gives information about the strength, stiffness, and ductility of a connector, but fails to provide a complete quantitative measure of the degree of composite action it will provide. Ultimately the role of a shear connector is to induce composite action, and a push test fails to assess a connector directly in this regard.

The main alternative to push tests is scaled beam tests, which are expensive and can be impractical. The research presented herein circumvents this issue by combining the use of push tests and basic composite beam finite element (FE) modeling to create a practical and accurate way to assess the performance of shear connectors under static loading, and conditions where the steel and concrete materials are behaving elastically. It can also be used to find accurate individual connector stress ranges to optimize fatigue design. The model presented has been validated through laboratory beam tests and a detailed 3D solid FE model.

[1]Department of Civil and Environmental Engineering, University of Waterloo, Waterloo, ON, Canada

**Corresponding Author:**
Address correspondence to Matthew Sjaarda: msjaarda@uwaterloo.ca

## Shear Connection and Composite Interaction Concepts

The degree of shear connection and the degree of composite interaction in a composite beam are concepts that are commonly misunderstood and frequently used interchangeably. It is necessary to define both of these terms clearly, as they are essential for evaluating shear connection performance. The degree of shear connection refers to the equilibrium of forces in a composite beam at the ultimate strength limit state (ULS), while the degree of composite interaction deals with the compatibility of displacements (and strains), namely the shape of the strain profile through the depth of a given section (*1*). The two terms are related to each other, but they are not interchangeable.

The degree of shear connection, $\eta$, is shown in Equation 1 and is defined as the ratio of the ultimate shear strength of the interface in a shear span, $F_i$, to the minimum strength necessary for the section to develop its full composite flexural capacity at the end of the shear span, $F_f$.

$$\eta = \frac{F_i}{F_f} \quad (1)$$

Figure 1*a* shows the longitudinal equilibrium of forces at the ULS for a full shear connection ($\eta = 1$), partial shear connection ($0 < \eta < 1$), and no shear connection ($\eta = 0$). In design codes, the shear strength of the interface is calculated using the ultimate strength of all shear connectors in the shear span ($A_{sc}f_u$), neglecting friction and bond. The section can reach its full flexural capacity when $F_i$ is equal to $F_f$, which is the lesser of the steel beam area multiplied by the yield strength of the material ($A_s f_y$) and the product of the stress block factor $\alpha_1$, the concrete area $A_c$, and the 28-day compressive strength of concrete, $f'_c$. It should be noted that the shear span is determined using a moment envelope for moving loads, and is independent of the load at any given time. As a result, a beam or girder can be described as having a constant degree of shear connection.

The degree of composite interaction, $\varphi$, is difficult to define completely without the use of differential calculus and compatibility equations. In 1951 Newmark et al. published a classic paper on linear elastic partial-interaction theory, which was later extended and simplified. The laborious theory development will not be shown here (see (*1, 2*)). Instead, since the calculation of the degree of composite interaction is not required in the design, a simplified definition shall suffice. This simplified degree of composite interaction is given by Equation 2 and is defined using the ratio of the neutral axis separation, $h$, to the maximum neutral axis separation, $h_0$. This maximum neutral axis separation is the distance between the steel and concrete neutral axes if there was no composite interaction (i.e., steel and concrete are bending on their own and share the same curvature). Figure 1*b* shows three beam strain profiles illustrating full or complete interaction ($\varphi = 1$), partial interaction ($0 < \varphi < 1$), and no interaction ($\varphi = 0$). It should be noted that all mechanical shear connectors require slip to resist force, and thus full interaction is only possible through a stiff chemical bond. The strain discontinuity in partial and non-interactive strain profiles is indicative of interfacial slip. This slip strain (or slip per unit of beam length) can be integrated over the length of the beam to find the total slip at any point.

$$\varphi = 1 - \frac{h}{h_0} \quad (2)$$

Unlike the degree of shear connection, the degree of composite interaction is not defined at a particular section along a composite beam, nor is it defined under a particular loading condition. As a result, a beam or girder cannot be described as having a constant degree of composite interaction. The degree of composite interaction changes with loading, over the length of the beam, and even over the life of a beam; it has been shown that the stiffness of a composite beam decreases when subjected to repeated loading (*3*). In contrast to the degree of shear connection, the degree of composite interaction is difficult to calculate by hand. Due to its simplicity and ease of calculation, the level of shear connection is commonly used to characterize composite action for design purposes. This is despite the fact that the degree of composite interaction is the actual measure of the performance of connectors along the shear interface, since it directly measures the ability of the connector to resist slip and transfer force into the concrete slab.

## Background

### *Design Assumptions*

To simplify the design and analysis of composite beams, the degree of composite interaction is often assumed to be complete (100%), particularly in design codes. Newmark et al. stated, in 1951, that this assumption is reasonable for a properly designed beam (*2*). Additional to the assumption of full interaction, design codes neglect the effects of friction along the steel-concrete interface, instead assuming that shear is carried only by the connectors. The Canadian steel design code for buildings places a limit of 40% on the degree of shear connection that must be provided, stating that below 40% shear connection, composite action cannot be assured (*4, 5*).

In contrast to the code for steel buildings, the Canadian Highway Bridge Design Code (CHBDC) (*6*) and the AASHTO Load and Resistance Factor Design (LRFD) Bridge Design Specifications (*7*) do not permit partial shear connections, stating that the full capacity of a composite section must be developed through shear connection regardless of the loading conditions (*6, 8*). Yura, Methvin, and Engelhardt contend that current bridge code design rules may require an excessive number of shear studs, to the point where it can be difficult to fit them all in the beam (*3*). They

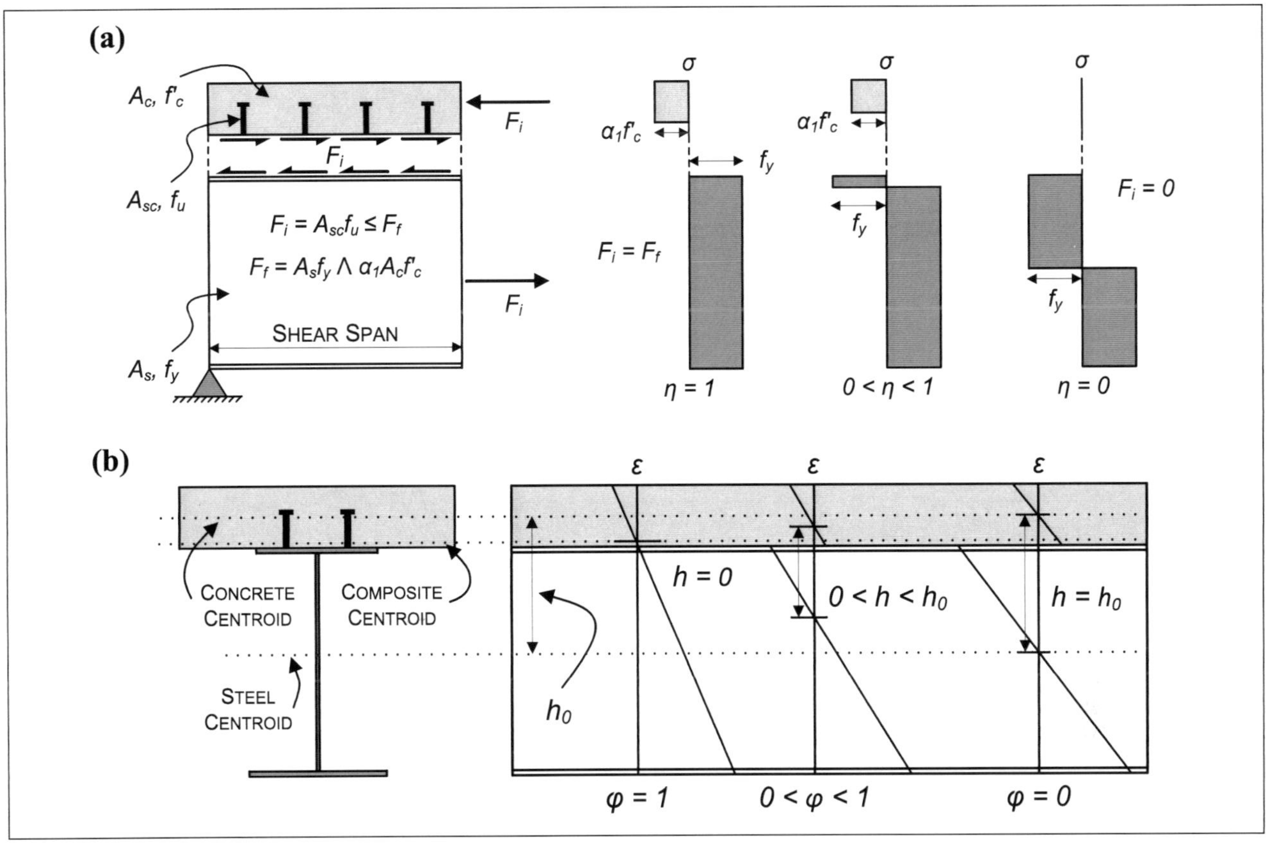


**Figure 1.** Degrees of (*a*) shear connection and (*b*) composite interaction.

recommend that partial composite action be considered for bridges, so that shear connection design is based on required strength and not on cross section properties, and that the lower limit for shear connection should be placed at 50% (*3*). However, one should keep in mind that in most bridges the number of shear connectors required is governed by the fatigue limit state, and this may require more studs than a full shear connection in any case.

## Partial Shear Connection and Strength

The basic effect of a partial shear connection on strength is that it theoretically changes the ultimate failure mode of the composite beam to failure of the shear connectors. Additionally, as the number of shear connectors decreases, the degree of composite interaction also decreases and interfacial slip increases; concerns regarding cyclic interfacial slip may be the primary motivation for requiring a full shear connection in bridges. According to Yura et al., partial shear connection construction has been used in the building industry since 1969, and the majority of composite beams designed today use a partial connection (*3*). The reason is that a partially shear-connected beam is very efficient; it can be shown that a beam with 50% shear connection will have approximately 75–85% of the moment capacity compared with the same beam with 100% shear connection. It may be thought by some that having a full shear connection carries significance when it comes to preventing interfacial slip. However, from the definitions laid out previously, it is clear that slip depends on the stiffness of the connection, not on the strength. As a result, there is no transition in the way a composite beam behaves as the level of shear connection reaches the "full" value. The only change that occurs is in the overall failure mode of the beam; with a partial shear connection the failure mode is no longer controlled by the cross section, but is instead initiated in the shear connection itself. This itself is dubious, since the magnitude of interfacial slip required to fail a shear connector (under static loading) is not likely to occur.

## Partial Composite Interaction and Fatigue

The assumption of complete interaction in composite beams is conservative from a fatigue perspective. Complete interaction assumes that shear connectors are rigid, and therefore attract or develop more connector shear force than they actually do. The elastic calculation of shear flow with a full interaction assumption leads to overestimated connector design shear forces. Despite not being accounted for, it is known that partial interaction is inevitable under service loads when mechanical shear

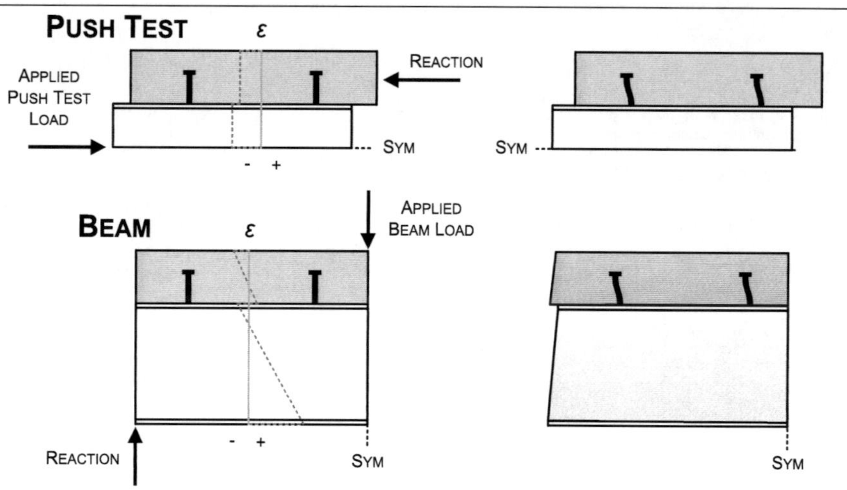


**Figure 2.** Strain profiles in push tests and beams.

connectors are used, and although the interfacial slip only has a small effect on strength, it can have a significant influence on the fatigue life of a composite beam or girder. In the past, rigorous hand calculations would have been needed to calculate the beneficial effects of partial interaction and interfacial friction. At present, however, commercially available FE analysis packages enable modeling that will take partial interaction into account based on known (input) connector stiffness or load-slip response. This stiffness data can be in the form of a single number (termed the "modulus of shear connectors" by Newark et al.), or a full load-slip curve obtained from shear connector push tests (*2*). Furthermore, Oehlers and Seracino (*9*) devised a simplified method to incorporate partial interaction into design or analysis, by using linear segments to approximate the shear flow force range envelope obtained from computer simulation and partial interaction theory. Even small reductions in the connector design force manifest themselves in significant extensions of fatigue life.

### *Comparison of Push Tests and Beams*

Mechanical shear connectors were first proposed for composite construction in the 1920s, but it was not until the 1960s that systematic testing began. The use of push tests began early on as both an ultimate and fatigue strength measurement tool. A push test is a relatively inexpensive test used to study shear across an interface, meant to replicate the longitudinal shear in a composite beam. Push tests remain standard practice today (see Eurocode 4 Annex B, (*10*)). Although they have been confirmed as a reliable method for determining shear capacity (*9*) they do not perfectly simulate the shear behavior in a beam. The value of push test results is questionable due to the difference in boundary conditions that results (*11*). In a push test, the shear forces on individual connectors remain constant relative to one another throughout the test. A beam test, on the other hand, features force redistribution; when one shear connector begins to crack and fail (in fatigue, for example), other neighboring connectors compensate and accumulate shear force, and the crack grows more slowly because the cracked connector now attracts less force. Additionally, the force being transferred to the slab does not remain constant as connectors begin to plastically deform. Furthermore, a beam test engages friction at the slab-beam interface, which lessens the force transferred through the connectors.

It is worth comparing the strain profile of the composite section in the push test and beam test cases, and discussing the interfacial slip. The key difference between beams and push tests is in the way the connector shear forces arise. In the beam, the externally applied load causes a strain gradient with a discontinuity at the interface (also called "slip strain"), and connectors are subjected to shear force in order to resist the accumulation of slip due to the strain discontinuity along the length of the beam. In a push test, however, connectors resist shear force because they are part of the load path between the applied load and the balancing reaction at the base of the specimen. A push test is determinate, while a composite beam is not. Figure 2 shows the strain profiles and final deformed shapes in each case.

A push test is useful for generating load-slip curves for shear connectors. These curves give information about connector strength, stiffness, and ductility. The drawback of the load-slip curve is that it does not give a quantitative indication of the composite action that may result from the presence of a connector in a beam. Since the induction of composite action is the primary function of a shear connector, a model is needed to evaluate connectors on this basis.

## Research Challenge and Objectives

The main objective of the modeling procedure developed and described here is to provide a tool to evaluate shear

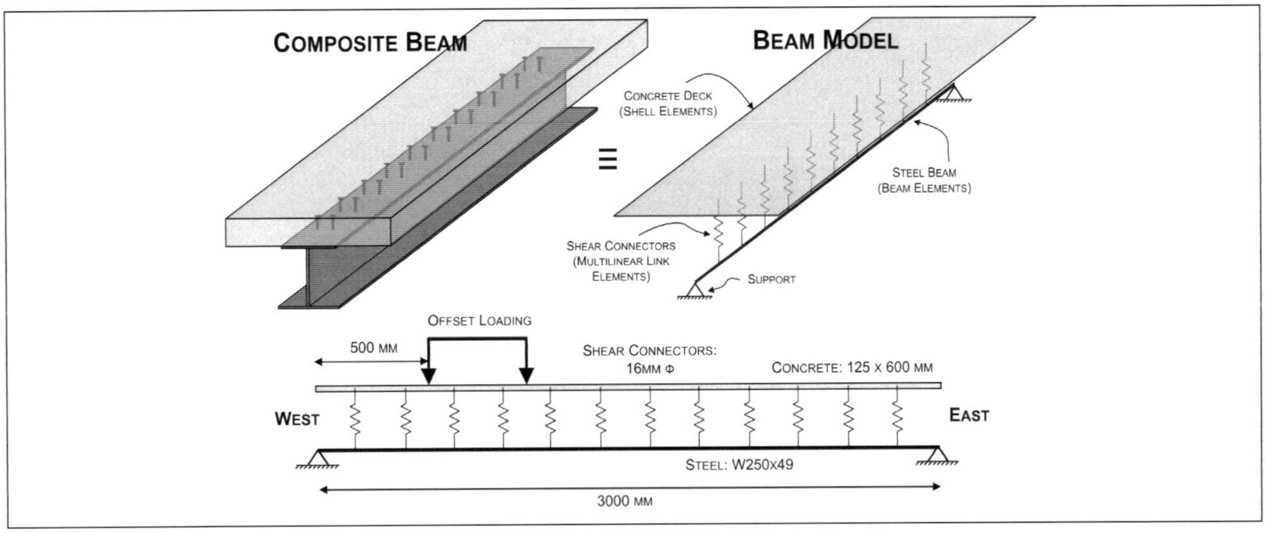


**Figure 3.** FE beam model showing element types, geometry, and loading setup.

connectors for their intended behavior. This behavior includes the degree of composite interaction that results from the shear connection, and by extension includes cross-sectional stresses, and beam deflections. The analysis tool will allow for the comparison between connector types with known load-slip curves, including studs and through-bolts, as well as other shear connection types. In the following sections, an FE model is presented, which takes a load-slip curve from a push test, and uses this curve to model the composite beam and slab behavior through the use of multilinear links. In addition, a shear stud force-deformation relationship obtained from a beam test will be used as the model input and the results will be compared with those obtained using data from push tests.

## Finite Element Analysis

### FE Beam Model Description

A SAP2000 model was programmed using the CSi Application Programming Interface (API) (*12*). The model works in conjunction with a spreadsheet, which contains necessary geometry and material data. In particular, the spreadsheet contains the load-slip properties of the shear connectors, represented as multilinear links with load-slip behavior defined by the user. This load-slip behavior is the key variable in the research presented. The model contains simplifications, including elastic material behavior for the steel and concrete, which allow it to run analyses quickly. As a result, validation of the model was necessary in order to assess the significance of these simplifications.

Figure 3 depicts the prototype composite beam to be modeled and a schematic of the model. The concrete deck is represented by shell elements, while the model uses beam elements for the steel beam. The two are connected using multilinear link elements representing shear connectors. The model uses a "division" size specified by the user, which splits the deck and steel beam into segments. The division size is recommended to be small (about 25 mm or 1 inch) to allow for versatility in accommodating different connector spacings and for more locations where data is output (link elements can only connect to shell and beam element ends and edges). In order to prevent ingress (overlap) between the concrete deck and the steel beam, gap link elements were used at every division and given rigid properties in the vertical direction. Friction is not included in the beam model at any point in this research, but is possible to implement by altering the gap link element properties.

The SAP model is set up for parametric analysis, allowing material properties, loading, and geometry (including connector spacing), to be specified by the user. However, for the purposes of this research the geometry and loading setup has been set to match beam specimens from an experimental program previously completed at the University of Waterloo (*13*) to allow for model validation. Validation was also completed using an ABAQUS 3D solid FE model, developed previously (*13*).

### Model Geometry and Loading

Twelve full-scale composite beam specimens were tested to investigate the fatigue of shear connectors (*13*). The static tests conducted prior to fatigue testing were used to validate the SAP FE model, along with results from an ABAQUS 3D solid FE model. The specimens were simply supported with a 3,000 mm (approximately 10') span and consisted of a W250×49 (imperial designation W10×33) steel section connected to a 600 mm (2') wide and 125 mm (5") thick concrete slab. The strength of the concrete was approximately 45 MPa (6,500 psi), and 24 welded shear studs per beam, each

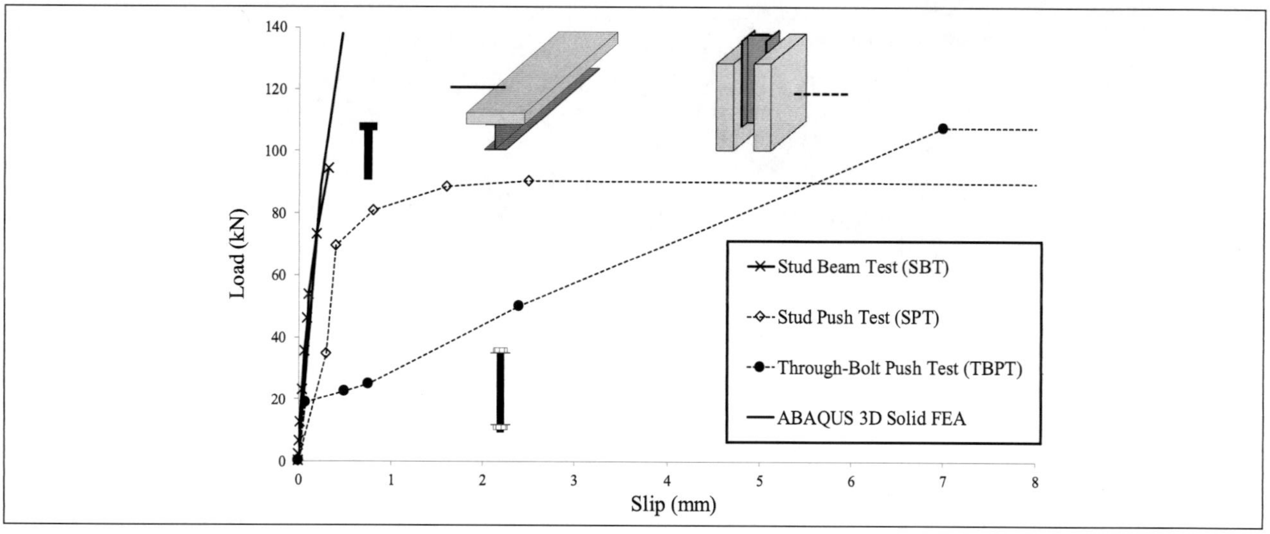


**Figure 4.** Force-deformation (or "load-slip") curves used as SAP model inputs.

16 mm (5/8") in diameter, were used at a spacing of 125 mm (5", two studs per row). The loading configuration consisted of two point loads that were offset from the beam midspan to create a varying shear diagram along the length of the beam. This geometry corresponds to a degree of shear connection of 33%. A full description of the FE model can be found in (*11*) and a schematic is shown in Figure 3.

### *Load-Slip Relationships*

Three load-slip relationships were used in the SAP model for the results that will be discussed. Two of these come from push tests completed by Chen et al. in 2014 (further description can be found in (*14*)), and the other comes from the beam tests described in the preceding section and detailed in (*13*). In order to obtain stud load-slip data from the beam tests, linear variable displacement transducers were used to obtain slip values before and after a row of studs, and strain gauges on the steel were used to calculate the force in the slab transferred by those studs. Each load-slip curve was approximated by linear segments, following the measured load and slip data. The curves, with loads normalized for a single stud, are shown in Figure 4, along with the force-deformation data from studs in the ABAQUS 3D solid beam model. The first three curves, which are used as the SAP model inputs, are given abbreviations for reference (shown in the Figure 4 legend). The stud beam test curve is referred to as "SBT," the stud push test curve as "SPT," and the through-bolt push test curve as "TBPT."

Note the strong correlation between the load-slip results from the stud beam test and the stud beam FE model; both curves are significantly stiffer than the stud push test curve, and both have higher ultimate strengths. These characteristics can be attributed to differing boundary conditions and adjacent geometry; neighboring studs prevent slip in beams, even as studs yield, and this does not occur in push tests. The early reduction in stiffness for the bolt connector is a result of the bolt slipping into bearing as friction is overcome. After this point, the stiffness is increased again and ductility and strength are achieved through dowel action of the bolt.

## Results and Discussion

### *Comparison of Model Connector Stress and Interfacial Slip Results*

A comparison of connector stresses and interfacial slips between the SAP FE results using the stud beam test (SBT), stud push test (SPT), and through-bolt push test (TBPT) load-slip curves will now be shown and discussed. Figure 5*a* shows connector stresses along the beam length for each analysis, along with the results of the stud beam FE results from ABAQUS and the stud stresses calculated from elastic theory assuming full composite interaction. These results correspond to a beam loading of 200 kN or 45 kips (100 kN at 500 mm and 1,000 mm from the west end, or 22.5 kips at approximately 20" and 40"). The stresses were computed using Equation 3, where $V$ is the transverse shear, $Q$ is the second moment of area of the transformed section, $s$ is the connector spacing, $I$ is the moment of inertia of the transformed section, and $A$ is the area of the shear connector.

$$\sigma = \frac{V \cdot Q \cdot s}{I \cdot A} \tag{3}$$

Note the high degree of correlation between the connector stresses predicted by Equation 3, the experimental measurements, the ABAQUS model, and the SAP model using the SBT load-slip curve as the multilinear link input.

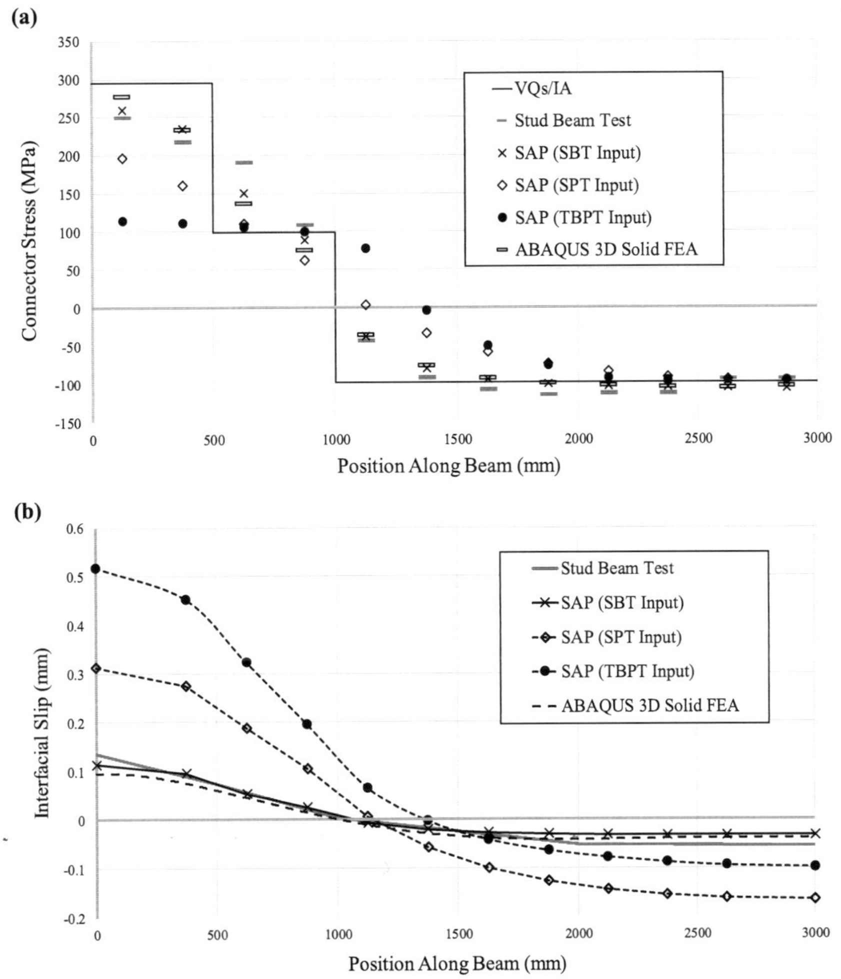


**Figure 5.** (*a*) Connector stresses and (*b*) interfacial slip along the beam for a load of 200 kN.

The SAP model results are within 5% of the ABAQUS FE results and 12% of the experimental measurements on average. However, using the SPT load-slip curve results in significantly lower stresses (on average 30% lower than when using the beam test load-slip curve). These results indicate that using push tests to approximate the interface shear resistance will result in an underestimation of connector stiffness and force.

In addition to examining stud connector stresses, the interfacial slip values along the length of the beam can also be compared to assess the connector performance. Figure 5*b* shows the predicted slip values for the three SAP model results and the ABAQUS model. Additionally, the results from the experimental beam test (*13*) are shown. It is apparent that the SAP model using the SBT load-slip curve gives slip values that are similar to the solid ABAQUS model and the experiment, but the SAP FE results using the push test load-slip curve give much larger slip values. The values using the through-bolt load-slip curve are even larger, owing to the reduced stiffness of that connection type.

## *Comparison of Model Degree of Composite Interaction Results*

While connector stress and slip are important metrics, the primary goal of the analyses was to assess the connectors in terms of the degree of composite interaction they induce. Figure 6 shows the degree of composite interaction along the beam length, according to the definition shown in Figure 1*b* and given in Equation 2. Interaction values along the beam length vary significantly, reflecting the restraint of slip around the leading face of each connector, and less restraint on the trailing face. The "leading" connector face is the one farthest from the maximum moment (in this beam the maximum moment is 1,000 mm from the west end, denoted as the zero mm position along the beam).

In general, the analysis results using the SBT load-slip curve to define the link elements show higher values of composite interaction, with an average value of 69.6%, whereas the beam using the SPT curve links averaged 50.4%. Both show a parabolic trend, with the degree of composite interaction peaking

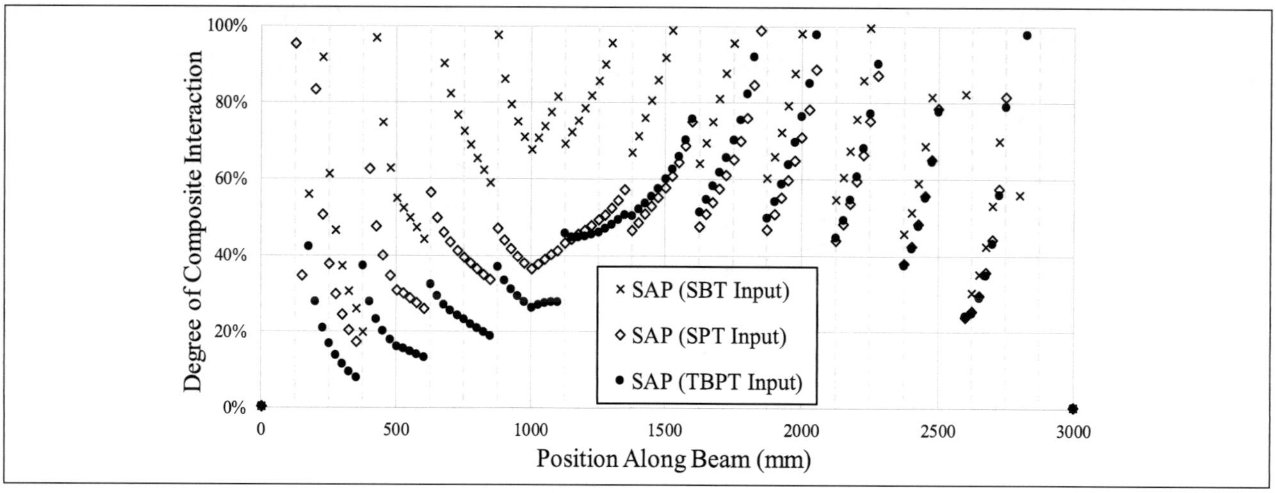


**Figure 6.** Degree of composite interaction along the beam length for a load of 200 kN.

**Table 1.** Summary of Beam Behavior Outcomes When Using Different Load-Slip Curves As Inputs

| Measurement | Stud beam test (experimental) | SAP model load-slip input: SBT | SPT | TBPT |
|---|---|---|---|---|
| Maximum deflection | 2.66 mm | 2.14 mm | 2.68 mm | 2.86 mm |
| Degree of composite interaction | – | 69.6% | 50.4% | 41.5% |
| Maximum steel stress | 104 MPa | 101.2 MPa | 109.1 MPa | 112.8 MPa |

around midspan. To explain the increased interaction near the point of maximum moment and the midspan one must look to Figure 5, and observe where there is the lowest amount of slip for the highest connector force. The amount of composite interaction at any point is completely driven by the compatibility of displacements, which is why the type of load-slip curve used to define the flexibility of the connector is so important to the results. All three results also exhibit larger variation toward both ends of the beam. This is because the percentage difference in the compressive force in the slab is larger on either side of the outside connectors (those toward both beam ends) than it is on the inside connectors.

The difference between the SPT results and the through-bolt results can again be seen primarily on the west end of the beam, where the slip is high (for slip values, see Figure 5). Prior to the initiation of slip in the through-bolt connection (at a connector load of about 19 kN or stress of 93 MPa, see Figure 4), the stiffness of a stud and a through-bolt connector are comparable. As a result, in the region of the beam where bolt slip has not occurred (from about 1,300 to 3,000 mm), similar levels of composite interaction develop.

## *Summary of Results*

Table 1 summarizes the outcomes of using each of the three shear connector load-slip inputs, while also showing the experimental SBT results. The average deflection from the tests and the maximum steel stress lie between the SAP model results using the SBT curve and the SPT curve. Comparing the model results using the link element properties based on the push test results, the through-bolts result in a 7% increase in maximum beam deflection, an 18% decrease in average degree of composite interaction, and a 3% increase in maximum stress in the steel beam bottom flange compared with studs. The increase in the maximum stress in the beam is small compared with the overall decrease in degree of composite interaction, because the maximum stress occurs where the beam moment is highest and shear is not. The reduction in degree of composite interaction is most affected in areas of high shear where slip has accumulated. The maximum moment is, by definition, the end of the shear span and the location where the interfacial slip vanishes.

## Conclusion

A simple FE model has been presented that allows for the input of a load-slip curve for a given shear connector type and outputs beam behavior including degree of composite interaction, deflection, material and connector stresses, and interfacial slip. The load-slip curve input can be in the form of force-deformation data from a beam test, or load-slip data from a push test. The work presented in this paper has led to the following conclusions:

- The degree of composite interaction is the proper measure of composite performance, while the degree of shear connection only applies at the ultimate limit state, and only refers to one location along a composite beam. Although the calculation for the degree of composite interaction is not simple, it is made easier through the use of FE models, such as the one described in this paper.
- Shear connector force-deformation results from beam tests are better indicators of beam behavior than results from push tests. The FE model results using the beam load-slip curve input (SBT) matched the experimental results well for connector stresses and interfacial slip.
- Bolt slip in through-bolt connectors significantly increases interfacial slip and decreases the degree of composite interaction in a beam. However, it influences maximum material stresses to a lesser degree because maximum stresses do not occur in areas of high shear and slip. For a proper comparison between stud and bolt connectors, load-slip data for through-bolts should be obtained from beam tests.
- Although the fatigue limit state (FLS) currently prevails over the ultimate limit state for the design of shear connectors, this may not be the case for connections using bolts. As a result, the current research may be useful for the design of full bridges to determine accurate stud stress ranges for FLS design, but it may be doubly useful for through-bolts or other novel connectors, affecting both FLS and ULS.

Additional questions about through-bolt shear connections remain, such as the effect of repeated loading on connectors above and below the bolt slip loads and how such loads may affect bolt pretension and overall connection integrity. This research is currently on-going at the University of Waterloo, and it is hoped that bolted beam tests (both static and fatigue) will contribute to the development of a viable alternative to shear studs in applications where deconstruction, fatigue resistance, and precast construction are factors.

## Acknowledgments

Financial support for this research provided by the Canadian Institute of Steel Construction (CISC), the Natural Sciences and Engineering Research Council of Canada (NSERC), and the Ministry of Transportation of Ontario (MTO) is gratefully acknowledged. Technical input was also provided by Taylor Porter and is much appreciated by the authors.

## Author Contributions

The authors confirm contribution to the paper as follows: study conception and design: Matthew Sjaarda (MS), Scott Walbridge (SW), Jeffrey West (JW); data collection: MS; analysis and interpretation of results: MS, SW, JW; draft manuscript preparation: MS. All authors reviewed the results and approved the final version of the manuscript.

## References

1. Oehlers, D. J., N. T. Nguyen, M. Ahmed, and M. A. Bradford. Partial Interaction in Composite Steel and Concrete Beams with Full Shear Connection. *Journal of Constructional Steel Research*, Vol. 41, No. 2/3, 1997, pp. 235–248.
2. Newmark, N. M., C. P. Siess, and I. M. Viest. Tests and Analysis of Composite Beams with Incomplete Interaction. *Proceedings of the Society for Experimental Stress Analysis*, Vol. 9, No. 1, 1951, pp. 75–92.
3. Yura, J. A., E. R. Methvin, and M. D. Engelhardt. *Design of Composite Steel Beams for Bridges*. Report No. FHWA/TX-08/0-4811-1. Center for Transportation Research, The University of Texas at Austin, 2008.
4. Canadian Standards Association. *CAN/CSA S16-09: Design of Steel Structures*. Canadian Standards Association, Mississauga, Ontario, 2009.
5. Canadian Institute of Steel Construction. *CISC Commentary on CSA S16-09*. Canadian Standards Association, Mississauga, Ontario, Canada, 2012.
6. Canadian Standards Association. *CAN/CSA S6-06: Canadian Highway Bridge Design Code*. Canadian Standards Association, Mississauga, Ontario, 2006.
7. American Association of State Highway and Transportation Officials. *LRFD Bridge Design Specifications*. American Association of State Highway and Transportation Officials, Washington, D.C., 2012.
8. Canadian Standards Association. *S6.1S1-10 Commentary on CAN/CSA S6-06*. Canadian Standards Association, Mississauga, Ontario, 2010.
9. Oehlers, D. J., and R. Seracino. A Tiered Approach to the Fatigue Assessment of Composite Steel and Concrete Bridge Beams. *Proceedings of the Institution of Civil Engineers – Structures and Buildings*, Vol. 152, No. 3, 2002, pp. 249–257.
10. CEN. *Eurocode 4, ENV 1994: Design of Composite Steel and Concrete Structures*. European Committee for Standardisation, 2005.
11. Porter, T. *The Fatigue Resistance of Headed Shear Stud Connectors in Steel-Precast Composite Girders*. Master of Applied Science in Civil Engineering thesis. University of Waterloo, 2016.
12. SAP 2000 Version 17. Computers and Structures, Inc., Berkeley, CA, 2014.
13. Sjaarda, M., T. Porter, J. S. West, and S. Walbridge. Fatigue Behaviour of Welded Shear Studs in Precast Composite Beams. *Journal of Bridge Engineering*, Vol. 21, No. 11, 2017, p. 04017089.
14. Chen, Y., Y. Zhao, J. S. West, and S. Walbridge. (2014). Behaviour of Steel – Precast Composite Girders with Through-Bolt Shear Connectors under Static Loading Through-Bolt Support. *Journal of Constructional Steel Research*, Vol. 103, 2014, pp. 168–178.

*The Standing Committee on Steel Bridges (AFF20) peer-reviewed this paper (18-06059).*

Article

TRR
JOURNAL OF THE TRANSPORTATION RESEARCH BOARD

Transportation Research Record
2018, Vol. 2672(41) 186–195

DOI: 10.1177/0361198118792126
journals.sagepub.com/home/trr
SAGE

# High Resolution Modeling and Modeling of Connections in Pony-Truss Bridges

**Martin A. Butler[1], James A. Swanson[1], Gian A. Rassati[1], and Eric F. Dues[2]**

**Abstract**

Pony-truss bridges are usually modeled using simplifying assumptions, such as pinned connections and analyzing the trusses separately from the rest of the structure. Pony-truss bridges are also generally designated as possessing many Fracture Critical Members (FCM), a characteristic that requires more rigorous inspections. Eliminating or reducing the number of FCM is desirable as it reduces inspection costs. A high-resolution model of a pony-truss bridge was assembled in ABAQUS CAE that explicitly models the behavior of the gusset plates in the connections and the angles in the framed stringer connections. This model indicated alternative load paths that are potential sources of redundancy useful for retrofit or reclassification of bridges with FCM designation. Higher-resolution models of framed stringer connections were also made in ABAQUS CAE using solid elements explicitly modeling the bolted connections. The connections were loaded in tension; what the stringers must hold if they supply an alternative load path for the truss in the event of fracture in the bottom chord. The primary objective of the work described is to provide high-resolution finite element analysis results that can be used to create general behavior for lower-resolution finite element models used in fracture critical analyses to reclassify FCM as non-fracture critical. The results of high-resolution modeling reveal the behavior that is necessary to include in these lower-resolution models, as well as potential alternative load paths. Lower-resolution models are assembled in SAP2000 and matched to the higher-resolution models.

Following the 1967 collapse of the Silver Bridge, federal legislation was passed in the 1968 Federal-Aid Highway Act establishing the National Bridge Inspection Standards (NBIS) that requires inspection of all bridges at least 20' long every 2 years (*1*). In Ohio, the Ohio Revised Code requires a more stringent annual inspection of all bridges 10' and longer (*2*). Bridges designated as having fracture critical members (FCM) require a costly inspection; inspectors must get within arm's reach of the FCM (*3*). Thus, reducing the number of FCM is valuable even if not all FCM are reclassified. There are several definitions of FCM. The National Bridge Inspection Standards defines FCM as "a steel member in tension, or with a tension element, whose failure would probably cause a portion of or the entire bridge to collapse" (*4*). The American Association of State Highway and Transportation Officials defines FCM in the Manual for Bridge Inspection (MBE) as "steel tension members or steel tension components of members whose failure would be expected to result in a partial or full collapse of the bridge" (*5*) and in the Bridge Design Specifications FCM is defined as "component in tension whose failure is expected to result in the collapse of the bridge or the inability of the bridge to perform its function" (*6*). These definitions share that some component of the member must be in tension and its loss must cause significant change to the behavior.

The assumption made in design is that a pony-truss will be non-redundant. Existing bridges have some redundancy resulting from design, detailing, and construction methods. Connections will be a gusset plate with dozens of rivets or bolts that will transmit some moment. Floor systems are assumed to transfer only traffic loads, however some floor systems may assist in reapportioning loads. Reconsideration of modeling techniques are suggested that will result in more complex models but may also reduce owner inspection requirements or even allow structures to be built with a targeted reliability index, the way connections and members are designed.

[1]Department of Civil and Architectural Engineering, University of Cincinnati, Cincinnati, OH
[2]Gannett Fleming, Inc., Columbus, OH

**Corresponding Author:**
Address correspondence to Martin A. Butler: butlerm3@mail.uc.edu

## Risk-Based Assessment

The demand for annual or biennial inspection results in an approximately uniform distribution of resources for assets that do not have uniform reliabilities. Risk-based approaches for determining inspection intervals are defined by the National Cooperative Highway Research Program (NCHRP) document 12-82 (*7*, *8*). This approach requires consideration of both the probability of an event and its consequence. Thus, probable but benign events do not require frequent inspections, whereas improbable catastrophic events may call for more frequent inspections. Rubrics determining the necessary frequency of inspection determine two factors: the probability occurrence (O) and the consequences of occurrence (C). The product of these two factors is the inspection priority number (IPN), a value that indicates the priority a failure state should have, determining the appropriate inspection interval (*9*).

A 2012 FHWA memo (*10*) states that a refined analysis may be used to reevaluate a fracture critical designation. Though guidelines for such analyses are not provided, the memo indicates that reclassified members must have been fabricated with a modern fracture control plan. These plans were developed in 1978 by AASHTO; thus a refined analysis of bridges constructed prior to 1978 cannot result in reclassification (*11*).

Behavior from high-resolution models may be used to help in the creation of lower-resolution models that are more feasible for small-scale bridges. The present work describes high-resolution modeling that was undertaken as part of a larger effort to develop a refined fracture critical analysis method for reclassifying FCM as not fracture critical that could be implemented using typical analysis software.

## Pony-Truss Bridges in Ohio

Pony-truss bridges are composed of two simply supported trusses supporting a floor system. Trusses are composed of a top chord in compression, a bottom chord in tension, and diagonal members that may experience either. A pony-truss has no members framing into the top of the trusses, so the flexural support of the floor beams framing into panel points of the truss is used to provide stability.

Floor systems of pony-truss bridges are decks supported by stringers which are supported by floor beams. Floor beams are connected to truss panel points with moment connections. Stringers may bear on top of floor beams in an underslung configuration, or they may frame into them. Either way the assumption is that they are simply supported without continuity. Decks bear on stringers and are often attached using clips for steel or timber decks, or shear studs for concrete decks. In Ohio, the lateral loads are usually resisted by either rods or angles framed into the floor beams.

Modern pony-truss bridges are usually made from light rolled sections, mostly in the range of 20–40 lb/ft. The members are oriented such that they are in weak axis bending, making connections much easier and allowing the compression members of the top chord to be laterally supported against weak axis buckling. Older bridges may be constructed from built-up members of riveted plates.

### *Alternate Load Paths and Potential Retrofits*

The trusses are usually analyzed on their own in two dimensions, precluding alternative load paths that use secondary systems. A three-dimensional model indicated several potential alternative load paths that should be included in models. Stringers hold tension, gusset plates carry moment, and decks redistribute loads from weakened trusses to stiffer portions of the structure and loads across broken floor beams. Care should be taken in the connections of these load paths, as in evaluating damaged structures, members and connections are often loaded in ways they were not designed for.

One alternative load path for damage given special consideration is through the stringers. This load path requires several load transfers; first loads go into the floor beams, which are put into weak axis shear and bending, into the stringer, and then out the same way onto the other end of the damaged panel. If damage to the truss occurs in the last bay, the stringer must be axially anchored to the abutments. The connections limits this path. However, if tensile continuity between stringers were added, the resulting load path could provide redundancy to the bottom chord. Additionally, moment continuity across stringers can assist in maintaining a bridge following damage to a floor beam. This is not easily achieved with framed stringer connections, but is common in underslung stringers, which may be continuous over multiple panels.

Decks and lateral load systems are often neglected during design, but their wide range of connections allow for several alternative load paths. Even fully composite with the stringers, decks provide little tensile stiffness. However, decks do provide flexural stiffness that can be an effective alternative load path for damage to both trusses and floor beams.

Lateral load systems are also ineffectual in providing an alternative load path. Models of the Campbell Avenue Bridge built in SAP2000 without the lateral load system and with larger lateral load system composed of C12x30 did not significantly increase the capacity of the damaged bridge. An alternative lateral system connecting the stringer connections to each other also did little to increase capacity. A more promising retrofit method than this is post-tensioned cables, either draped or along the tension chord (*12*).

Finally, truss connections are assumed to be frictionless pins, but there is some resistance to moment available. The moment capacity of neither the connections nor the members is great, as members are in weak axis bending and the gussets often yield before members, providing little resistance to rotation. The small amount of stiffness added by the consideration of the moment capacity may be useful, though even damaged trusses are dominated by axial loads.

## High-Resolution Modeling

A high-resolution model was built in ABAQUS/CAE Version 6.13 of an existing pony-truss bridge, the Campbell Avenue Bridge. It has an overall span of 160' 0" broken into 10 truss panels, each with a length of 16' 0". The trusses are spaced laterally at 31' 4" center to center. W36 × 135 floor beams support eight stringer lines (W16 × 45) spaced at 6' 0" in a framed configuration with a seat angle. The deck is 28' 0" wide and is made up of an 8 ½" thick reinforced concrete deck made composite with the stringers. Two element types were used in the creation of this model. Rolled steel members, gusset plates, connecting angles, and decking were modeled using a general-purpose shell element S4R, a four-node doubly curved shell with reduced integration, hourglass control and finite membrane strains. The lateral rods were modeled using B31, four-node beam element. Bolts were modeled using mesh-independent fastener multipoint constraint using rigid beam interactions projected with a 1" diameter. These multipoint constraints used the structural distribution of force, necessary for the prying present in some of the connections, particularly among the angles connecting the stringers. No mesh sensitivity analysis was performed; a mesh size of 1" was used for rolled steel shapes and gusset plates as suggested by Stains (*13*). The high-resolution model is shown in Figure 1 with the deck hidden for clarity; the framed stringer connections are shown in Figure 1, displaying the mesh.

The deck was made composite with the stringers using mesh-independent-fastener type multipoint constraints. These constraints do not allow for pullout, but the concrete damaged plasticity model used for the deck material allows for the concrete to lose stiffness if it is sufficiently loaded. The interaction between floor beam and deck was defined as a contact interaction, with hard normal behavior and tangential behavior defined by a friction coefficient of 0.2. Steel members and gusset plates were also given this contact interaction if it was possible for them to touch.

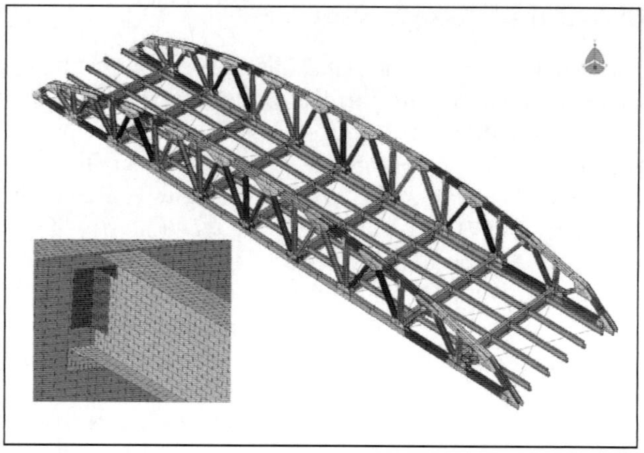

**Figure 1.** High-resolution model and close-up of the high-resolution model's stringer connection.

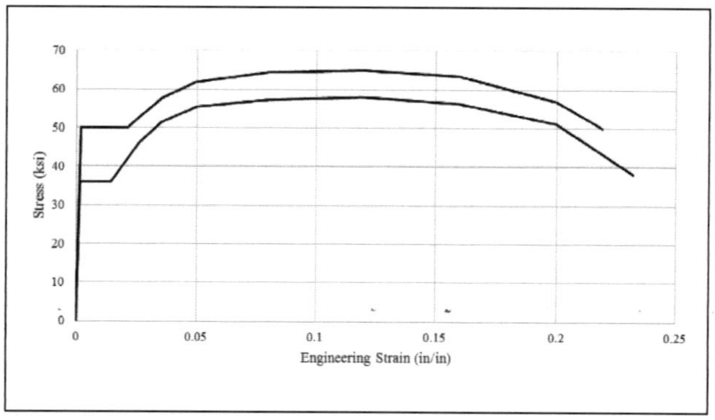


**Figure 2.** A572 grade 50 stress–strain behavior.

All bolts in truss, stringer, and floor beam connections were modeled using mesh independent fasteners that constrain nodes in a specified region to nodes in another region with rigid beams. For the bolts the area is nodes within a 1" diameter. This method provides good initial behavior, but it does not allow for net section fracture or bolt failure. It also does not allow for friction to contribute to the connection behavior.

Rolled section members were modeled using shell elements with 1" square mesh. These models ignore the fillet at corners, reducing the cross-sectional area. They also neglect residual stresses and initial out of straightness, so that yielding of members will initialize differently and buckling will occur more suddenly than in actuality.

Two material properties were used in the high-resolution model, one for A572 grade 50 steel and one for 4 ksi concrete. As demanded by the bridge plans, all steel portions of the bridge were made of this steel, including angles. The nonlinear behavior of the steel is shown in Figure 2. It is the nominal behavior for A572 grade 50 steel. The plans also called for a 4 ksi reinforced concrete deck, which is uncommon; the majority of pony-truss bridge decks are made of corrugated steel filled with asphalt. An alternative steel deck was also considered which ignored the strength contributions of asphalt. Finally, a version of this model was produced with no deck, where the traffic and deck dead loads were applied directly to the stringers.

The concrete was modeled using the Concrete Damaged Plasticity material behavior available in ABAQUS. The values used for this concrete were scaled from the study by Jankowiak and Lodygowski (*14*) who calibrated values for this method. The values were scaled in the same manner as was done by McConnell (*15*) where the modification gave good results for different concrete strengths used in bridge decks. The strength of the concrete modeled by Jankowiak and Lodygowski was higher than that studied by McConnell, 7.25 ksi compared with 3.1 ksi, so the 4.0 ksi concrete was within that range. The concrete damaged plasticity model is used to capture the tension cracking and compression crushing that occurs as concrete fails as well as the permanent deformations that also occurs.

The lateral rod anchor plate was modeled as a part of the floor beam rather than a separate part welded to the floor beam with the lateral rods tied to the anchor plate portion of the floor beam part. Framed stringer connections were modeled with angle geometry considered explicitly. Each of the truss supports were constrained to a single point using a kinematic coupling constraint, and this point restrained. Two of them at one end of the bridge were pinned, with all directions of translation restrained. The other two at the opposite end of the bridge as well as the ends of the stringers, were defined as rollers, allowing translation only in the direction of the roadway.

Loads were applied in two steps. In the first step, dead loads were applied using the body force loads applied to every part. The weight of the roadway was considered in the density of the concrete or steel deck. For the case with no deck, the weight of the deck and bearing surface were distributed to the top flange of the stringers. The second step of loading considered an HS-20 truck load applied to the deck. The truck load was applied as four pressure loads over 10" × 20" tire patches. The centers of these tire patches were placed at the critical location for the member considered along the length of the bridge, with the tire patch beginning 12" from the edge of the deck. The pressures were determined based on the axle loads of an HS-20 truck, but the loads were incremented until the model failed to converge, indicating collapse. This procedure is outlined more explicitly in Swanson et al. (*16*). In practice, the procedure closely resembles the ASCE 41 (*17*) nonlinear static procedure (NSP), with the major difference being that the loading is directed downward. Damage was modeled by fully removing a member from the model.

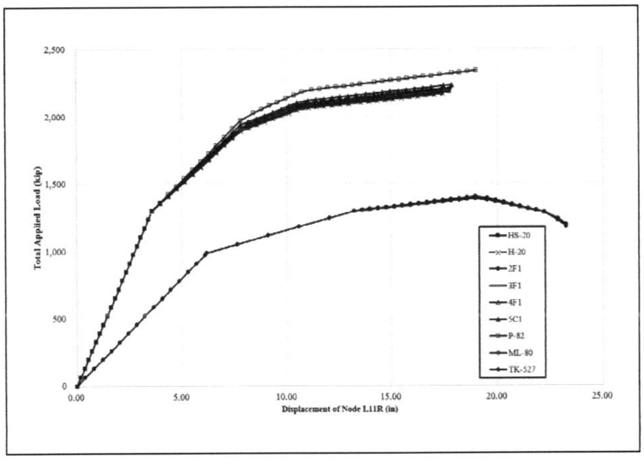


**Figure 3.** Truck loads on member B5-B6, both lanes.

## *Loads on Models*

*Truck loading.* As the behavior for these high-resolution collapse models are highly nonlinear, the nature of the loading has a significant effect on the collapse progression. It may then appear that a wide range of loadings must be considered. However, when many different truck loadings are compared, the damaged behavior is quite similar. A nonlinear model of the Campbell Avenue Bridge was built in SAP2000 and subject to the redundancy evaluation procedure proposed in NCHRP Report 406. The locations of fractured members are shown in Figure 10 and include members B1-B2, U2-B2, B5-B6, and U10-B6, which were removed one at a time. The pushover curves for both damaged and intact bridges for the loss of B5-B6 are shown in Figure 3. Similar curves were obtained for B1-B3 and U2-B3, the first of these gave results similar to those shown, the latter collapsed before dead load was fully applied. These curves suggest the truck loading is unimportant for the collapse behavior; only the P-82 and the 5C1 truck loadings give noticeably stiffer results when comparing the deformation and the base reactions, and these only for the intact structure. These two loadings are wide and they give higher capacity values so should be avoided. For the rest of this work the H-20 load is used for its easy application. H-20 loads were also used in NCHRP Report 406, which obtained ratios of intact to damaged live load factors for a variety of bridge types. Obtaining a ratio similar to that of the four girder bridges indicates a sufficiently redundant structure.

*Load Factors.* Load factors for dead load and live load are important considerations for nonlinear analysis, particularly as the proposed analysis requires two steps, applying the dead load before incrementing the live load and the live load effect is incremented until failure. High-resolution modeling of FCM fracture in loaded bridges performed by Cha et al. indicates that the impact factor for live loads in the event of member fracture is 1.2–1.3 for trusses (*18*). Upon first glance it may appear necessary to apply some impact load factor to the live loads, and depending on the analysis procedure this may be appropriate.

One method of determining the redundancy of a bridge is to compare its post-damaged capacity with its intact capacity, as is done in the method proposed in NCHRP Report 406. This procedure compares the ratio of damaged live load capacity to undamaged live load capacity with that ratio for four-girder bridges. Given that it is generally agreed that four-girder bridges are redundant, having comparable damaged strength ratios to this bridge type indicates a redundant structure. As the live load factors would be linearly scaled by dynamic effects, and the live load factors of the four-girder bridges would also be scaled by the same value, these effects do not affect the comparison. If an analysis is being performed to evaluate the behavior in the event of a fracture divorced from the NCHRP Report 406 context, impact factors do need to be included.

## Results of High-Resolution Modeling

High-resolution modeling may be used to directly measure the reliability of the bridge. However, a high-resolution model is tedious to produce and requires software not readily availible to most. A lower-resolution model made in more readily available software such as SAP2000 is desirable for implementing reevaluations on a wider scale. Behavior exhibited by the high definition model must be captured by the lower definition models. The high-resolution modeling also identifies potential alternate load paths, which may suggest modes of retrofit or different methods of construction that will result in future structures having fewer FCM. Some notable results of the high-resolution model are that while stringers possess potential as an alternative load path for tension chord members, decks hold essentially no tension in the event of bottom chord loss, and gusset plate connections often yield before members do.

In all cases, even for an intact bridge with a concrete deck, the first yielding to be exhibited was in the angles connecting the stringers to the floor beams. This yielding is a result of tension in the connection, with angles yielding as they are pried from the floor beam. The assumption that these will yield first does not reflect belief that these connections are consistently under-designed in practice, rather it is a reflection of these connections being designed as simple shear connections. These framed connections are quite soft in tension, as is shown in the high-resolution connection modeling. Further, since the multipoint constraints do not allow slip that is available in bolted connections, the stringer connections in the high-resolution model are initially stiffer than they ought to be, especially if oversized holes or slots are used. These connections are designed to be simply supported and assume no tensile demand at all. However, this is not the case even with relatively small levels of overloading. In extreme cases, such as the loss of a bottom chord member, this continuity may be useful, particularly for retrofit.

Decks and lateral systems are largely ignored in stability design. The high-resolution models reveal that in the event of loss of a bottom chord member a concrete deck does not take that load. In fact, along the panel missing a bottom chord member, the membrane stress in the concrete deck is actually reduced relative to the load it experiences in the rest of the bridge. The type of deck had almost no impact on the high-resolution model performance. Decks are not useless for alternative load paths though, they are useful for the event of fracture in the floor beam, modeled here as a moment release in the floor beam. Comparisons between the high-resolution models with concrete and corrugated steel decks are presented in Figure 4 for intact bridges as well as those experiencing the loss of a bottom chord member. This does not indicate that deck does nothing to enhance the stability of damaged structures.

Another result of the high-resolution modeling is that truss members may yield well after gusset connections. Gusset plate yielding and concrete cracking often led to collapse before members yielding. The gusset plate

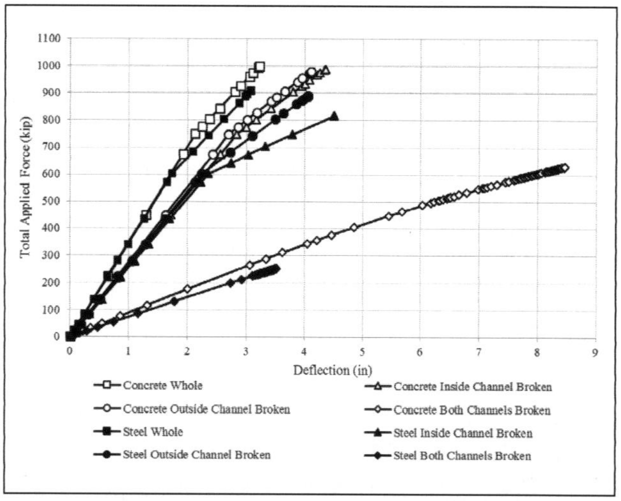


**Figure 4.** Comparison of corrugated steel and concrete deck high-resolution models for damage to truss member B5-B6.

yielding is the result of interaction of bending and axial loads. However, as soon as yielding is initiated due to the combination of moments and axial loads the axial loads dominate. Plastic hinges form in the gusset between the last row of bolts of one member and the first row of bolts on the next. The presence of axial loads is not surprising in trusses, and the addition of bending in the plane of the truss is intuitive in the event of lost members. Fortunately, a change to the geometry of a structure great enough to cause significant bending loads in what should be a truss, is also sufficient to remove the truss forces. This creates the situation that the connections of a damaged truss will generally be loaded axially if the structure is relatively intact, or by moment if it is greatly changed. Thus, a situation where both axial and bending loads are concurrently significant is unlikely. Out-of-plane bending can also occur due to a reduction in stiffness of one truss relative to the other. In the case of internally redundant members (two channels used as a single bottom chord member, for example) the loss of one of the member elements will also cause out-of-plane bending in the gusset plates, assuming the damage is sufficiently far from the gusset. The out-of-plane bending changes the initialization of yielding but not the overall behavior, which remains localized to the middle of the gusset connections.

## High-Resolution Connection Modeling

There has already been research into high-resolution modeling of bolted and welded steel connection behavior (*19*, *20*). A deeper understanding of the nonlinear behavior of these connections does not necessarily suggest a change in connection design, because connections in regular use should not be pushed to their nonlinear limits. However, when considering a damaged structure, as is done when attempts are made to perform refined analyses to reevaluate a structure's fracture critical status, this behavior becomes important. One of the results of the high-resolution model was the realization that stringer connections initiate yielding earlier than other portions of the model, and that stringers will carry tension.

The high-resolution model of the full bridge indicated that even under normal loading on an intact structure, stringers of a pony-truss bridge carry tension. Higher-resolution models of just the framed connections were produced in ABAQUS CAE to determine the tensile behavior of the framed stringer connections. The connection considered was that of the Campbell Avenue Bridge, connecting a W16x45 stringer to a W36x135 floor beam using three L5" × 3 ½" × 3/8" angles. The web angles were 9" long while the seat angle was 8" long, bolts were 7/8" in diameter. The monitored displacement for the force displacement curves shown below was a node at the end of the stringers cope.

These connections were modeled using the method suggested by Ruffley (*19*). Bolts were explicitly modeled with simplified geometry, with circular nuts and heads, and were fully pretensioned by reducing the temperature of the shank of the bolt. The bolts, stringers, floor beams, and angles were all modeled using C3D8R elements, an eight-node solid element with hourglass control. An alternative model of one of the connections was produced using C3D20R elements, a 20-node quadratic solid element. This gave the same results as the model using C3D8R elements, so the C3D8R were used for the rest of the connection modeling.

Three common sizes of angles are often used for the framed stringer connections, L5" × 3" × ¼", L5" × 3" × 3/8", and L5" × 3" × ½". Two lengths of angles were considered, 8" and 9", and models were created incorporating both two and three bolts in the shear angles of the stringer. Angles were modeled using nominal material values for both A36 and as A572-50 steel. Only the thickness of the angles and the type of steel used in the angles were significant in determining connection behavior. The location of a slip plateau for the ½" thick angles, the only angles to have a clear slip plateau before yielding, is dependent on the number of bolts in the stringer. After the slip plateau, the curves swiftly converged again, so this is of little importance in directly modeling nonlinear behavior. Figure 5 shows the mesh of the high-resolution connection model using 9" long angles that are 3/8" thick with three bolts. Models of the

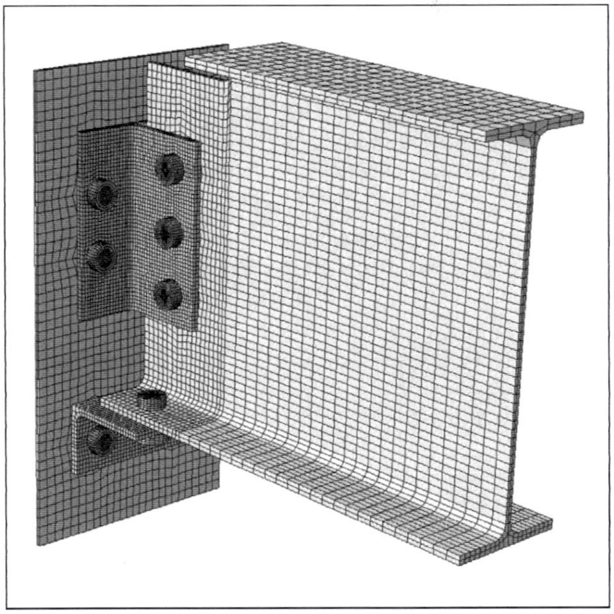

**Figure 5.** Mesh of a high-resolution connection model.

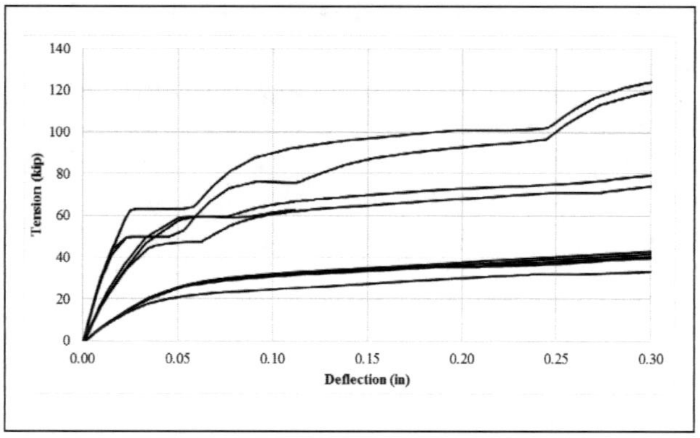


**Figure 6.** Force–deflection behavior of 50 ksi stringer connections with seat angles.

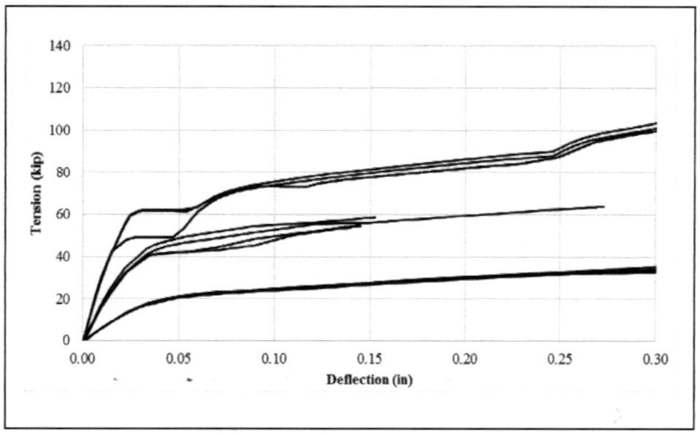


**Figure 7.** Force–deflection behavior of 36 ksi stringer connections with seat angles.

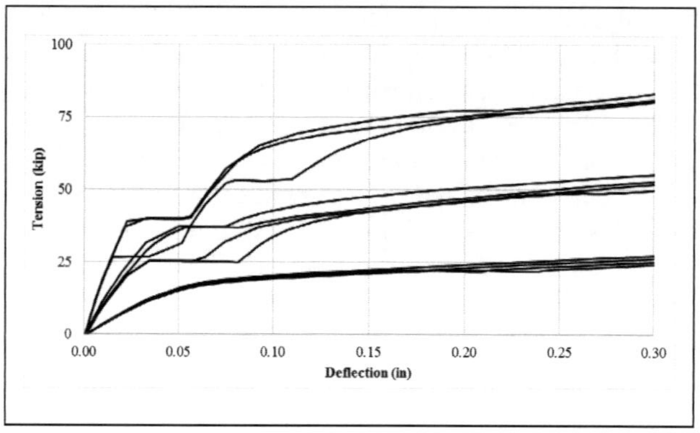


**Figure 8.** Force–deflection behavior of 50 ksi stringer connections without seat angles.

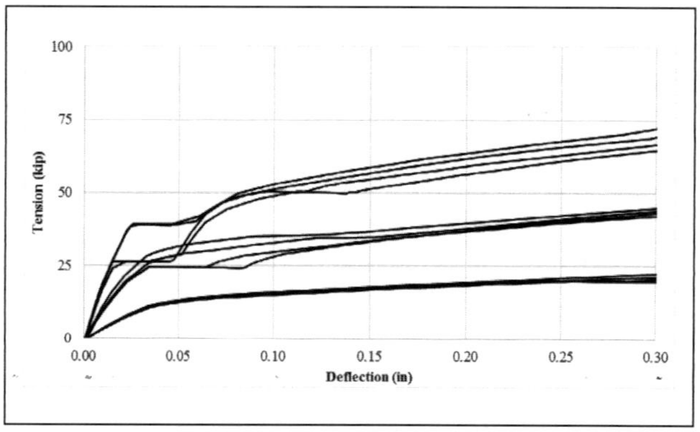


**Figure 9.** Force–deflection behavior of 36 ksi stringer connections without seat angles.

connections were produced both with and without a seat angle. The force-deflection curves for connections with seat angles are shown in Figures 6 and 7 and force-deflection curves for connections without seat angles are shown in Figures 8 and 9. The lower group of curves in each of the figures represents ¼" thick angles, the middle groups of curves represent 3/8" thick angles, and the upper groups represent ½" thick angles. As the non-linear behavior of the stringers is nearly entirely determined by the thickness of the angles, the nonlinear behavior in lower-resolution models may be taken directly from these curves.

## Lower-Resolution Models

Lower-resolution finite element models were made in SAP2000 version 14 to match the high-resolution model of the same bridge, shown in Figure 10. Instead of using shells for rolled shaped, beam members were used. Gusset plates were not explicitly modeled, hinges based on the yielding pattern of the gusset plates were included on the ends of the truss members. The deck was included as shells, with a mesh size of 24", and was fully composite with the stringers, though it was not constrained to the floor beam nodes. This is important to ensure that the deck does not carry the stringers load.

These models were loaded similarly with an unfactored dead load followed by live loads in the ratio of an HS-20 truck load incremented until collapse. These lower-resolution models are similar to what engineers generally have access to when evaluating small bridges. Damage to the concrete deck was neglected and the corrugated steel deck was modeled as a flat shell with modifications to the stiffness for bending and in-plane tension. The nonlinear behaviors of the lower-resolution models are defined based on the behavior of the high-resolution model, with discrete plastic hinges defined in the members to capture yielding of the gusset plates. In addition, the stringers were modeled as having tensile continuity based on the responses of the high-resolution connection modeling, specifically, that of 50 ksi 3/8" thick angles with a seat angle.

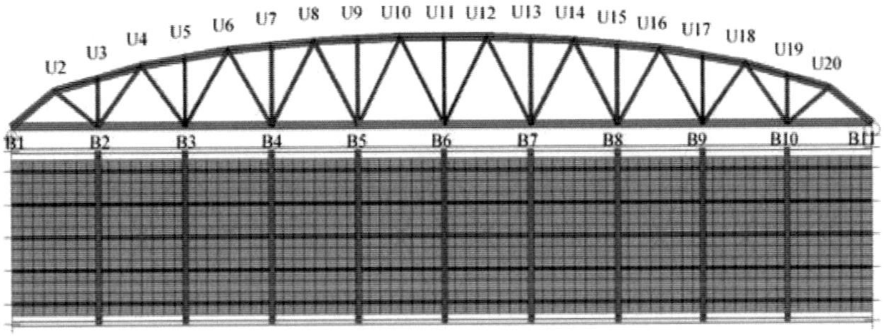


**Figure 10.** Side and top view of the lower-resolution model.

Although the lower-resolution model is of the same bridge as the higher-resolution model, the principles used in establishing it are applicable to other truss structures. The nonlinear behavior of the connections of the trusses are not based on the high-resolution model, but on the geometry and material behavior of the gussets and members. The nonlinear behavior of the stringers connections was taken directly from the high-resolution connection models. Thus, the reevaluation procedure may be reproduced for other bridges using the force deformation curves for stringer connections.

### *Plastic Hinge Definition*

The high-resolution model and connection model reveal some patterns in how yielding initializes and the shape of the yielding in the gusset plate connections. The yielding in the gusset plates initializes along the width of the flange and between the last row of bolts on one member and the first row of bolts on the next member. This gives a sectional area based on the thickness of the gusset plates and the geometry of the members and a hinge length based on the connection geometry. Depending on the thickness of the gusset this may also initialize nonlinear bending material as well. Nonlinear behavior of the lateral system was also considered, with an axial hinge based on the bending of the rods' anchor plates.

The high-resolution connection modeling shows that the behaviors of framed connections are highly predictable based on the gage of the angle used in framing them and the material behavior of the angles; they are relatively unchanged by the number of bolts in the angles or the length of the angles. The resultant force–displacement curve is readily used to define nonlinear behavior of the stringer connections. For the lower-resolution models, the rest of the simply supported assumptions are maintained, including releases for torsion at one end and moment releases at both ends. The high-resolution connection models were not subjected to moment loadings to determine that these typical connections fully lack rotational stiffness, but low resolution linear elastic models suggest that even if fully stiff, they carry little end moment if there is a deck present.

Framed stringer connections are typical enough that a few high-resolution models can give force-deflection curves that are representative of the vast majority of them. The wide range of geometries and bolt layout among gusset plate connections precludes this exhaustive approach. Fortunately, the yielding in the gusset plates shown in the high-resolution modeling is consistent and limited in scope. The yielding initializes between the last row of bolts in one chord member and the first row of bolts in the next member. Member yielding occurs along the length of the member but does not extend into the gusset connection. Thus, discreet plastic hinges of the smaller gusset sections as well as of the members can sufficiently capture the nonlinear behavior. These observations are used in defining the discrete plastic hinges in the lower-resolution models described by Swanson et al. (*16*) and elsewhere that are representative of the behavior observed in the high-resolution models.

### *Comparison of High- and Low-Resolution Models*

A comparison of the collapse curves for the high-resolution model and the low-resolution model of the same bridge is shown in Figure 11 giving the push-down for the center node of the damaged bottom chord. States shown are the intact state, a damaged state with one half of a bottom chord member removed, and for a damage state with the full bottom chord member removed. Two partial damage states are presented for the high-resolution model, one for the outer channel removed and another for the inner channel removed. Only one partial damage state is presented for the lower-resolution SAP2000 model, as the connections were not explicitly modeled and the complexities of that load path are absent. The partial damage matches closely between the

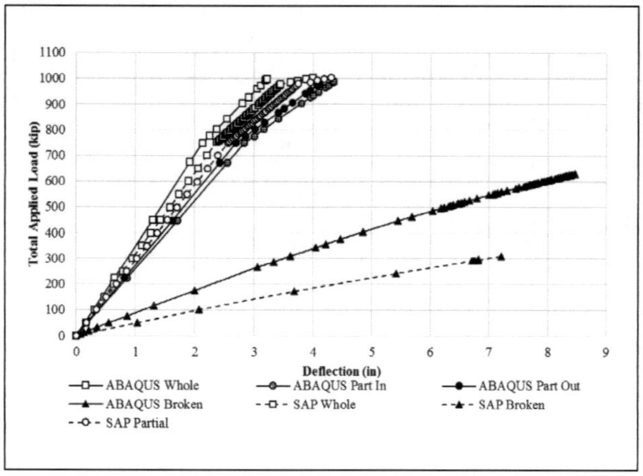


**Figure 11.** Total applied load versus deflection from high- and low-resolution models.

models, but the intact and fully damaged lower-resolution models have noticeably lower stiffnesses. In the case of the intact state, this is not significant, and the collapse occurs at the appropriate load. For the loss of the full bottom chord member there is a substantial difference in stiffness. One modeling aspect that may cause this is the implementation of the fully composite deck. In the high-resolution model the deck is constrained to the stringers over the entire length of the stringers. In the lower-resolution model, however, the coarser mesh means that for the deck to be noncomposite with the floor beams it is also noncomposite with the stringers for two feet on either side of the floor beams since each deck element is roughly 2' × 2'.

## Conclusion

Models of truss bridges are generally conservative assuming a lack of moment continuity that is present in gusset connections and ignoring support from the floor and lateral systems. Consideration of the contribution of the floor system, including connections in the floor system and the deck itself, as well as the limited moment resistance present in gusset connections is instrumental in reevaluating a member's fracture critical status. High-resolution models of a pony-truss bridge indicate that simple methods may be used for modeling the nonlinear behavior of gusset connections that may yield before members themselves do. Framed stringer connections are not designed for axial continuity, but high-resolution models of common framed stringer connections reveal a limited and consistent tensile capacity that could assist as an alternative load path for damaged trusses. The curves presented for framed stringer connections are suitable for defining nonlinear hinge properties in lower-resolution finite element models that would be commonly available to engineers. Whereas high-resolution modeling described is time intensive, it is an option for unique structures and was extremely useful in characterizing behavior and determining important sources of alternative load paths as part of an effort to develop a refined analysis procedure based on lower-resolution modeling.

## Author Contributions

The authors confirm contribution to the paper as follows: study conception and design: MAB, JAS, GAR, EFD; data collection: MAB, JAS, EFD; analysis and interpretation of results: MAB, JAS, EFD; draft manuscript preparation: MAB. All authors reviewed the results and approved the final version of the manuscript.

## References

1. Federal Highway Administration. *Highways, National Bridge Inspections Standards.* Code of Federal Regulations, Title 23, Part 650. U.S. Government Printing Office, Washington, D.C., 2011.
2. ORC 5501.47. *Ohio Revised Code Section 5501.47: Bridge Inspections.* 2010. http://codes.ohio.gov/orc/5501.47. Accessed March, 2014.
3. Connor, R. J., R. Dexter, and H. Mahmoud. *NCHRP Report 354: Inspection and Management of Bridges with Fracture-Critical Details.* TRB, National Research Council, Washington, D.C., 2005.
4. Federal Highway Administration. *Nation Bridge Inspection Standards.* Federal Register 69, No. 239. Federal Highway Administration, Washington D.C., 2004.
5. American Association of State Highway and Transportation Officials. *Manual for Bridge Evaluation, 2nd Edition, with 2011, 2013, 2014, 2015, and 2016 Interim Revisions.* American Association of State Highway and Transportation Officials, Washington D.C., 2016.
6. American Association of State Highway and Transportation Officials. *LRFD Bridge Design Specifications 7th Ed with 2015 Interim Revisions.* American Association of State Highway and Transportation Officials, Washington D.C., 2015.
7. Ghosn, M., F. Moses, and D. M. Frangopol. Redundancy and Robustness of Highway Bridge Superstructures and Substructures. *Structure and Infrastructure Engineering,* Vol. 6, No. 1–2, 2010, pp. 257–278.
8. Ghosn, M., and F. Moses. *NCHRP Report 406: Redundancy in Highway Bridge Superstructures.* TRB, National Research Council, Washington D.C., 1998.
9. Applebury, C. R. *Reliability-Based Methodology for Bridge Inspection Planning.* MS thesis. University of Missouri, Columbia, 2011.
10. Federal Highway Administration. *Clarification of Requirements for Fracture Critical Members.* Memorandum from Federal Highway Administration HIBT-10, 2012.
11. American Association of State Highway and Transportation Officials. *Guide Specifications for Fracture Critical Non-Redundant Steel Bridge Members.* American

Association of State Highway and Transportation Officials, Washington D.C., 1978.

12. Ayyub, B., and A. Ibrahim. Posttensioned Trusses: Reliability and Redundancy. *Journal of Structural Engineering*, Vol. 116, No. 6, 1990, pp. 1507–1521.
13. Stains, J. M. *Towards the Development of Redundancy Assessment Protocols for Steel Truss Bridges*. MS thesis. Benjamin M. Statler College of Engineering and Mineral Resources, West Virginia University, Morgantown, 2012.
14. Jankowiak, T., and T. Lodygowski. Identification of Parameters of Concrete Damage Plasticity Constitutive Model. *Foundations of Civil and Environmental Engineering*, Vol. 56, No. 6, 2005, pp. 53–69.
15. McConnell, J., D. Wurst, G. McCarthy, and M. Sparacino. *Investigation of Load-Path Redundancy in Aging Steel Bridges*. A report submitted to the University of Delaware University Transportation Center, 2013.
16. Swanson, J. A., G. A. Rassati, and E. F. Dues. *Fracture Critical Analysis Procedures and Design and Retrofit Approaches for Pony Truss Bridges in Ohio*. No. FHWA/OH-2016/13. Ohio's Research Initiative for Locals, the Ohio Department of Transportation, Office of Statewide Planning & Research, 2016.
17. American Society of Civil Engineers. *Seismic Evaluation and Retrofit of Existing Buildings*. ASCE 41-13. American Society of Civil Engineers, 2013.
18. Cha, H., L. Lyrenmann, R. J. Connor, and A. H. Varma. Experimental and Numerical Evaluation of the Postfracture Redundancy of a Simple Span Truss Bridge. *Journal of Bridge Engineering*, Vol. 19, No. 11, 2014, p. 04014048.
19. Ruffley, D. J. *A Finite Element Approach for Modeling Bolted Top-and-Seat Angle Components and Moment Connections*. MS thesis. University of Cincinnati, 2011.
20. Adkins, K. A. *A Model for Prediction of Fracture Initiation in Finite Element Analyses of Welded Steel Connections*. MS thesis. University of Cincinnati, 2014.

*The Standing Committee on Steel Bridges (AFF20) peer-reviewed this paper (18-03496).*

Article

TRR
JOURNAL OF THE TRANSPORTATION RESEARCH BOARD

Transportation Research Record
2018, Vol. 2672(41) 196–206

DOI: 10.1177/0361198118781653
journals.sagepub.com/home/trr
SAGE

# Longitudinal Joint Performance of a Concrete Hollow Core Slab Bridge

**Diego Maria Barbieri[1,2], Yuechi Chen[3], Enrico Mazzarolo[3], Bruno Briseghella[3], and Angelo Marcello Tarantino[1]**

**Abstract**

Hollow core slab bridges are constructed by placing prefabricated or prestressed box beams adjacent to each other, grouting the small longitudinal space (hinge-joint) between the slabs and casting a reinforced concrete deck. The longitudinal cracking appearing at hinge-joint locations leads to a premature deterioration of the deck. This paper presents a theoretical and experimental study of a hollow core slab bridge composed of three beams and a cast-in-place deck. A real-size specimen was built according to Chinese code specifications. The behavior of the longitudinal joints was investigated by applying the standard vehicle load. The tests do not highlight any longitudinal cracks. A finite element model was created from the experimental data. A finite element parametric analysis revealed some practical design indications regarding the following inputs: deck thickness, concrete strength, and hinge-joint steel bars. Furthermore, these analyses testify that C-shape and X-shape stirrups do not play an active role in preventing the joint longitudinal cracks. This research confirms the reliability of the design method, at least for static loads, while further studies are needed to investigate the effect of both periodical loadings and different temperatures on upper and lower surfaces of the beams.

A hollow core beam is a precast slab of prestressed or ordinary concrete used for small- and medium-span bridges (prestressed bridge elements and systems). Slabs have tubular voids extending the full length of the bridge, typically with a diameter equal to 2/3 ÷ 3/4 of the thickness.

Box beams are widespread because of the ease of construction, reduced weight, and transportation costs. They are convenient for accelerated bridge construction processes (*1*, *2*, *3*).

When the box beams are placed adjacent to each other, there is no continuity enabling them to function as a single structure. The small longitudinal joint between hollow core slabs, typically called a hinge joint, is normally filled with grout or an appropriate non-structural sealant to prevent water leakage and moisture penetration between beams, as well as to support load transfer (*4*). The deck is built with reinforced cast-in-place concrete. The cure of this layer may take some days; however, very little deck forming is needed. Concrete compressive strength and minimum thickness may be required (*3*, *4*, *5*).

The lack of available analyses and models for the design of the connection between precast members represents an important issue in the construction of these systems. Empirical relations form the basis for some design guidelines (*5*); these, however, have been unable to provide the durability demands on the connection.

Consequently, the major drawback in the construction of these systems is the longitudinal cracking visible on the top of the cast-in-place layer at hinge-joint locations; this is the leading cause of premature deterioration of the structure (*1*). A single slab may eventually carry the entire load when the lateral connection is completely damaged; this condition is known as single plate load effect (*6*).

The bridge can be analyzed in its entirety: a plate with opposite edges simply supported and the other two edges free under uniformly distributed load; this is a macromechanical model of a precast orthotropic bridge (*1, 4*). The formation process of the longitudinal cracks can be studied with the bonding relationships between cast-in-place and precast concrete (*6*, *7*, *8*, *9*, *10*). The research developed so far has not succeeded in defying the longitudinal cracking issue definitively.

[1]Modena and Reggio Emilia University, Modena, Emilia-Romagna, Italy
[2]Norwegian University of Science and Technology, Department of Civil and Environmental Engineering, Trondheim, Sør-Trøndelag, Norway
[3]Fuzhou University, College of Civil Engineering, Fuzhou, Fujian, China

**Corresponding Author:**
Address correspondence to Diego Maria Barbieri:
diegomb271@gmail.com; diego.barbieri@ntnu.no

The current Chinese infrastructure system widely employs simple span hollow core slab bridges with lengths ranging from 6 to 25 m. The spreading of the longitudinal cracking has become a great concern. The causes can vary. The current design approach (hinge-plate method) assumes only shear passes through the hinge joints and ignores the complex stress state; in addition, roads are crossed by overloaded vehicles, and questionable efficiency in asset maintenance is an ongoing problem. Furthermore, practice shows the cracking issue involves hollow core slab bridges independently of the age, length, and beam numbers.

The national Chinese code JTG D-62 defines the full-depth longitudinal joint which is now in use (*11*). It specifies longitudinal steel bars and C-shape, X-shape stirrups in the hinge joints; the stirrups would strengthen the bond performance between cast-in-place and precast concrete. The code does not require any calculation checks.

For this paper, an experimental test was performed to achieve a better understanding of the issue by evaluating the load that could bring about the formation of longitudinal cracking. A laboratory test was carried out on a full-scale specimen composed of three hollow core slabs and a 10-cm cast-in-place concrete slab. A finite element model (FEM) was then built from experimental data, and a parametric analysis shows the main outcomes considering different deck thickness, concrete strength, and joint stirrup reinforcements.

## Methodology

### Laboratory Test

A full-scale hollow core slab concrete deck was built to investigate the response of the longitudinal joints between the precast beams. The span length was 8 m and the width almost 4 m. Geometrical dimensions, steel bar diameters, and quantities were defined by standard drawings (*12*). The bridge was composed of three hollow core slabs and a cast-in-place layer (Figure 1*a*); there are four contact surfaces between the slabs and the hinge joints. Figure 1*b* and *c* show the geometrical dimensions, steel bars, and stirrups in the slab and hinge joints. PVC tubes form the holes inside the beams. Each beam rests on four laminated rubber bearings, which fulfil a simple support constraint.

The aim of the test was to investigate the possible development of longitudinal cracks on the deck by gradually applying the standard truck vehicle load specified by the national Chinese code JTG D-60 (*13*). Since its total dimension exceeds the bridge span length, the test applied the two most loaded axles highlighted in Figure 2*a*: each consists of two tires (Figure 2*b*) placed in the middle of the span (Figure 2*c* and *d*).

Figure 3*a* shows a general view of the bridge. Two hydraulic jacks apply controlled force by means of two reaction beams connected with four 0.6 x 0.2 m mortar pads positioned on the deck; each jack exerts its action on two contact areas (Figure 3*b*). Each contact area is subjected to a load ranging from 0 to 180 kN with increments of 5 kN, then from 180 to 300 kN with increments of 10 kN; the average time between two consecutive steps is two minutes. The hydraulic jacks eventually apply a force approximately four times the standard value, being 70 kN, the standard load specified by code JTG D-60 (Figure 2*a*).

Strain gauges monitor the strain of the longitudinal steel bars (Figure 3*c*). Dial indicators positioned

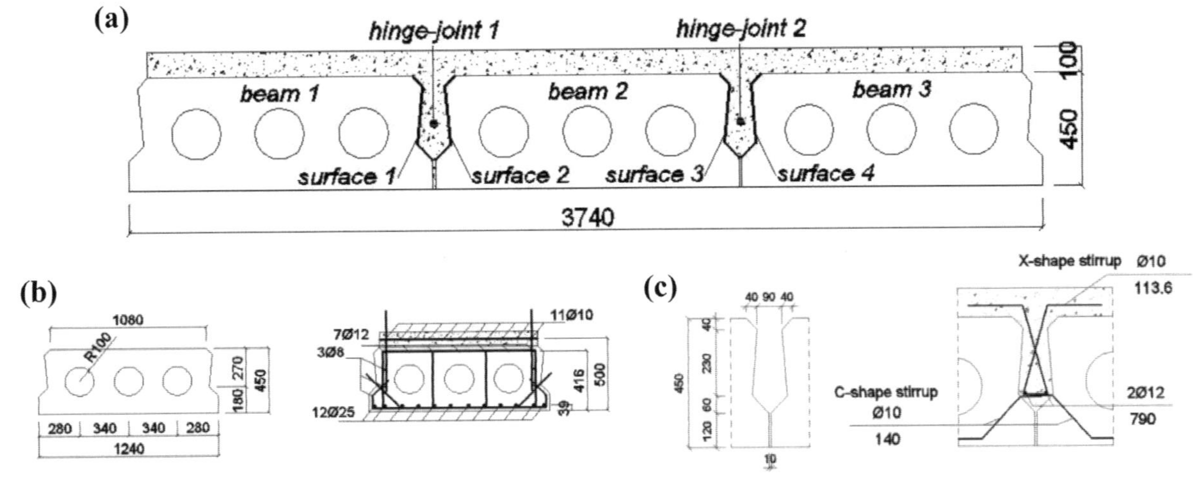


**Figure 1.** Bridge cross-section layout: (*a*) geometrical dimensions and bars in a slab, (*b*) geometrical dimensions and bars in a hinge-joint, and (*c*) dimensions in mm.

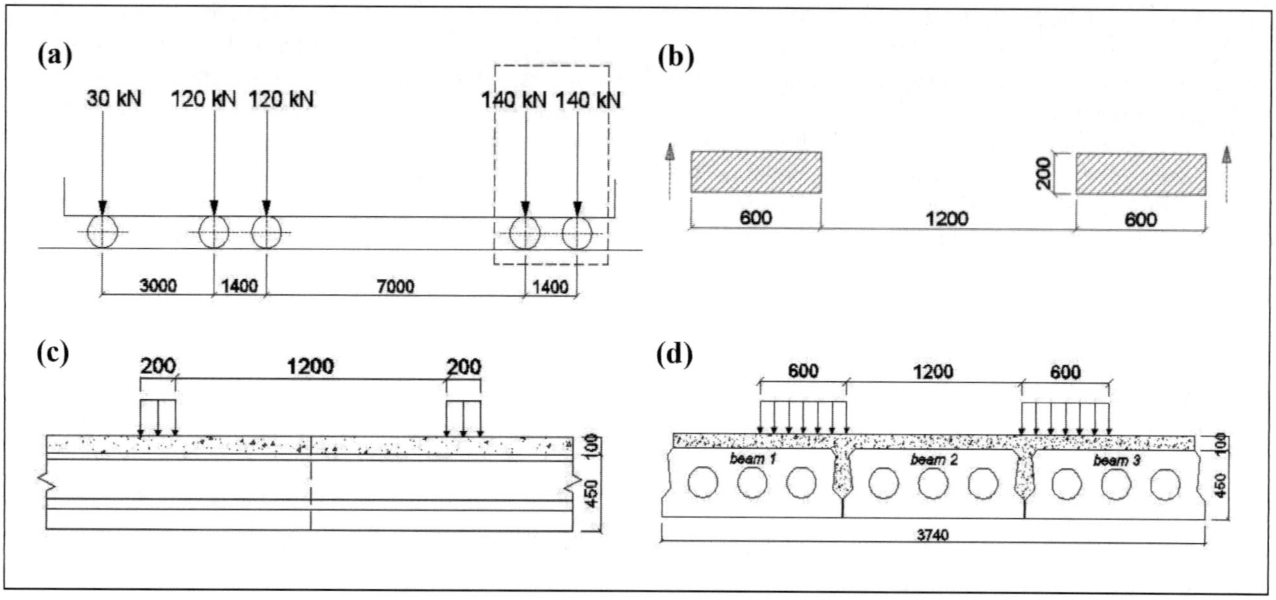


**Figure 2.** Standard truck vehicle load after JTG D-60, dimensions in mm: (*a*) schematic standard truck vehicle load, (*b*) contact areas for one axle, (*c*) load position, longitudinal view, and (*d*) load position, cross-sectional view.

vertically (Figure 3*d*) assess the deflection of the beams $\Delta_{v1}$, $\Delta_{v2}$ (Figure 3*e*). Dial indicators placed horizontally (Figure 3*d*) measure the relative displacement of the edges at the bottom of the slabs $\Delta_{h1} + \Delta_{h2}$ (Figure 3*e*). These data were correlated with the possible occurrence of longitudinal cracking. Figure 3*f* shows the position of the strain gauges, represented by black dots, and the vertical and horizontal dial indicators. The equipment was set both at 1/2 and at 3/4 of the span length. The test employed 6 strain gauges and 14 dial indicators.

Concrete specimens were tested to evaluate the actual compression strength and elastic modulus according to code GB/T 50081-200218 (*14*). The slabs were composed of C30 concrete with initial elastic tangent modulus E = 28,200 MPa. The deck was made of C40 concrete with initial elastic tangent modulus E = 30,700 MPa. The HRB 335 steel bars had characteristic yield strength of 335 MPa and elastic modulus of 210 GPa.

## Finite Element Analysis

### *Concrete and Steel Constitutive Relationships*

Test results were compared with a numerical simulation performed with ABAQUS software. A concrete damaged plasticity model simulated the concrete behavior of both the slab and the cast-in-place deck (*15*). The model assumes that the two main failure mechanisms are tensile cracking and compressive crushing of the material.

The uniaxial compressive response of concrete is characterized by damaged plasticity (Figure 4*a*). The response in uniaxial compression is linear until the value of initial yield $\sigma_{c0}$. In the plastic regime, the response is characterized by stress hardening followed by strain softening beyond the ultimate stress $\sigma_{cu}$. The stress–strain response follows a linear elastic relationship in uniaxial tension; the value of failure stress is $\sigma_{t0}$ (Figure 4*b*).

The post-failure behavior for direct straining is modelled with tension stiffening, which allows definition of the strain-softening behavior for cracked concrete. Tension stiffening is modelled thanks to Hillerborg's fracture criterion; $G_f$ is the energy required to open a crack of unit area (*16*). The brittle behavior of concrete is characterized by a stress-displacement response rather than a stress-strain response. This model assumes a linear loss of strength after cracking which begins when the tension reaches the associated fracture energy. $\sigma_{t0}$; $u_{t0}$ is the displacement at which the complete loss of strength takes place (Figure 4*c*).

The following parameters were used for the slab concrete: density $\rho$ = 2,500 kg/m$^3$; Young's modulus E = 30,700 MPa; Poisson's ratio $\nu$ = 0.23; dilatation angle $\psi$ = 36°; eccentricity $\varepsilon$ = 0.1; ratio $f_{b0}/_{c0}$ = 1.16; parameter K = 0.666; viscosity parameter $\mu$ = 0.01; tensile strength $f_{t0}$ = 2.01 MPa; fracture energy $G_f$ = 80 N/m.

The following parameters were used for the cast-in-place concrete: density $\rho$ = 2,500 kg/m$^3$; Young's modulus E = 28,230 MPa; Poisson's ratio $\nu$ = 0.18; dilatation angle $\psi$ = 36°; eccentricity $\varepsilon$ = 0.1; ratio $f_{b0}/_{c0}$ = 1.16; parameter K = 0.666; viscosity parameter $\mu$ = 0.01; tensile strength $f_{t0}$ = 2.40 MPa; fracture energy $G_f$ = 120 N/m.

Concrete components are meshed using fully integrated, 8-node linear bricks (C3D8I element); there are 25,392 linear hexahedral elements (Figure 5).

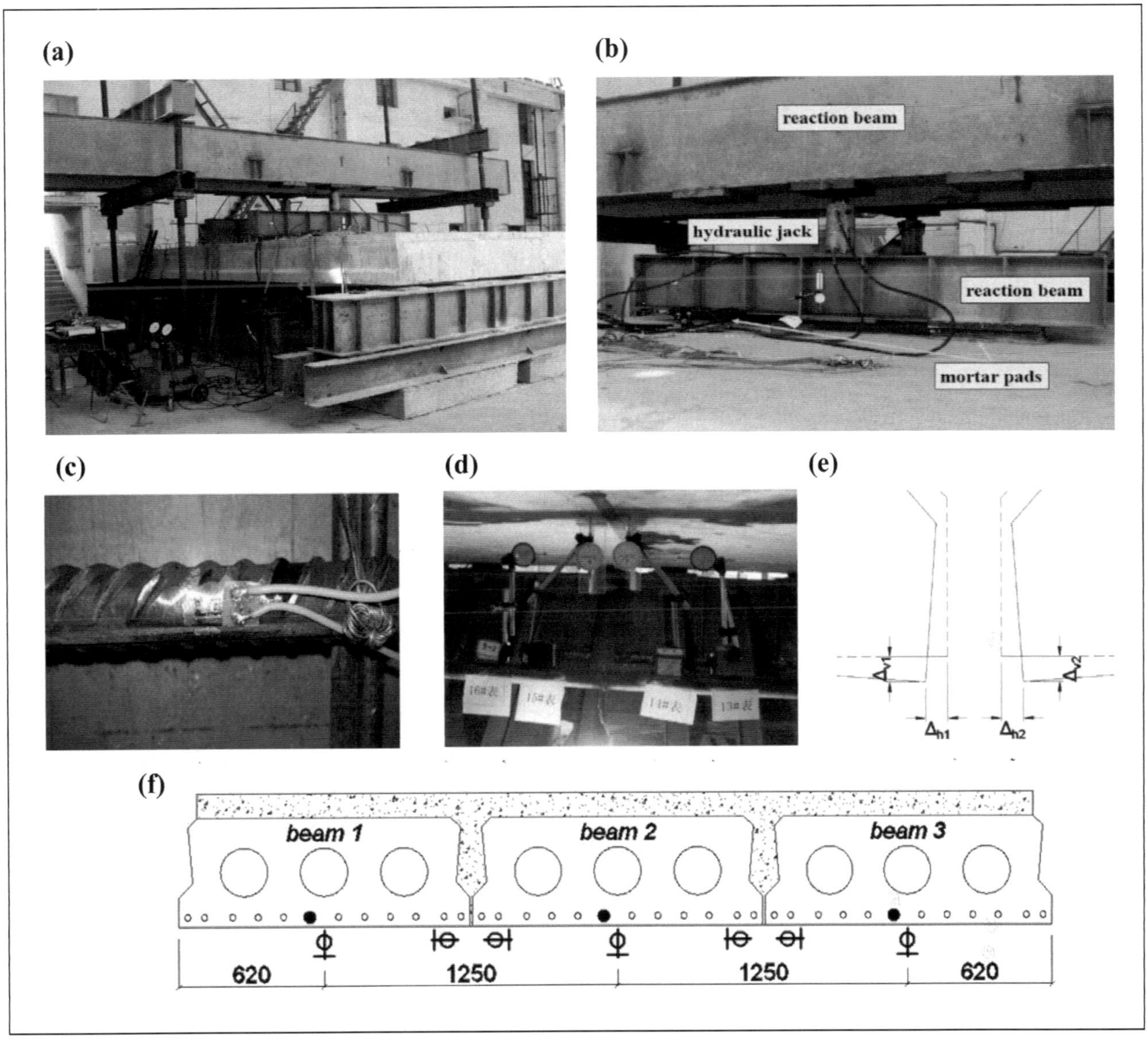


**Figure 3.** Bridge test equipment, dimensions in mm.

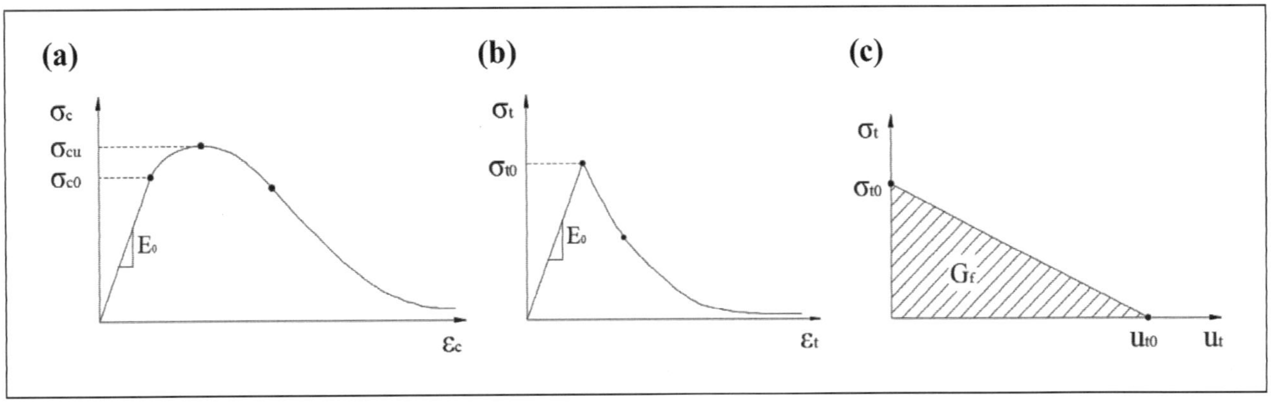


**Figure 4.** Concrete damage plasticity response to uniaxial compression, tension, and post-failure stress-fracture energy curve.

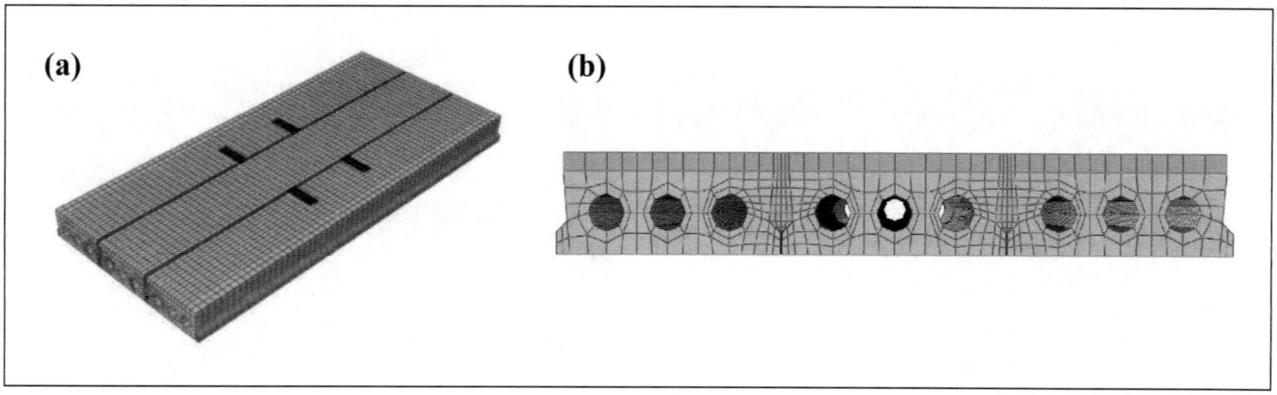


**Figure 5.** FEM with parts and mesh in evidence: (*a*) perspective view and (*b*) frontal view.

Longitudinal steel bars at the bottom of the slabs and inside hinge joints, C-shape, and X-shape stirrups were modelled by means of one dimensional rods embedded in oriented surfaces. Bars were implemented with classical metal plasticity model and were superposed on the mesh of the concrete parts.

The following parameters were used for the steel bars: density $\rho$ = 7,850 kg/m$^3$; Young's modulus E = 200 GPa; Poisson's ratio $\nu$ = 0.30; yield stress $f_y$ = 335 MPa. Steel components were meshed using 2-node linear three-dimensional trusses (T3D2 element) for a total number of 3,760 elements.

Simple supported boundary conditions were specified at both edges of the bridge. The loads taken into consideration were the self-weight and the vehicle force, which were gradually applied as in the experiment. Three steps form the analysis: the first initial step establishes boundary conditions and contacts between the surfaces; the second stage is a static, general step, which applies the structure self-load (gravity); the third stage is a static, general step, which exerts the vehicle load on the contact surfaces (pressure).

## *Contact Constitutive Relationships*

Surface-based cohesive behavior with generalized traction-separation model describes the contact between the precast concrete and the cast-in-place concrete. The relationship assumes a linear elastic separation law before damage; the progressive degradation of the cohesive stiffness drives the failure of the cohesive bond.

The failure mechanism consists of two parts: a damage initiation criterion and a damage evolution law. Damage begins to occur when the maximum contact stress ratio at a point reaches the peak value of the contact stress. The variable D represents the overall damage at a contact point; it is initially equal to 0. The scalar damage quantity D monotonically evolves from 0 to 1 on further loading after the initial damage is met. For linear softening ABAQUS uses an evolution of the damage variable that reduces to the expression

$$D = \frac{\delta_m^f(\delta_m^{max} - \delta_m^0)}{\delta_m^f(\delta_m^f - \delta_m^0)}, \tag{1}$$

$\delta_m^0$ is the displacement at the stress peak value; $\delta_m^f$ is the last displacement; and $\delta_m^{max}$ represents the maximum value of the effective separation attained during the loading history. A bilinear model needs to determine the slip stiffness $K$ (MPa/mm), the final slip value $S_u$ (mm), and the peak shear stress $f_v$ (MPa). Stiffness values typically range from 5.0 MPa/mm to 15.0 MPa/mm; the bond-slip values $S_u$ vary from $S_u/S_0 = 1.5$ (brittle fracture) to $S_u/S_0 = 3.0$ (extended rupture).

Surface roughness, interface agent, and the shape of the contact surface are the critical parameters in the definition of $f_n$ and $f_v^0$, respectively, the pull strength along the axis normal to the contact surface (Figure 6*a*), and the shear strength (Figure 6*b*). The following expressions define the curves implemented in the model regarding bonding mechanism between older (slab) and newer (cast-in-place) concrete (*10*).

$$f_n = 0.7 f_{ct}, \tag{2a}$$

$$f_v^0 = 0.01 f_c^0, \tag{2b}$$

$f_{ct}$ and $f_c^0$ are respectively the minimum values of tensile and compressive strength of the concrete between hollow slabs and hinge-joints.

## *FEM Parametric Analysis*

Starting from the calibrated FEM, a parametric analysis was performed to study the influence of the main parameters on the joints' response. The analysis focuses on deck thickness, deck concrete strength, and the presence of C-shape and X-shape stirrups. The analyzed outcomes are the horizontal displacements of slabs bottom edges

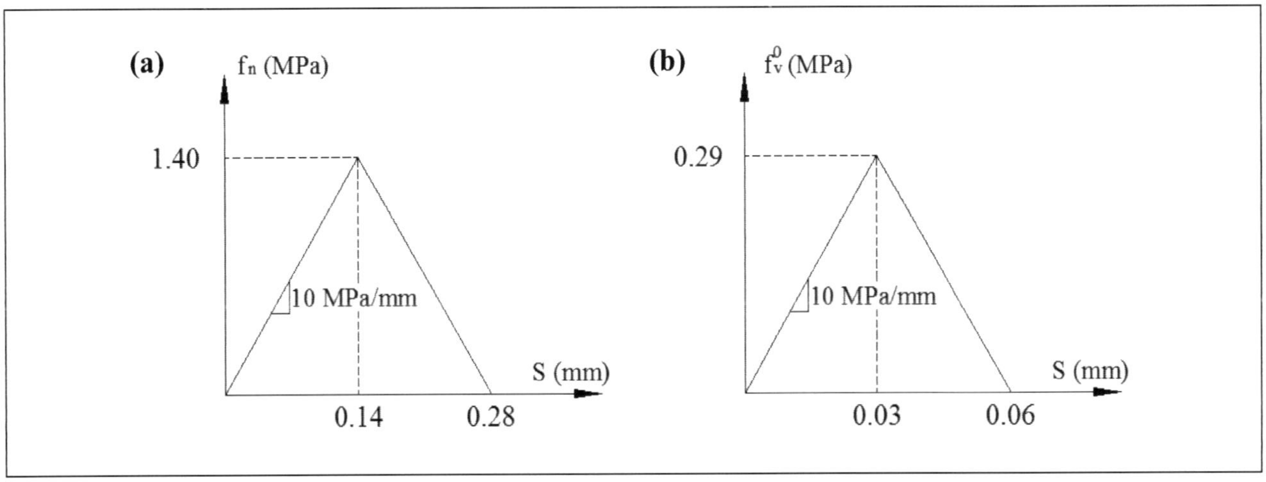


**Figure 6.** Bond-slip implemented relationships: (*a*) pull strength and (*b*) shear strength.

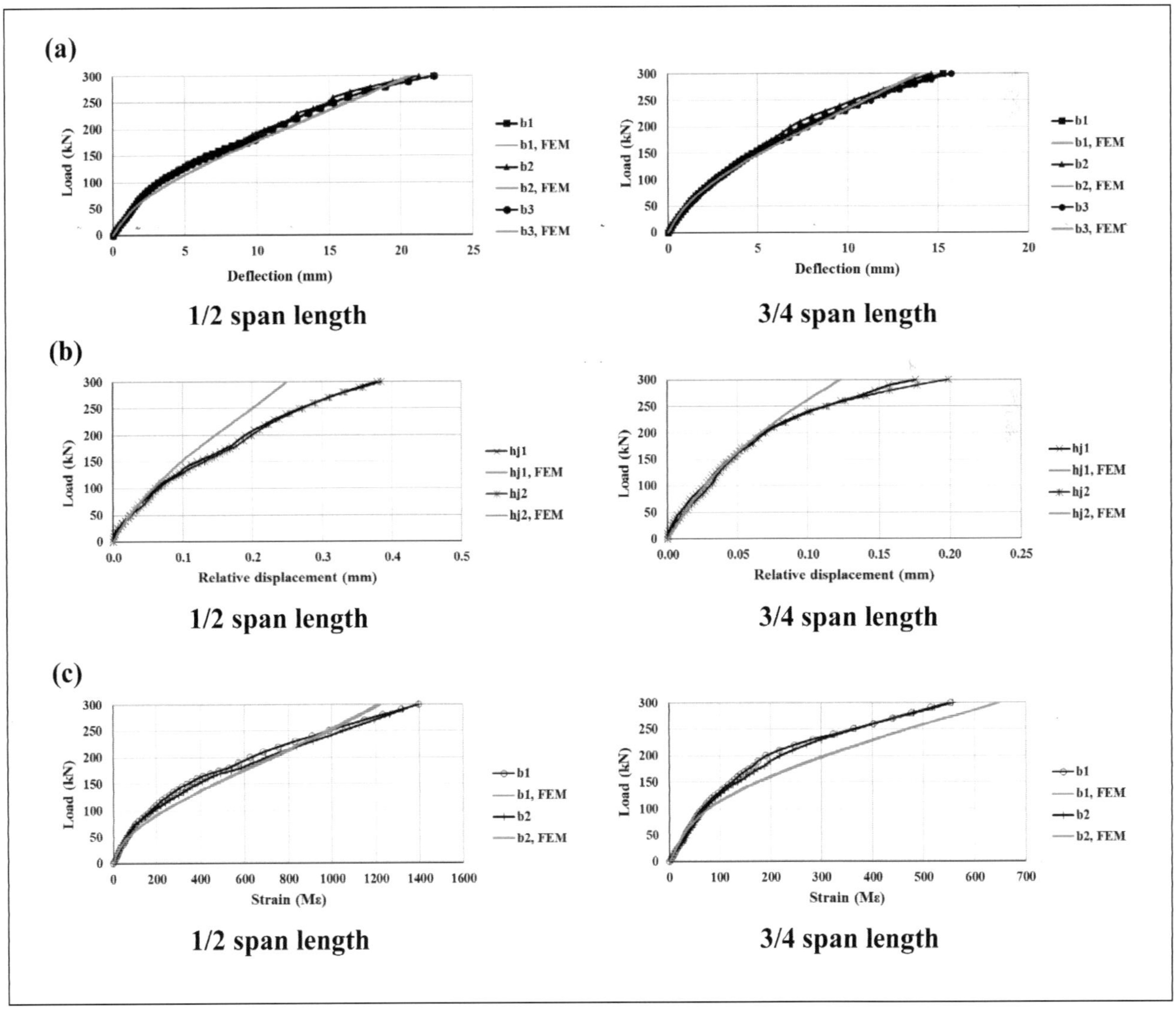


**Figure 7.** Comparison between experimental data and FEM results: (*a*) deflection of the beams, (*b*) horizontal displacement of the slab bottom edges, and (*c*) bottom bar strain.

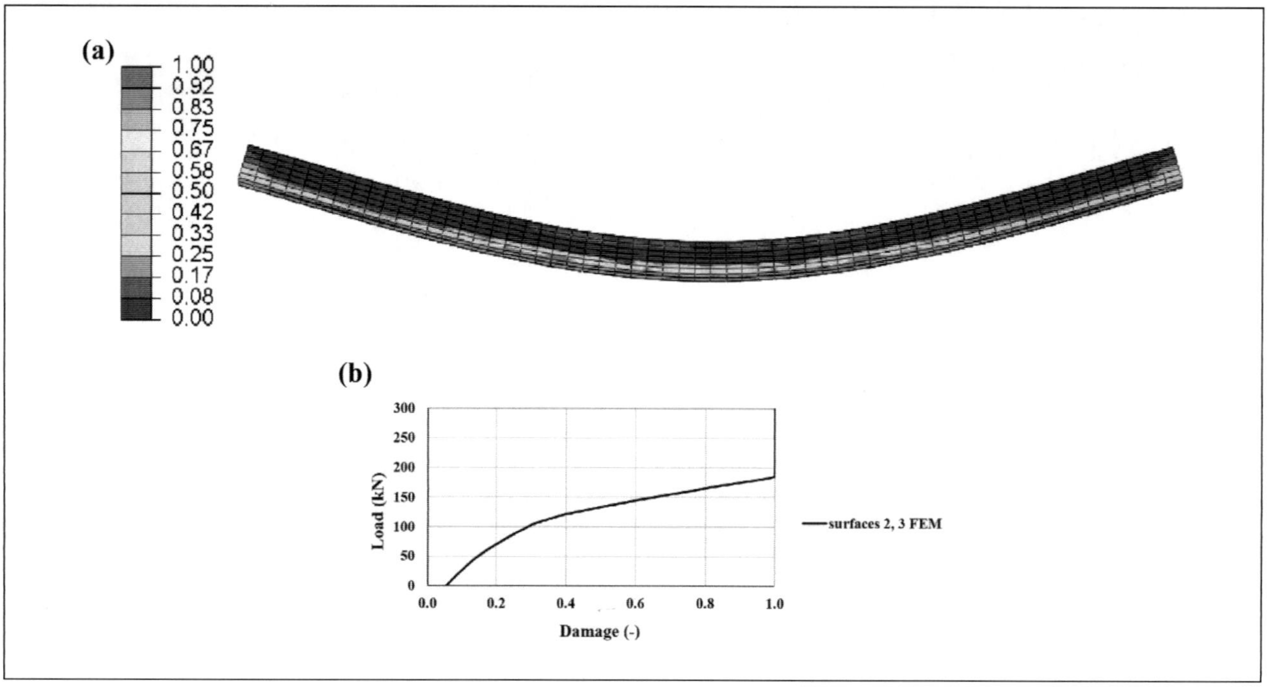


**Figure 8.** De-bonding shear stress damage criterion: (*a*) hinge-joint surfaces 2 and 3 at final load and (*b*) first de-bonded point damage evolution.

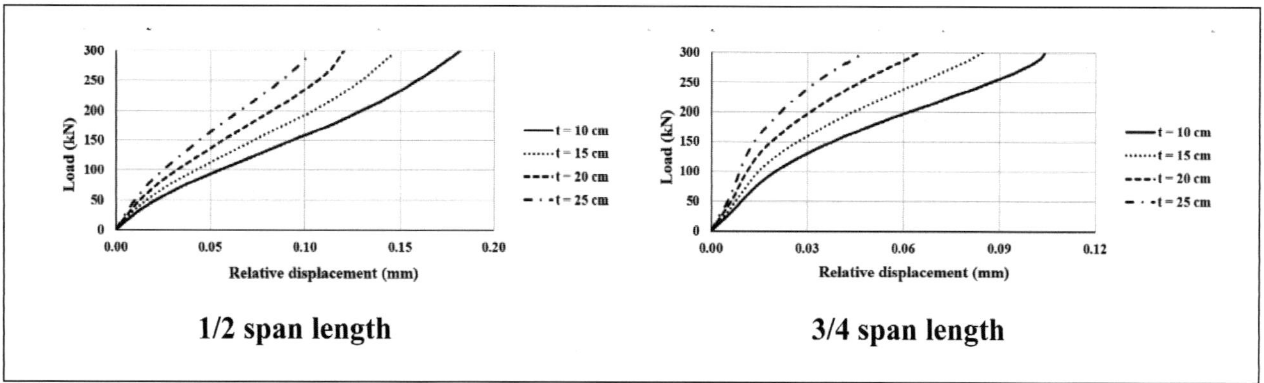


**Figure 9.** Horizontal relative displacements of the slab bottom edges.

and the extension of the de-bonded areas. The results obtained for slabs 1 and 2 are the same as for slabs 2 and 3 due to the symmetry of the problem.

## Results and Discussion

### *Comparison Between Experimental and FEM Results*

Figure 7 compares the test results and the finite element analysis outcomes; the Y-axis refers to the load applied to a single wheel contact area. Figure 7*a* shows the vertical displacement of the beams; the loading process can be roughly divided into three stages ("b" stands for beam in the legend). The load deflection initially increases according to a linear trend. There is a first inflection point when the load reaches approximately 30 kN (0.43 times the standard vehicle load); the curve starts following a non-linear path and some transversal cracks appear at the bottom of the beams. When the load is close to 200 kN (2.86 times the standard vehicle load) another inflection point occurs and the stiffness of the bridge is reduced further.

Figure 7*b* shows the horizontal relative displacement of the bottom edges of the slabs near the hinge joints ("hj" stands for hinge joint in the legend). In the elastic stage the relative displacement is almost 0 mm. Cracks begin and the growth of the curve becomes non-linear

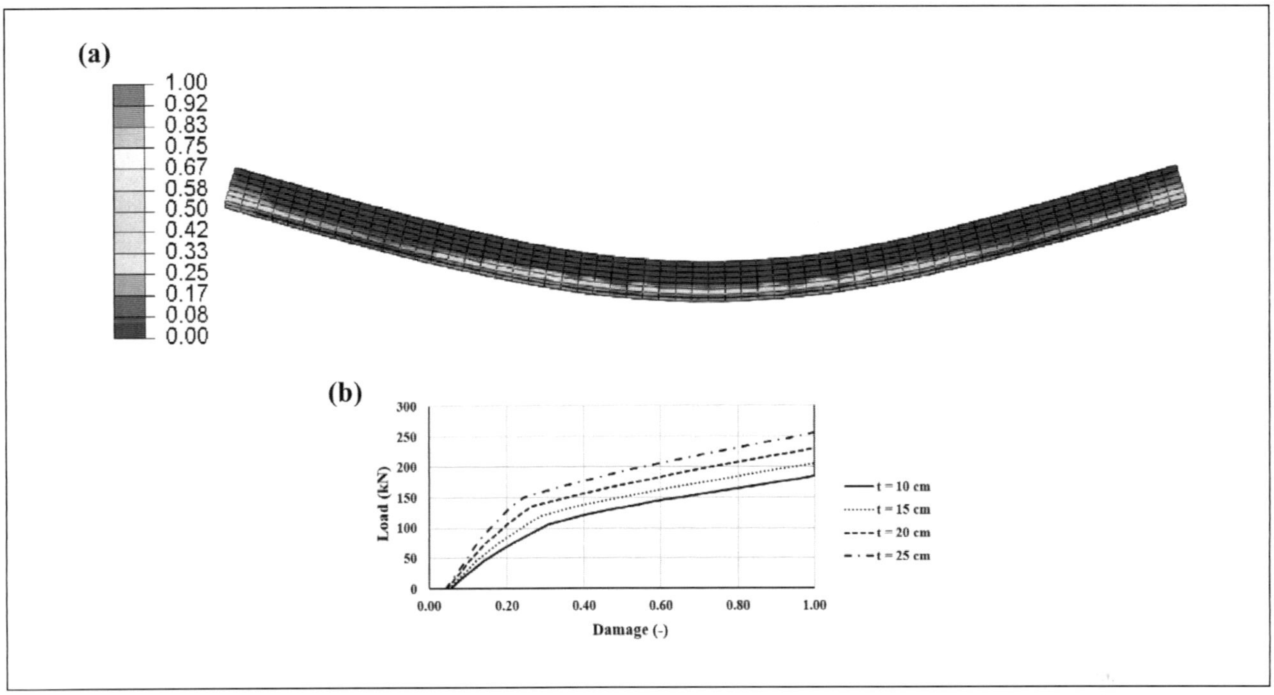


**Figure 10.** De-bonding shear stress damage criterion: (*a*) hinge-joint surfaces 2 and 3 at final load, deck thickness = 25 cm and (*b*) first de-bonded point damage evolution for different deck thicknesses.

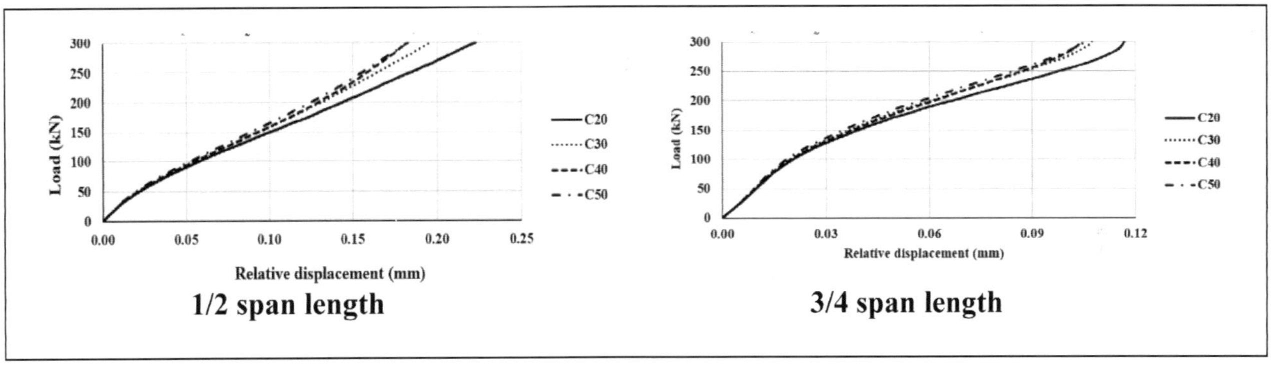


**Figure 11.** Horizontal relative displacements of the slab bottom edges.

when the load reaches 30 kN. Relative displacement becomes significantly larger after 200 kN.

Figure 7*c* depicts the strain of the steel bars at the bottom of the slabs. The longitudinal strain increases linearly until the load is higher than 100 kN. When the load reaches the final value of 300 kN, the strain is about 1,400 Mε; this value corresponds to a stress of 290 MPa, very close to the steel yield strength. The general trends of both experimental data and FEM results correspond well with each other.

De-bonding between hinge-joints and slabs occurs when tangential stress reaches the peak value $f_v^0$ (Figure 6*b*). First de-bonding occurs near the middle of the span when the load applied on a single contact area is 185 kN (2.64 times the standard vehicle load). At the end of the test, the de-bonded area at the bottom of symmetrical surfaces 2 and 3 was roughly 3.5 m long, as highlighted in red in Figure 8*a*. Figure 8*b* shows the shear stress damage criterion evolution for the first de-bonded point; the area belonging to the hinge-joint separates from the surface of the slab when the curve reaches the unit value. Symmetrical surfaces 1 and 4 have negligible extension of de-bonded areas.

Notwithstanding the high load applied, the test did not stimulate the formation of the longitudinal cracks on the cast-in-place deck. This could be the consequence of

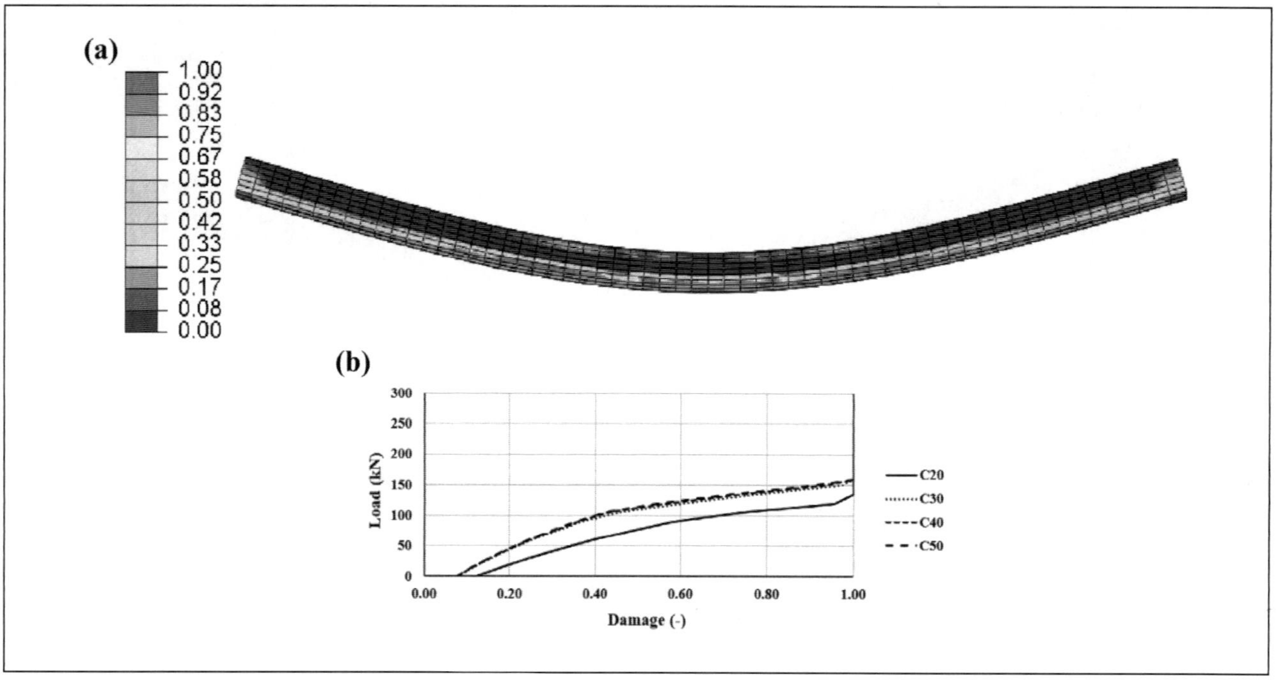


**Figure 12.** De-bonding shear stress damage criterion: (*a*) hinge-joint surfaces 2 and 3 at final load, concrete C50 and (*b*) first de-bonded point damage evolution for different concrete strength.

a static test. The cracking issue appears to have been caused by both temperature gradients and live loads and then propagated because of live loads (*17*, *18*).

## *FEM Parametric Analysis Results*

*Deck Thickness.* Four different deck thicknesses "t" were modelled: 10 cm, 15 cm, 20 cm, and 25 cm; this change significantly affects the horizontal relative displacement of the beams (Figure 9).

Figure 10*a* shows the de-bonded area with the 25-cm deck. The extension is considerably reduced if compared with the test results (Figure 8*a*). Increasing the thickness of the pavement is an effective method to reduce the shear stress of hinge joints and delay the arising of the de-bonding phenomenon (Figure 10*b*).

*Deck Concrete Strength.* Four different concrete strengths were used for the cast-in-place deck: C20, C30, C40, and C50. The difference in concrete quality does not bring about any considerable changes as the results are very close. Therefore, a possible special reinforced concrete deck seems to play a non-effective role (Figure 11).

De-bonded areas do not undergo any substantial changes, as their extensions are very close to the test results for all types of concrete strength (Figure 12*a*). According to the adopted de-bonding formulation (Equation 2), the separation between the hinge-joint and the slab surface is governed by the weaker concrete; this is the reason why the C20 concrete damage curve is outdistanced from the others (Figure 12*b*).

*C-shape and X-shape Stirrups' Influence.* The considerable amount of time required for the positioning of C-shape and X-shape stirrups makes it interesting to understand their actual role. The results obtained, considering the model with or without, are very similar (Figure 13).

The extension of the de-bonded areas into the configuration without stirrups (Figure 14*a*) is very close to the test result (Figure 8*a*); moreover, the shear stress damage criterion trends are very close (Figure 14*b*). The position of the stirrups does not seem to play an important role.

## Discussion and Conclusions

Cracking of the deck near the longitudinal joints during the service life of hollow core slab bridges is a recurring problem.

To investigate the structural behavior of the longitudinal joints, a theoretical, experimental, and numerical study was developed. A real-size specimen was built according to Chinese code specifications and the behavior of the longitudinal joints was investigated. The quasi-static load did not stimulate the longitudinal cracking, even if the final load value was much higher than the one designated by code JTG D-60. Applying a four-time design load eventually brought about the yielding condition of longitudinal bottom steel bars.

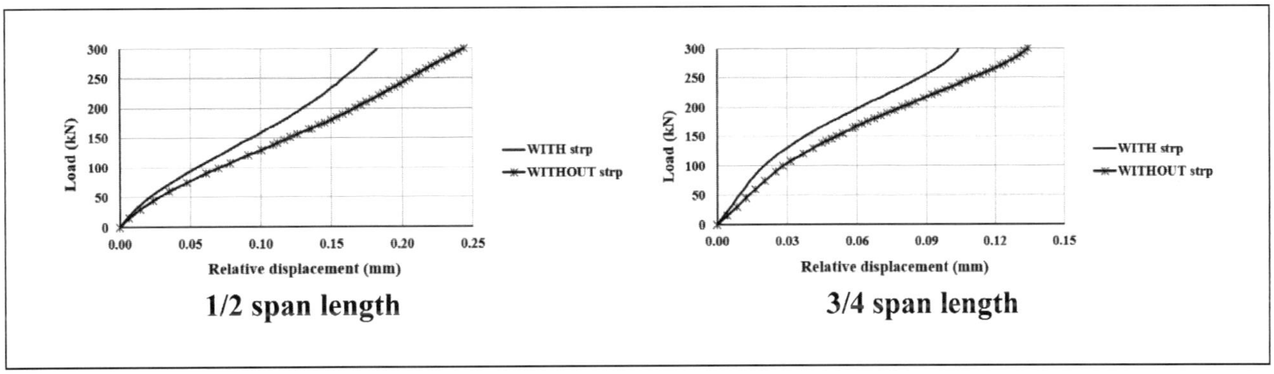


**Figure 13.** Horizontal relative displacements of the slab bottom edges.

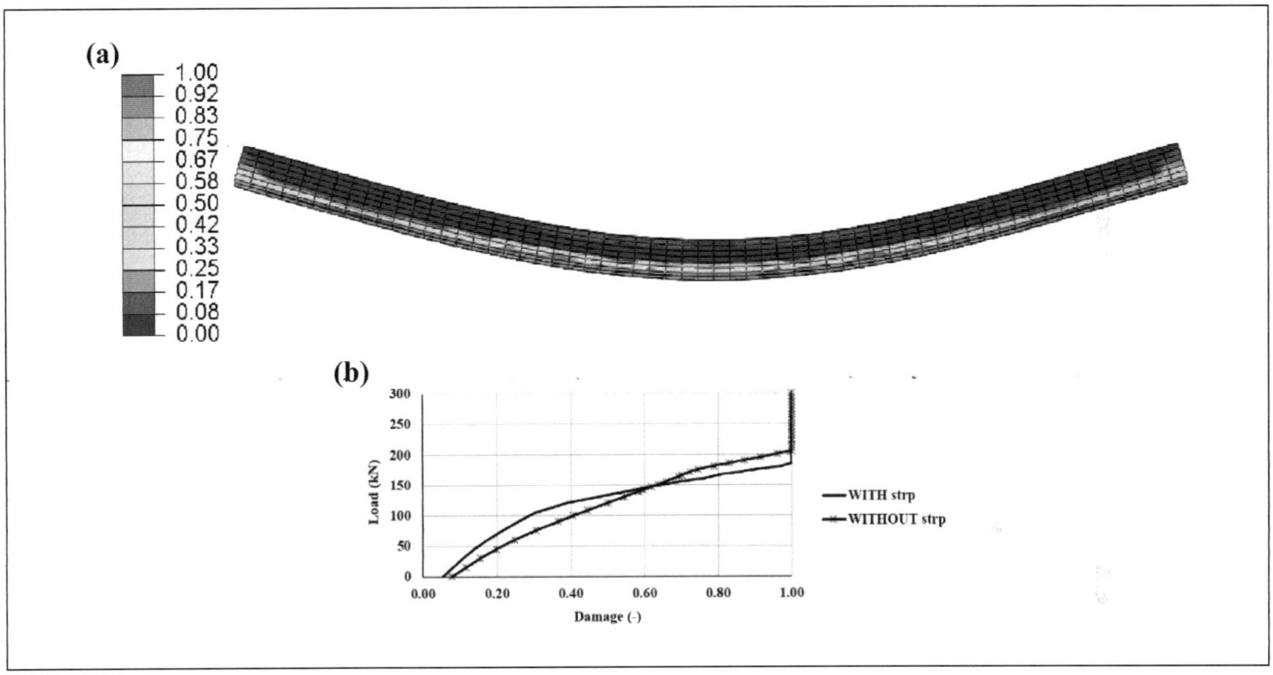


**Figure 14.** De-bonding shear stress damage criterion: (*a*) hinge-joint surfaces 2 and 3 at final load, without stirrup and (*b*) first de-bonded point damage evolution with and without stirrups.

After defining the bond-slip relationship between precast and cast-in-place concrete, the de-bonding phenomenon between the two first occurred when the load reached about 2.5 times the design value.

Furthermore, the parametric analyses showed that increasing the thickness of the cast-in-place deck had a greater impact on reducing the horizontal displacement of slab bottom edges and the correlated de-bonding phenomenon; on the other hand, this action necessarily implies an increase in the structural mass. These analyses also showed that C-shape stirrups didn't seem to play an important role in preventing the longitudinal cracking, as the horizontal displacement and de-bonded areas were the same obtained with the stirrups.

The research presented in this paper confirms the reliability of the design method at least for static loads, while further studies could investigate the effect of periodical loadings and different temperatures on the upper and lower surfaces of the beams.

## Author Contributions

The authors confirm contributing to the paper as follows: study conception and design: Barbieri, Chen, Mazzarolo, Briseghella, Tarantino; data collection: Barbieri, Chen; analysis and interpretation of results: Barbieri, Chen, Mazzarolo, Briseghella, Tarantino; draft manuscript preparation: Barbieri, Mazzarolo, Briseghella. All authors reviewed the results and approved the final version of the manuscript.

## References


1. Attanayake, U., and H. Aktan. First-Generation ABC System, Evolving Design, and Half a Century of Performance: Michigan Side-By-Side Box-Beam Bridges. *Journal of Performance of Constructed Facilities*, Vol. 29, No. 3, 2015. https://doi.org/10.1061/(ASCE)CF.1943-5509.0000526
2. El-Remaily, A., M. K. Tadros, T. Yamane, and G. Krause. Transverse Design of Adjacent Precast Prestressed Concrete Box Girder Bridges. *PCI Journal*, Vol. 41, No. 4, 1996, pp. 96–113.
3. Khaleghi, B. Use of Precast Concrete Members for Accelerated Bridge Construction in Washington State. *Transportation Research Record: Journal of the Transportation Research Board*, 2005. 11S: 187–196.
4. Ulku, E., U. Attanayake, and H. M. Aktan. Rationally Designed Staged Posttensioning to Abate Reflective Cracking on Side-By-Side Box-Beam Bridge Decks. *Transportation Research Record: Journal of the Transportation Research Board*, 2010. 2172: 87–95.
5. Russell, H. G. *NCHRP Synthesis 393: Adjacent Precast Concrete Box Beam Bridges: Connection Details*. Transportation Research Board of the National Academies, Washington, D.C., 2009.
6. Zhang, J., Y. Leng, S. Cheng, and H. He. Numerical Analysis on Changing Regularity of Transverse Load Distributions with Different Crack Lengthens of Plate-Girder Hinge Joint. *Journal of Highway Transportation Research and Development*, Vol. 8, 2010. 2172: 40–44.
7. Li, S., and S. Yang. Study on the Hinge-Joined Plate Method. *MATEC Web of Conferences*, Vol. 22, 2015. http://dx.doi.org/10.1051/matecconf/20152204020
8. Xiong, R., A. Liu, and Q. Yu. A Study about Transverse Load Distribution of the Hinged Joint Cellular Slab Bridge with Beam Grid Method. *Mechanic Automation and Control Engineering (MACE)*, 2010, pp. 4233–4236. http://dx.doi.org/10.1109/MACE.2010.5536049
9. Yuan, J., and B. Graybeal. Full-Scale Testing of Shear Key Details for Precast Concrete Box-Beam Bridges. *Journal of Bridge Engineering*, Vol. 21, No. 9, 2016. http://dx.doi.org/10.1061/(ASCE)BE.1943-5592.0000906
10. Ye, J., J. Liu, Y. Bo, and Y. Fu. Experiment on Shear Property of Hinge-Joints of Concrete Hollow Slab. *Journal of Highway and Transportation Research and Development*, Vol. 201330, No. 6, 2013, pp. 1–7.
11. China Ministry of Transport. *JTG D62-2004. Code for Design of Highway Reinforced Concrete and Prestressed Concrete*. People's communication press, 2004 (in Chinese).
12. Fujian Communications Planning and Design Institute (Fuzhou sheng jiaotong guihua sheji yuan). *Standard Drawings*. Fuzhou, Fujian, China. 2013.
13. China Ministry of Transport. *JTG D60-2004. General Code for Design of Highway Bridges and Culverts*. People's communication press, 2004 (in Chinese).
14. China Ministry of Construction. *GB/T 50081-2002. Standard Method of Mechanical Properties on Ordinary Concrete*. People's communication press, 2002 (in Chinese).
15. Kmiecik, P., and M. Kamińsky. Modelling of Reinforced Concrete Structures and Composite Structures with Concrete Strength Degradation Taken into Consideration. *Archives of Civil and Mechanical Engineering*, Vol. 11, No. 3, 2011, pp. 623–636.
16. Hillerborg, A., M. Modéer, and P.-E. Petersson. Analysis of Crack Formation and Crack Growth in Concrete by Means of Fracture Mechanics and Finite Elements. *Cement and Concrete Research*, Vol. 6, No. 6, 1976, pp. 773–781. https://doi.org/10.1016/0008-8846(76)90007-7
17. Dong, X. *Traffic Forces and Temperature Effects on Shear Key Connection for Adjacent Box Girder Bridge*. PhD dissertation. University of Cincinnati, Ohio, 2002.
18. Huckelbridge, A. A. Jr., H. El-Asnawi, and F. Moses. Shear Key Performance in Multibeam Box Girder Bridges. *Journal of Performance of Constructed Facilities*, Vol. 9, No. 4, 1995, pp. 271–285. http://dx.doi.org/10.1061/(ASCE)0887-3828(1995)9:4(271)


*The Structures Section (AFF00) peer-reviewed this paper (18-00258).*

Article

TRR
JOURNAL OF THE TRANSPORTATION RESEARCH BOARD

Transportation Research Record
2018, Vol. 2672(41) 207–220

DOI: 10.1177/0361198118792766
journals.sagepub.com/home/trr
SAGE

# New Connection Detail to Connect Precast Column to Cap Beam using Ultra-High-Performance Concrete in Accelerated Bridge Construction Applications

**Mohamadreza Shafieifar[1], Mahsa Farzad[1], and Atorod Azizinamini[2]**

**Abstract**

Accelerated bridge construction (ABC) is a paradigm change in delivery of bridges. ABC minimizes the traffic interruption, enhances safety to public and workers by significantly reducing on-site construction activities, and results in longer-lasting bridges. The use of precast elements is gaining attention owing to inherent benefits of accelerated construction. Designing an economical connection is one of the main concerns for these structures. New improved materials such as ultra-high-performance concrete (UHPC) with superior characteristics can provide solutions for joining precast concrete elements. In this paper two types of column to cap beam connection using UHPC are proposed for seismic and non-seismic regions. Among the merits of the proposed details, large tolerances in construction and simplicity of the connection can be highlighted which facilitates and accelerates the on-site construction time. The experimental program was carried out to evaluate the performance and structural behavior of the proposed connections. Four specimens were subjected to constant axial compressive loads and cyclic lateral loading. Results of the experiment showed that the displacement ductility of the specimens, incorporating suggested details, demonstrated adequate levels of displacement ductility. More importantly, the proposed connections prevented the damage into capacity protected element—in this case the cap beam. Analytical and nonlinear finite element analysis on the specimens was carried out to better comprehend the behavior of the proposed connections.

Rapid construction is in high demand in the bridge construction industry to minimize on-site construction time, reduce traffic interruption, and increase work zone safety. Accelerated bridge construction (ABC) offers a new solution for reducing construction time while building, replacing, or retrofitting bridges. The most common form of ABC uses prefabricated modular bridge elements. These elements require joints between elements and there are concerns regarding the durability and structural behavior of the connections (*1*). The behavior of the structure with precast concrete elements is ruled by the connections and the performance of the connection is influenced by various parameters. Therefore, simple connection detail results in predictable structural behavior. A simple detail can increase construction speed and precision of fabrication.

Various studies on connection between precast concrete elements in the bridges have been carried out (*2–6*). The Federal Highway Administration (FHWA) summarizes common connection details for prefabricated bridge elements (*7*). The column to cap beam or footing connection plays a critical role in the behavior of the bridge in seismic regions. Several details for joining column to cap beam have been proposed, but none of these methods have entirely solved construction issues. Among the methods are pocket, bar couplers, member socket, hybrid, integral, and mechanical connections.

In light of the challenges inherent in connecting precast concrete elements, exploring new materials such as ultra-high-performance concrete (UHPC) can be

[1]Department of Civil and Environmental Engineering, Florida International University, Miami, FL
[2]Department of Civil and Environmental Engineering, Accelerated Bridge Construction University Transportation Center, Miami, FL

**Corresponding Author:**
Address correspondence to Mohamadreza Shafieifar: mshaf017@fiu.edu

beneficial. UHPC is a cementitious material with a typical compressive strength exceeding 21 ksi (150 MPa), improved toughness and increased damage tolerance making it well-suited for use in heavily loaded structural components (*8*). UHPC includes a mix of Portland cement, fine sand, silica fume, ground quartz, superplasticizer accelerator, steel fibers, and water that provides properties such as high strength and durability. UHPC exhibits considerably large strain and tension capacity, compared to the regular concrete, which can be employed for ductile structures in seismic regions. Superior characteristics of UHPC, such as the short development length and the ability to gain strength during a short time can assist ABC construction. Several studies are available for UHPC material properties and its application in bridges (*9–11*).

Ductility of a structure is expressed as the capacity of the structure to absorb and dissipate energy during earthquake, by slight loss of strength and serviceability. Formation of plastic hinge in the columns can provide this inelastic behavior of the bridges. The properly detailed and closely spaced transverse reinforcement in conventional reinforced concrete columns can ensure ductile behavior during earthquakes. Transverse reinforcement plays an important role in confining the concrete within the core region of columns, resulting in improvements of strength and ductility. Consequently, to ensure ductile behavior, modern design codes impose stringent requirements for the detailing of transverse reinforcement in columns. According to the current seismic design methods, forming the plastic hinge in the column should provide the ductility to the bridge in seismic regions.

The objective of this research is to develop a new UHPC-based connection between the cap beam and column which can potentially be used in ABC application for both seismic and non-seismic regions. The connection, made primarily with UHPC, is designed for a target plastic hinge location. The feasibility study of this project was done in Shafieifa et al.(*12*). In the current research, four scaled columns with the proposed detail were tested under cyclic loading and the structural behavior of the specimens was highlighted. The results of the tests and numerical studies will be used in future for development of design guidelines.

## Description of the Proposed ABC Connection

In the proposed connection, UHPC is used to join the precast elements in the field. Two details are proposed: seismic and non-seismic details. In the seismic detail, two layers of UHPC are employed. As shown in Figure 1*a*, in the absence of top layer of UHPC, it was anticipated that the major damage will occur near the cap beam. Therefore, to prevent spreading of the damage to the cap beam another layer of UHPC was used as a column capital. The use of UHPC in two layers guaranteed the formation of the plastic hinge at the desired location of the column. The use of UHPC in the splice region allows the development of reinforcing bars over a short length. Therefore, the length of the gap to be filled by UHPC in the field is relatively small. For the non-seismic detail (Figure 1*b*), the cap beam and the column are simply joined with a layer of UHPC.

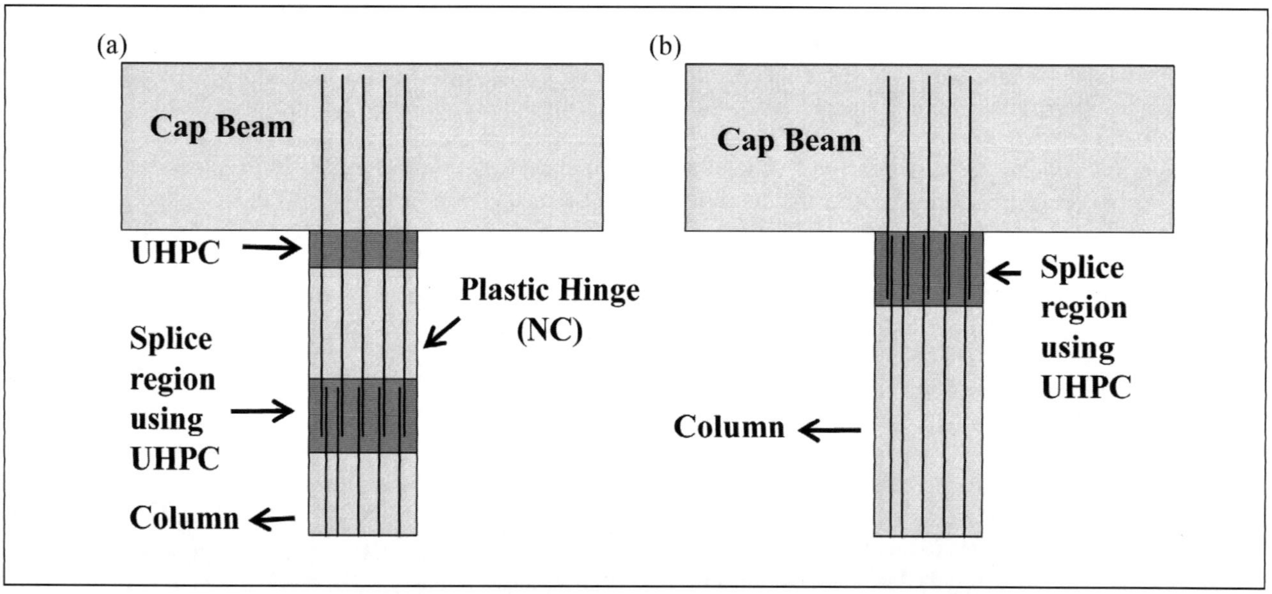


**Figure 1.** Details of the concept of the proposed connection: (*a*) seismic detail, and (*b*) non-seismic detail.

**Table 1.** Details of the Specimens

| Specimen ID | Geometry detail | Transverse reinforcement detail | Axial load ratio |
|---|---|---|---|
| S-2.5-10 (reference) | As shown in Figure 2 (seismic detail) | #3@2.5 in. in plastic hinge and splice region | 10% |
| S-4-10 | As shown in Figure 2 (seismic detail) | #3@4 in. in plastic hinge and no stirrups at splice region | 10% |
| S-2.5-20 | As shown in Figure 2 (seismic detail) | #3@2.5 in. in plastic hinge and one stirrup at splice region | 20% |
| NS-2.5-10 | As shown in Figure 3 (non-seismic detail) | #3@2.5 in. in plastic hinge and no stirrups at splice region | 10% |

For the seismic detail, the system consists of two parts. First part includes the cap beam, UHPC column capital, and a part of column where the plastic hinge forms, while the second part is the column. These two parts will be connected in the field using UHPC. The connection is proposed for precast column to cap beam or footing connections located in high seismic zones. Gaining the strength of UHPC over a short time and large tolerances for construction of the proposed details can provide a rapid method of bridge construction. The merits of the proposed connection system are large tolerance for construction, confinement of damage within plastic hinge, and prevention of rebar yield in the cap beam in the case of a large seismic event. In order to investigate the structural behavior of this connection, a set of experiments were conducted.

## Experimental Program

In this study, four scaled connections between a precast column and cap beam were constructed and tested under combined axial and reversed cyclic loading. Three seismic details and one non-seismic detail were investigated. The column and cap beam were designed based on common provisions for seismic regions (*13–15*) with the ductility ratio of 5. The length of the splice region was large enough to ensure that slip failure in this region would not occur.

### *Description of the Test Specimens*

Four connections of column to cap beam were designed for testing under lateral loading and constant axial load, designated as specimen S-2.5-10, S-4-10, S-2.5-20 and NS-2.5-10. Details of test variables and axial load of all specimens are given in Table 1. S-2.5-10, S-4-10, S-2.5-20 had a seismic proposed detail with different axial load rate and transverse rebar ratio in splice and plastic hinge zone. The axial load varied from 10%–20% of its pure axial capacity. In addition, specimen NS-2.5-10 was designed with non-seismic detail. Previous research indicates that sufficient steel fibers in concrete may increase shear capacity, section confinement, and provide more ductility (*16*). Therefore, to study the confinement effect of UHPC, different transverse reinforcement ratios in the splice region were used.

Figures 2 and 3 show the dimensions of the specimens in detail. The column height was 61 in. (155 cm) with a circular cross-section having diameter of 12 in. (30.5 cm). The column was typically reinforced longitudinally with eight No. 5 rebars (M16). Longitudinal reinforcement ratio was 2.19% for all specimens. Clear cover of the rebars was 1.5 in. (38 mm). The axial load of the reference specimen (S-2.5-10) was 56 kips (249 kN) which resulted in approximately 10% of the pure axial load capacity of the column section. The lap splice length of rebar in UHPC was 8 in. (200 mm), equal to 13 times the rebar diameter; although previous studies have shown that shorter lap splice lengths can be used if adequate cover concrete is provided (*17*).

### *Construction of the Test Specimens*

For the seismic details, the first step of the construction procedure was casting of the cap beam (support) followed by casting the first layer of UHPC and plastic hinge part. Casting splice region and column part was carried out 11 days later. In order to minimize the cold joint between the normal concrete (NC) and UHPC, the surface of the concrete was textured. Moreover, around 1 in. (25 mm) of the first layer of UHPC was embedded in the support to prevent any local damage at the concrete support. See Figure 4.

### *Test Setup and Loading Procedure*

The experimental test was carried out six months after casting the specimens. Quasi-static test was considered to evaluate the cyclic performance of the connection. A 150-kip (490 kN) hydraulic ram was used for cyclic testing of the cantilever column. A constant predetermined axial load was applied using two hydraulic rams. Initially, low displacement cycles were applied to the column to estimate the idealized yield displacement of the column ($\Delta_y$). See Figure 5.

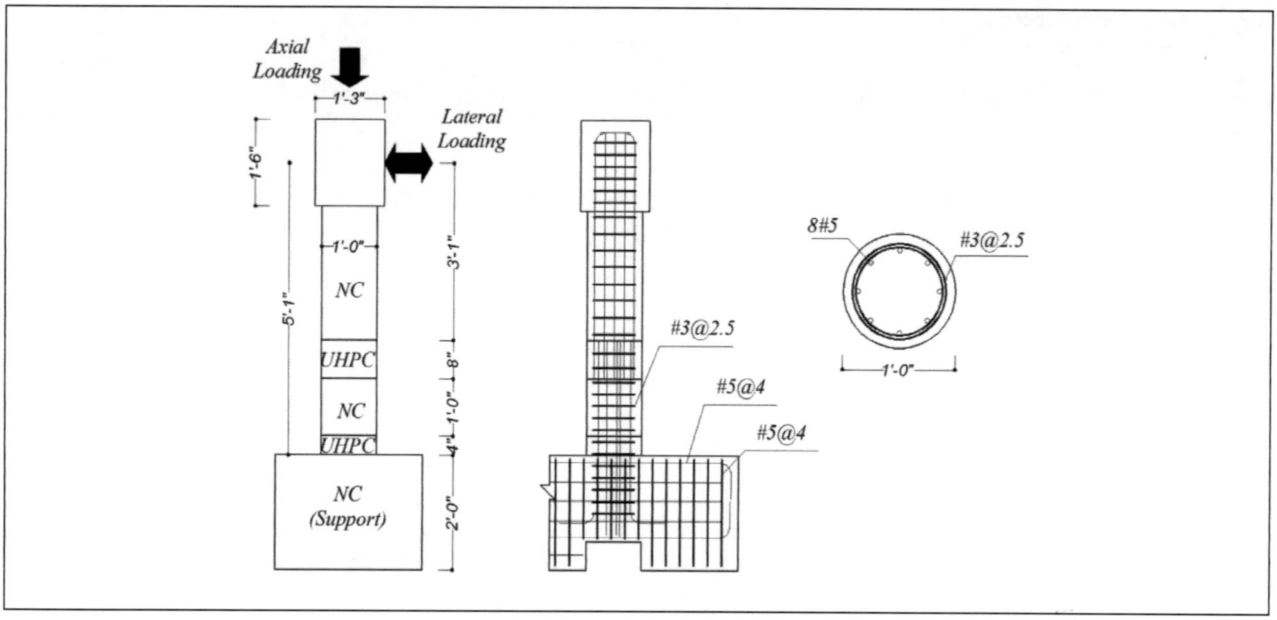


**Figure 2.** Specimens S-2.5-10, S-4-10, S-2.5-20 typical dimensions and reinforcement details.

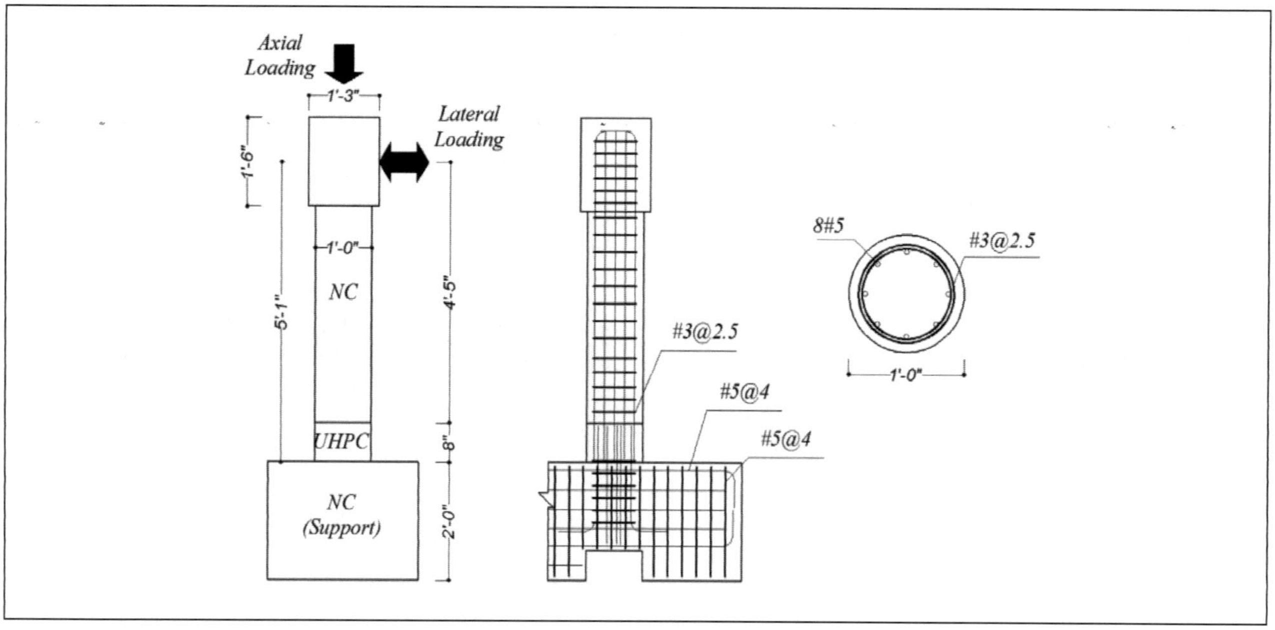


**Figure 3.** Specimen NS-2.5-10 (non-seismic detail) dimensions and reinforcement details (no stirrups at splice region).

The yielding displacement ($\Delta_y$) was defined by assuming the bilinear model, equivalent elastoplastic system with the same elastic stiffness and ultimate load as the real system, by following the procedure suggested by R. Park (*18*). After obtaining $\Delta_y$, the column was subjected to three cycles of $2\Delta_y$, $3\Delta_y$, $4\Delta_y$, etc. At the end of each cycle, the displacement was paused to observe the damages and trace the cracks. Yielding displacement of the seismic specimens (S-2.5-10, S-4-10 and S-2.5-20) was 0.65 in. (17 mm). The yielding displacement of non-seismic specimen (NS-2.5-10) was calculated 0.8 in. (20 mm). To evaluate the behavior of the specimens, the test setup was instrumented with string pots, load cells, strain gauges, potentiometers and pressure transducers. To calculate the curvature of the column, potentiometers were employed at four levels in plastic hinge regions.

**Figure 4.** Specimen construction procedures: (*a*) formwork of the support, (*b*) casting support, (*c*) casting first layer of UHPC, (*d*) casting plastic hinge, (*e*) reinforcement of the column, (*f*) casting splice region with UHPC and (*g*) casting column with NC.

**Figure 5.** Loading setup overview.

**Table 2.** Actual Material Properties of the Material

| Concrete material | |
|---|---|
| Concrete of supports and plastic hinge location | $f_c = 7.1$ ksi (49 MPa) |
| Concrete of the columns | $f_c = 6.4$ ksi (44 MPa) |
| UHPC of the first layer | $f_c = 20.1$ ksi (139 MPa) |
| UHPC of the splice region | $f_c = 24.9$ ksi (172 MPa) |
| Steel bars | |
| $f_y = 68$ ksi (468 MPa) | $f_u = 113$ ksi (780 MPa) |

Cylinder test was used to measure the compressive strength of normal concrete and UHPC a day after the experimental test. The actual material properties are reported in Table 2.

## Experimental Results

### *Observed Damage*

The development of cracks and crushing of the concrete in the plastic hinge region at different displacement levels are shown in Figure 6 for each specimen. For all seismic details (S-2.5-10, S-4-10 and S-2.5-20), the first cracks formed in the plastic hinge area followed by development

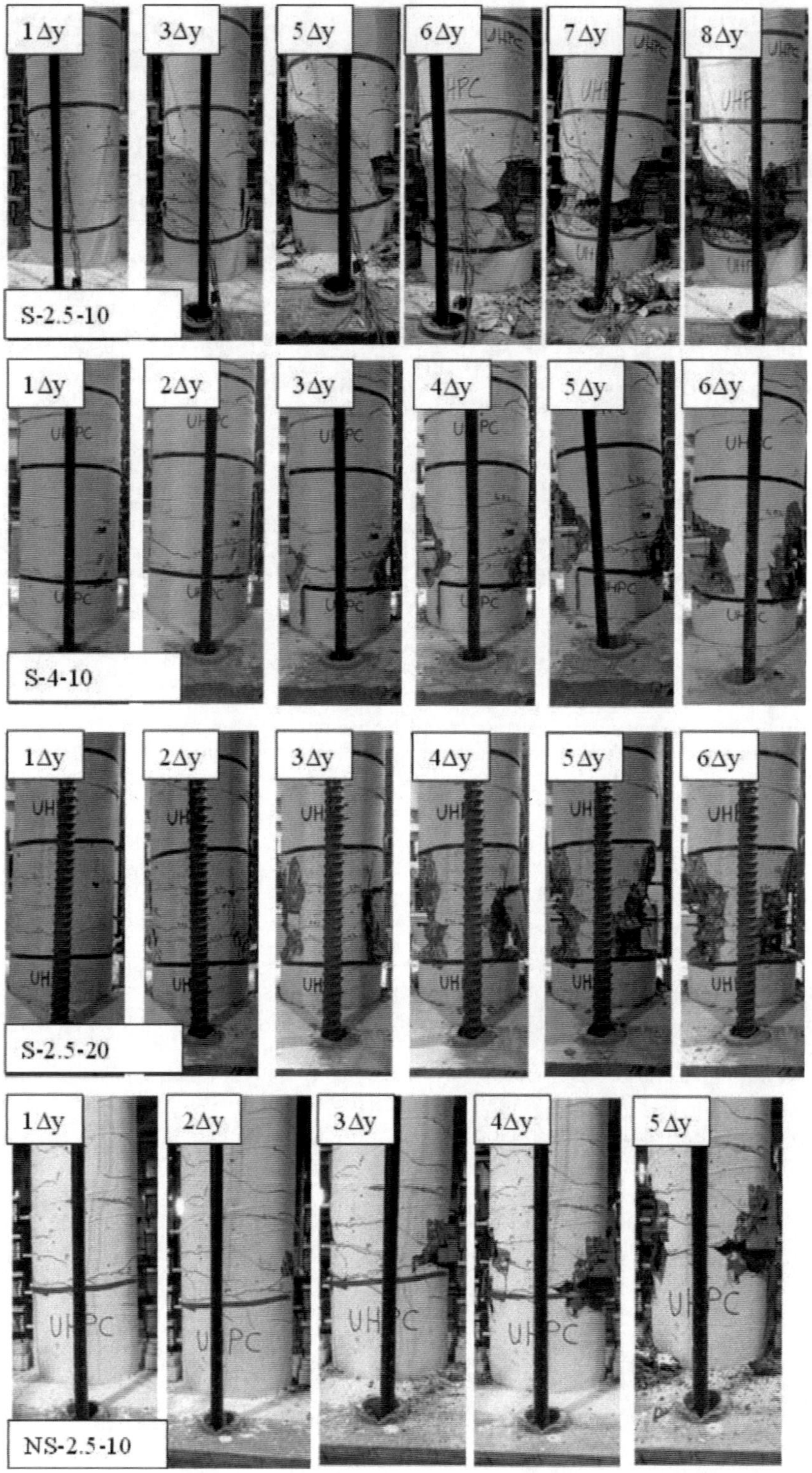


**Figure 6.** Plastic hinge damage at different displacement levels for all specimens.

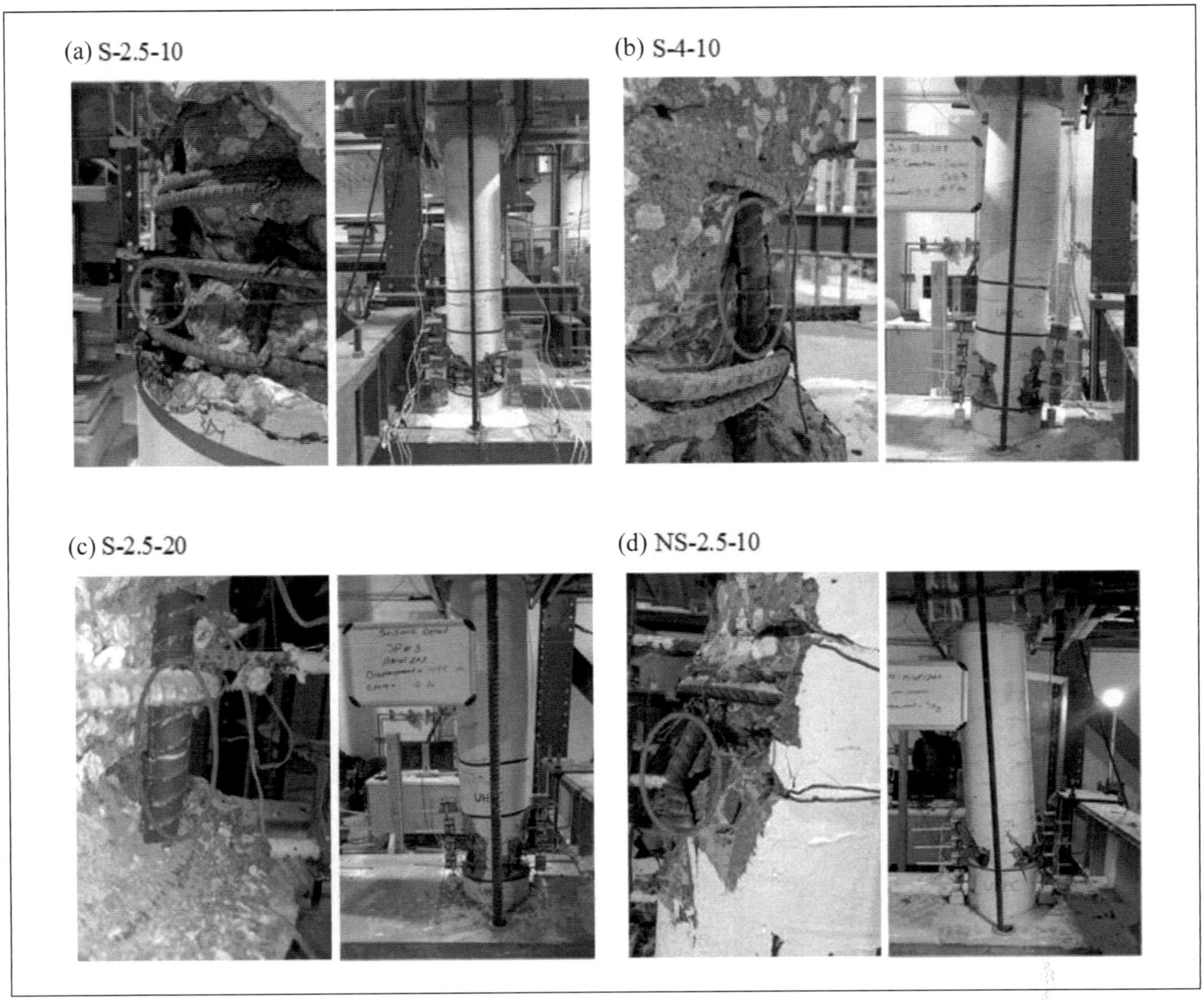


**Figure 7.** Failure pattern, longitudinal rebar fracture and location of the plastic hinge in the specimens.

of limited cracks in the column. For these specimens the spalling of the cover concrete was first observed at $2\Delta_y$ and damage was limited to the plastic hinge zone, which consisted of normal strength concrete. No spalling or cracking was observed in UHPC portions of all specimens. No bond failure between normal concrete and UHPC was observed, which indicates the sufficient bond between UHPC and normal concrete. The spalling of the cover concrete of non-seismic specimen (NS-2.5-10) was first observed at the first cycle of $2\Delta_y$. Limited concrete cracking was found on the surface of the cap beam of the specimen.

## *Mode of Failure*

Specimens failed when one of the longitudinal bars ruptured after development of buckling between stirrups due to spalling of the concrete cover. The concrete cover in the plastic hinge zone spalled off and then the core concrete crushed (Figure 7). Fracture of the rebar was located around 2 in. above the UHPC layer for seismic details. The displacement ductility of the specimens is shown in Table 3 indicating their performance in seismic zones based on common seismic provisions. Results of the test show that by increasing the stirrup spacing or axial load the ductility will decrease. By comparing reference specimen (S-2.5-10) with specimens S-4-2.5 (with less transverse reinforcement ratio) and S-2.5-20 (higher axial load), it was concluded that the ductility of the specimens was most influenced by the space between the stirrups in plastic hinge region.

## *Load Displacement Curve*

Table 3 shows the maximum displacement for each specimen. Even though the non-seismic detail was not

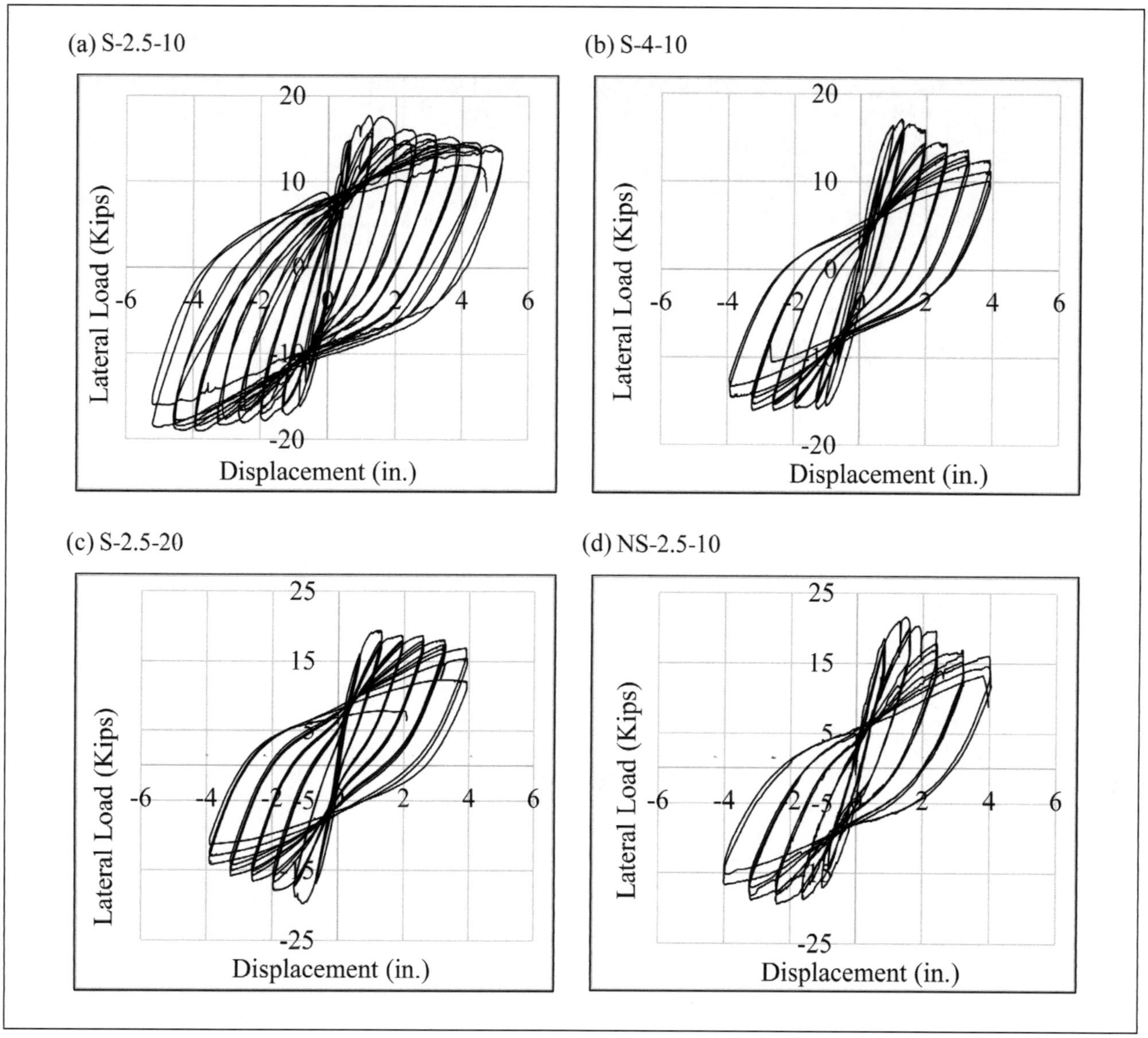


**Figure 8.** Hysteric load deflection curve of the specimens.

**Table 3.** Ductility and Maximum Drift of the Specimens

| Specimen ID | Maximum drift | Displacement ductility |
|---|---|---|
| S-2.5-10 | 8.5% | 8 |
| S-4-10 | 5.3% | 5 |
| S-2.5-20 | 6.4% | 6 |
| NS-2.5-10 | 6.5% | 5 |

designed for seismic loading, the specimen was subjected to the cyclic loading to show its capability for seismic area as well. Compared with the monotonic loading, cyclic loading can provide severe condition. Figure 8 shows the hysteresis load deflection curve of the specimens. In all specimens, rebar cages were shifted approximately ½ in. during the casting and asymmetry of the curves can be attributed to the casting imperfection of the specimen. The lateral load decreased gradually after the peak load, indicating the spall off of cover concrete and p-$\Delta$ effects. Lateral load reduction after peak load of specimen S-4-10 (with a larger distance between stirrups) and non-seismic detail (NS-2.5-10), were more than other specimens. Specimens with higher ductility ratio had more capacity to absorb and dissipate energy under extreme loads, such as earthquakes, by showing a limited loss of serviceability. However, it should be noted that the residual drifts of the specimen S-2.5-20 (with 20% axial

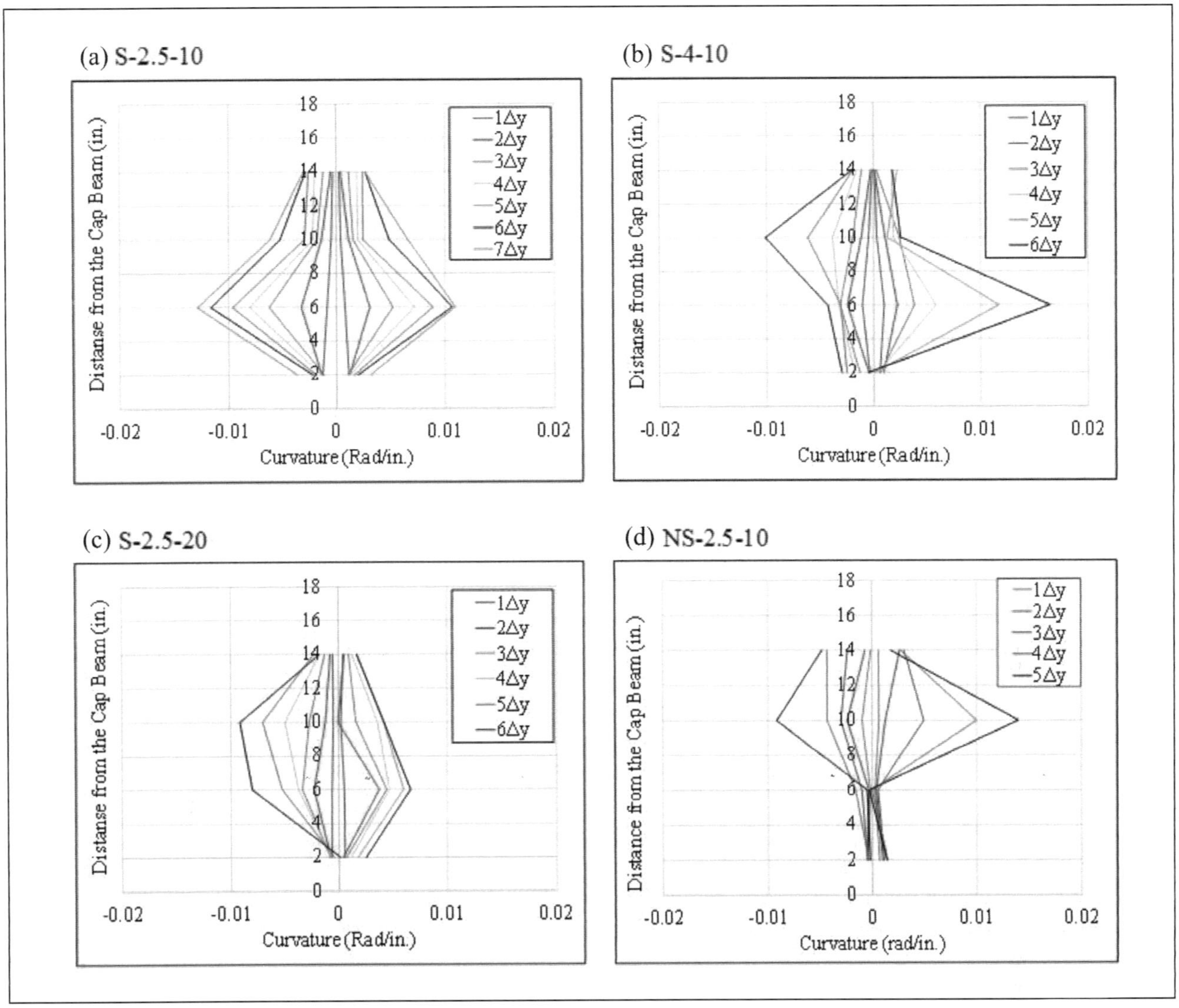


**Figure 9.** Plastic hinge curvature profiles for each displacement.

load) was smaller than those of columns with 10% axial load.

Due to the applied axial load, specimen S-2.5-20 had a higher load capacity compared to other seismic details. It is also worth noting, as the distance between load and critical section in the non-seismic detail (NS-2.5-10) was less than other specimens, the specimen could carry out a higher level of lateral load.

## *Measured Curvatures*

The specimens were instrumented with 8 potentiometers to measure the curvature of the column. The potentiometers were mounted on steel bars that passed through the column at four different distances (4 in. space) from the cap beam face. Some potentiometers stopped working at high displacement levels and further measurement of curvature was not possible. The maximum average curvature profile at each displacement level versus the column height is shown in Figure 9. Results show that in the seismic details, the plastic hinge was placed at the desired location (between two layers of UHPC). For the non-seismic detail (NS-2.5-10) the plastic hinge formed above the splice region and the results show that the use of UHPC in splice region prevented development of plastic hinge to the cap beam as well. This shows the capability of this specimen for seismic area as well.

By comparing the curves plotted in Figure 9, the effects of axial load in the plastic hinge can be investigated. In the specimen S-2.5-20 with higher axial load ratio the maximum curvature was less than other specimens. It is worthwhile to notice that the UHPC splice

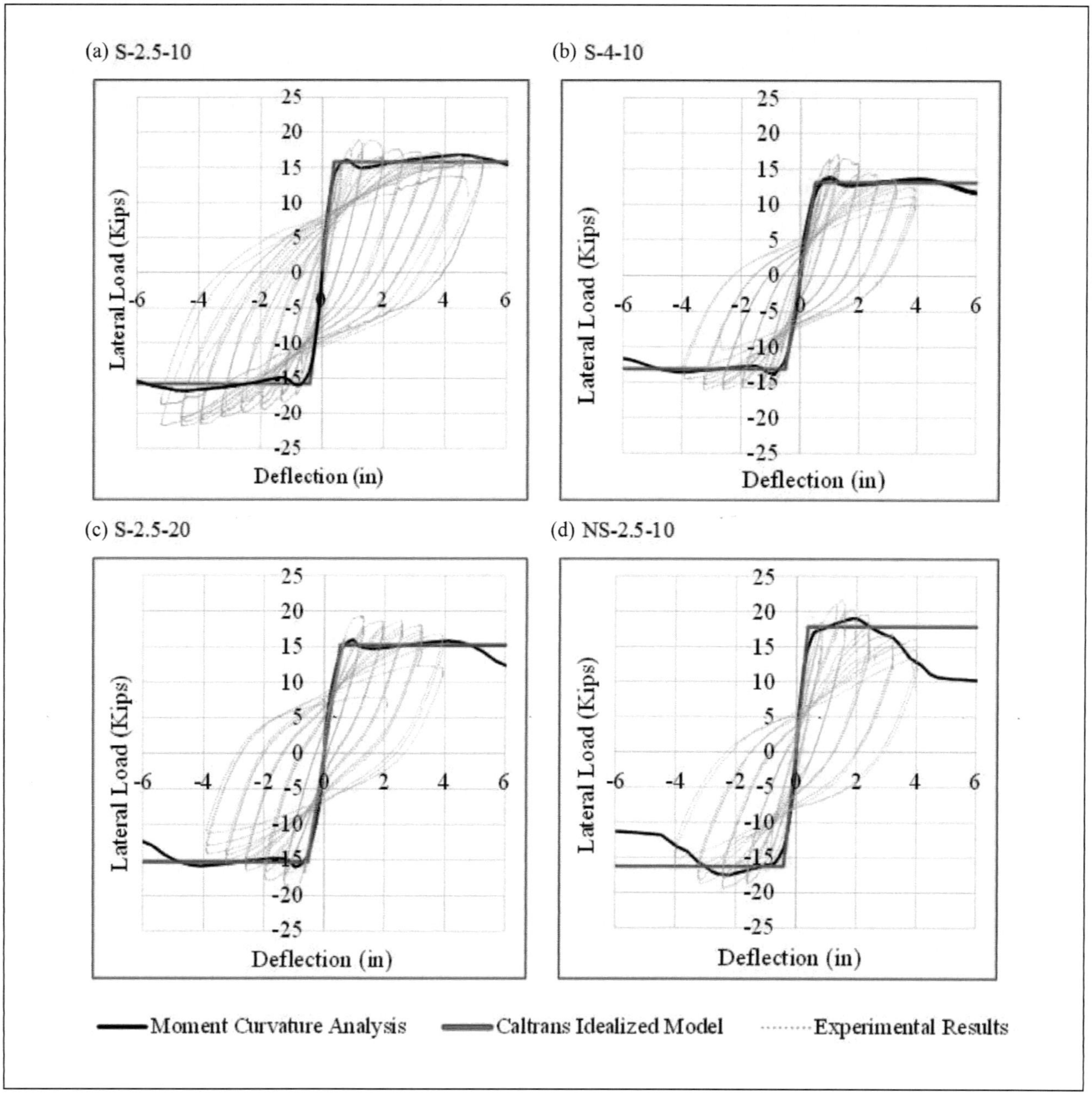


**Figure 10.** Load-displacement curves of the specimens, based on moment-curvature analysis and idealized Caltrans curve.

region in non-seismic detail prevented forming the plastic hinge in the cap beam.

## Moment-Curvature Analysis

The moment capacity of the specimens was calculated using moment-curvature section analysis software. For this analysis, the curves were obtained with related axial load and the idealized curves were derived according to Caltrans (*14*). Considering the equivalent analytical plastic hinge length and the length of the column, the local displacement capacity of a member can be calculated based on its moment-curvature analysis which includes idealized yield and plastic displacement due to rotation of the plastic hinge.

The effect of expected properties of the rebar, confinement of concrete and shifting the rebar due to construction imperfection were considered in the analysis. The idealized yield displacement and moment capacity of the reference specimen (S-2.5-10) calculated 0.54 in. (13 mm)

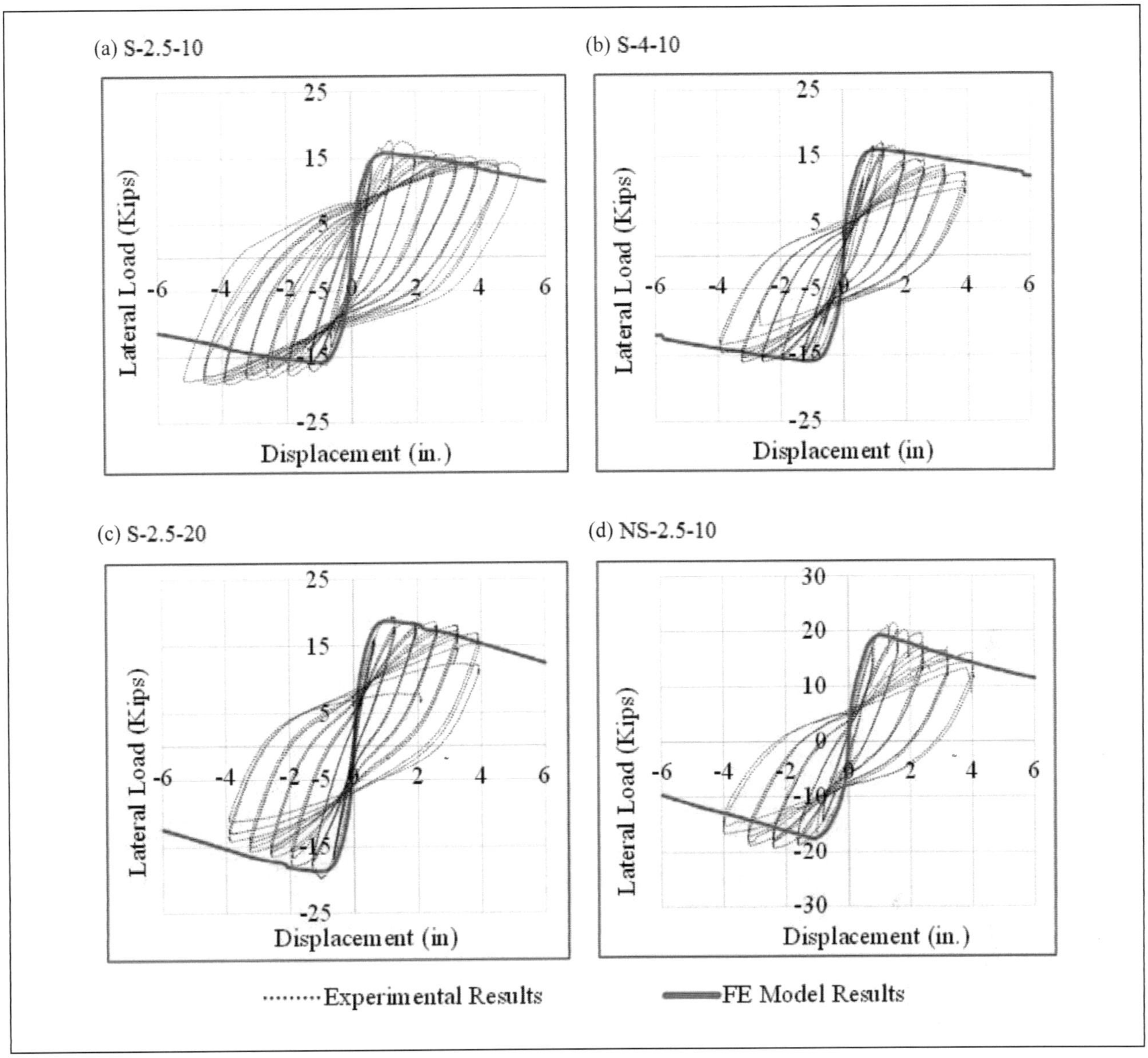


**Figure 11.** Measured hysteretic curves and numerical results of the specimens.

and 836 kip.in. (87 kN.m) respectively. The ductility of the reference specimen (S-2.5-10), based on Caltrans, was 9. Figure 10 shows the relation between experimental and analytical load-displacement curves of the column using Caltrans and moment-curvature analysis.

## Numerical Investigation

Many parameters can affect the load pattern on the bridges (*19*) and developing an experimental test for all cases is uneconomical, therefore developing a general finite element (FE) model can help the researchers for next steps. The numerical analyses were performed using Abaqus FE software. Dimensionally, the simulated models were the same as the test specimens. An eight-node solid element (C3D8) was selected to model the regular concrete and UHPC, while the truss elements (T3D2) were selected for embedded rebars. Concrete damage plasticity (CDP) model was employed for modeling the regular concrete and UHPC materials. The material strength was based on the measured compressive strength of regular concrete and UHPC. A quasi-static analysis with small load increments was used for numerical study. The models were subjected to a monotonic displacement control loading.

The input values used for uniaxial compressive strength of concrete materials in models were the actual

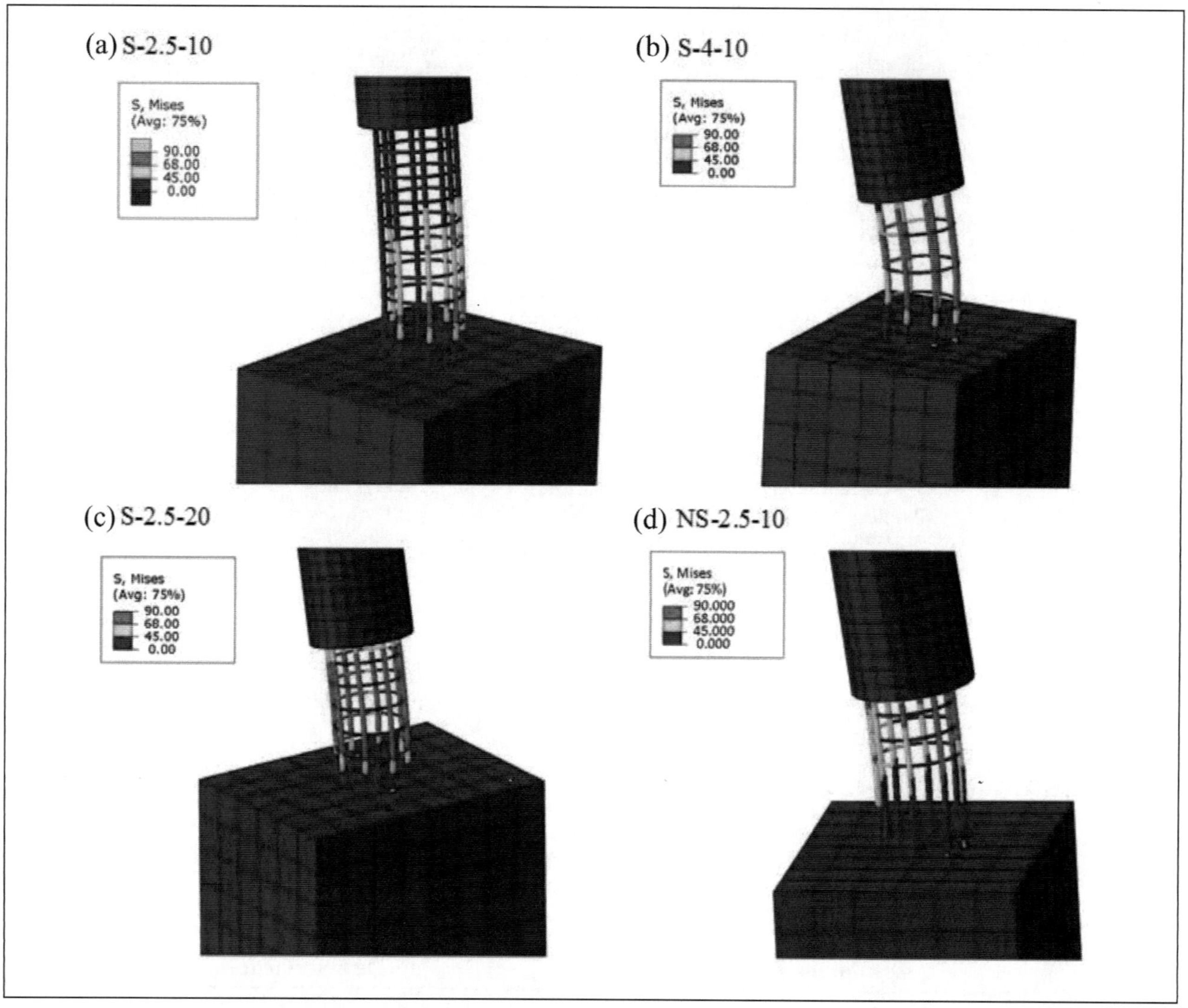


**Figure 12.** Rebar stresses in the numerical models.

values taken from cylinder test at the day after the experimental test. Therefore, the uniaxial compressive strength was 6.4 ksi (44 MPa) for normal concrete and 20.1 ksi (139 MPa) for UHPC. Defining the UHPC material properties was based on previous studies (*20*, *21*). The bars material was considered to be elastic-perfectly plastic material in both tension and compression. Steel material input data in the model for all rebars were taken from actual tensile test. Yield and ultimate stresses were 68 ksi (486 Mpa) and 113 ksi (780 MPa) respectively. Modulus of elasticity and Poisson's ratio of steel were assumed 29000 ksi (200 GPa) and 0.2.

## *Numerical Simulation Results*

A comparison of the results of the simulated specimens and experimental load–displacement curve is plotted in Figure 11. The difference between the FE model and experimental results can be attributed to the material properties, boundary conditions and construction imperfection. Despite the difference, a good correlation of the maximum load capacity was observed between FE model and experiments.

The lateral load capacities of all modeled specimens were similar to the analytical (moment-curvature analysis) results. The numerical results for the bars stresses are depicted in Figure 12. The model was able to simulate the column behavior with reasonable accuracy. As illustrated, the results of FE analysis could specify critical sections, and yielding development in rebars. The results of FE model show no rebar yielding or spalling in the cap beam.

The numerical results are generally in good agreement with the experimental observation justifying the modeling of the proposed detail. Based on developed FE model, parametric studies will be conducted to investigate the influences of other parameters in future studies.

## Summary and Conclusions

Two types of connection for precast concrete column to cap beam for seismic and non-seismic region are proposed. A layer of UHPC was used to connect precast elements. In order to force forming plastic hinge in the desired location for seismic detail, another layer of UHPC was added below the cap beam. Four specimens in cantilever configuration were tested under combined constant axial load and cyclic lateral load. Various responses of the connection were highlighted. Influence of transverse reinforcement and axial load on the behavior of the connections were investigated. The results from the experiments were used to develop a calibrated numerical model which can lead to developing a design guideline for proposed connection. Results of the study have led to the following conclusions:

- All the specimens with seismic detail showed ductile behavior and the plastic hinge formed in the desired location. High workability of the UHPC and large tolerance of bars in the proposed detail can facilitate and accelerate the on-site construction. Designers should consider the shifting of the plastic hinge which can affect the behavior of the structure.
- Judging from the observed ductility and failure mode of specimens, the main characteristic of the proposed connection is influenced by transverse rebar ratio. The distance between the stirrups plays a major role in preventing longitudinal bars buckling. Results show that increasing axial load can provide more capacity but less ductility for the specimen.
- No major crack was observed in the cap beam for the proposed seismic and non-seismic details. Therefore, the non-seismic detail, with a seismic design consideration can be an alternative detail even for seismic regions. This detail is preferred as it is easier to build and less expensive.
- No significant damage was found in splice region even in the absence of the transverse reinforcement in this region. Observing no failure in splice region indicates that UHPC can provide a good confinement and shear capacity. Eliminating the need for stirrups in the splice region can facilitate the construction procedure in the field. Moreover, in comparison with conventional concrete, even a short lap splice of bars in UHPC can transfer forces between spliced bars, which leads to a decrease in casting volume and saves time in the field.

### Acknowledgments

The research study, results of which are reported in this paper, was partially sponsored by Accelerated Bridge Construction University Transportation Center (ABC-UTC) at Florida International University. ABC-UTC is a Tier 1 UTC funded by U.S. DOT. The authors acknowledge and thank the sponsors for their support. The authors are also thankful to Ductal® by LafargeHolcim, for providing the UHPC material.

### Author Contributions

The authors confirm contribution to the paper as follows: study conception and design: M. Shafieifar, A. Azizinamini; data collection: M. Shafieifar, M. Farzad; analysis and interpretation of results: M. Shafieifar; draft manuscript preparation: M. Shafieifar, M. Farzad. All authors reviewed the results and approved the final version of the manuscript.

### References

1. Culmo, M. P. *Accelerated Bridge Construction-experience in Design, Fabrication and Erection of Prefabricated Bridge Elements and Systems*. No. FHWA-HIF-12-013. 2011.
2. Tullini, N., and F. Minghini. Grouted Sleeve Connections Used in Precast Reinforced Concrete Construction–Experimental Investigation of a Column-to-column Joint. *Engineering Structures*, Vol. 127, 2016, pp. 784–803.
3. Kim, T.-H., H.-M. Lee, Y.-J. Kim, and H. M. Shin. Performance Assessment of Precast Concrete Segmental Bridge Columns with a Shear Resistant Connecting Structure. *Engineering Structures*, Vol. 32, No. 5, 2010, pp. 1292–1303.
4. Stephens, M. T., D. E. Lehman, and C. W. Roeder. Design of CFST Column-to-foundation/cap Beam Connections for Moderate and High Seismic Regions. *Engineering Structures*, Vol. 122, 2016, pp. 323–337.
5. Shim, C., S. Lee, S. Park, and C. Koem. Experiments on Prefabricated Segmental Bridge Piers with Continuous Longitudinal Reinforcing Bars. *Engineering Structures*, Vol. 132, 2017, pp. 671–683.
6. Shim, C. S., C.-H. Chung, and H. H. Kim. Experimental Evaluation of Seismic Performance of Precast Segmental Bridge Piers with a Circular Solid Section. *Engineering Structures*, Vol. 30, No. 12, 2008, pp. 3782–3792.
7. Culmo, M. P. *Connection Details for Prefabricated Bridge Elements and Systems*. No. FHWA-IF-09-010. 2009.
8. Graybeal, B. A. *Material Property Characterization of Ultra-high Performance Concrete*. No. FHWA-HRT-06-103. 2006.
9. Farzad, M., A. Mohammadi, M. Shafieifar, H. Pham, and A. Azizinamini. *Development of Innovative Bridge Systems Utilizing Steel-Concrete-Steel Sandwich System*. No. 17-02229. 2017.
10. Baqersad, M., E. A. Sayyafi, and H. M. Bak. State of the Art: Mechanical Properties of Ultra-High Performance

Concrete. *Civil Engineering Journal*, Vol. 3, No. 3, 2017, pp. 190–198.
11. Farzad, M., M. Shafieifar, and A. Azizinamini. *Accelerated Retrofitting of Bridge Elements Subjected to Predominantly Axial Load Using UHPC Shell*. No. 18-05067. 2018.
12. Shafieifar, M., M. Farzad, and A. Azizinamini. *Alternative ABC Connection Utilizing UHPC*. No. 17-03398. 2017.
13. American Association of State Highway and Transportation Officials. *AASHTO Guide Specifications for LRFD Seismic Bridge Design*. American Association of State Highway and Transportation Officials, Washington, D.C., 2012.
14. Caltrans. *Seismic Design Criteria (SDC)*. California Department of Transportation, Sacramento, CA, 2010.
15. *AASHTO LRFD Bridge Design Specifications*. 7th ed. 1998, pp. 5–138.
16. Paultre, P., R. Eid, Y. Langlois, and Y. Lévesque. Behavior of Steel Fiber-reinforced High-strength Concrete Columns Under Uniaxial Compression. *Journal of Structural Engineering*, Vol. 136, No. 10, 2010, pp. 1225–1235.
17. Yuan, J., and B. A. Graybeal. *Bond Behavior of Reinforcing Steel in Ultra-high Performance Concrete*. No. FHWA-HRT-14-090. 2014.
18. Park, R. Evaluation of Ductility of Structures and Structural Assemblages from Laboratory Testing. *Bulletin of the New Zealand National Society for Earthquake Engineering*, Vol. 22, No. 3, 1989, pp. 155–166.
19. Sadeghvaziri, E., M. B. Rojas, IV, and X. Jin. Exploring the Potential of Mobile Phone Data in Travel Pattern Analysis. *Transportation Research Record: Journal of the Transportation Research Board*, 2016. 2594: 27–34.
20. Shafieifar, M., M. Farzad, and A. Azizinamini. Experimental and numerical study on mechanical properties of Ultra High Performance Concrete (UHPC). *Construction and Building Materials*, Vol. 156, 2017, pp. 402–411.
21. Shafieifar, M., M. Farzad, and A. Azizinamini. A Comparison of Existing Analytical Methods to Predict the Flexural Capacity of Ultra High Performance Concrete (UHPC) Beams. *Construction and Building Materials* Vol. 172, 2018, pp. 10–18.

*The Structures Section (AFF00) peer-reviewed this paper (18-04892).*

*Opinions and conclusions expressed in this paper are of the authors and do not necessarily represent those of sponsors.*

**Sage Publications USA**
sagepub.com

CURRAN ASSOCIATES INC.
proceedings.com

ISBN 978-1-5108-8602-5